Eric Höweler

Design for Construction

Tectonic Imagination in Contemporary Architecture

Eric Höweler

Design for Construction

Tectonic Imagination in Contemporary Architecture

Routledge
Taylor & Francis Group
LONDON AND NEW YORK

Cover image: Courtesy of Amateur Architecture Studio. Photograph by Laksana Studio.

First published 2025
by Routledge
4 Park Square, Milton Park,
Abingdon, Oxon OX14 4RN

and by Routledge
605 Third Avenue,
New York, NY 10158

Routledge is an imprint of the Taylor & Francis Group, an informa business

British Library Cataloguing-in-Publication Data
A catalogue record for this book is available from the British Library

Library of Congress Cataloging-in-Publication Data
A catalog record has been requested for this book

ISBN: 9781032562339 (hbk)
ISBN: 9781032562315 (pbk)
ISBN: 9781003434559 (ebk)

DOI: 10.4324/9781003434559

Typeset in Neue Haas Unica and Helvetica Neue

Book Design: Common Name
Layout Assistance: Alex Xiaotan Yang, Chandler Caserta
Copy Editing: Edith Fikes

Publisher's Note:
This book has been prepared from camera-ready copy provided by the author.

For Product Safety Concerns and Information please contact our EU representative: GPSR@taylorandfrancis.com
Taylor & Francis Verlag GmbH, Kaufingerstraße 24, 80331 München, Germany

Table of Contents

Design for Construction

Introduction

Agency and Imagination

"I've solved it!" architect Jørn Utzon exclaimed during a 1961 phone call with structural engineer Ove Arup.[1] Utzon was calling to explain his proposed solution for the construction of the iconic sail-like shells of the Sydney Opera House. According to Utzon, all the shells would be segments cut from a single sphere. This principle, which came to be known as the "spherical solution," would define the roof's geometry and provide the rigor needed to make the design constructible.

Utzon was awarded the commission in 1957 following an international competition held by the state government of New South Wales, Australia. The entry fee was 10 Australian pounds. His award-winning scheme showed several nested shells, evoking the petals of a flower or the profiles of sailboats. [3] The design's bold silhouette captivated the jury, which included Eero Saarinen, convincing them to award the project and the prize of 5,000 Australian pounds to Utzon.

The competition drawings were highly schematic, and the roof shells were drawn with delicate linework representative of a thin-shell concrete structure. The jury proposed that Utzon work with Arup, a renowned structural engineer, to realize the complex project. Studying the proposal, Arup soon concluded that the "thin-shell concrete" concept would not work at the intended scale, and the design team set to work to translate the early sketches into a geometric logic and structural system that could be engineered and constructed.

After three years of work, Utzon had a breakthrough. He is said to have used an orange to demonstrate his theoretical solution of segmenting a sphere to create a viable structure for the shells. He standardized the geometry of each segment — each would have the same radius — and arranged them with a high ridgeline that approximates a gothic arch. Eventually, wood models were produced to explain his spherical solution in a more didactic way. [1]

Utzon argued that his solution would require only one mold, as each curve would be the same in any direction.[2] This approach allowed the design team to define the outer geometry of the shells but left the structural and constructability questions unresolved. Utzon's "solution" for the design had only addressed part of the problem. The shells still needed to be engineered in a way that was ultimately constructible.

Before the building was completed, and amidst disagreements with Australian authorities and a falling out with Arup, Utzon resigned, and Australian architect Peter Hall was tasked with completing the design of the halls. The cost of construction skyrocketed to 10 times the initial amount, and politicians used the project to point fingers and assign blame for cost and schedule overruns.

The Sydney Opera House is one of the most storied structures of the 20th century. It is often

[1] Peter Murray, *The Saga of the Sydney Opera House: The Dramatic Story of the Design and Construction of the Icon of Modern Australia* (London; New York, NY: Spon Press, 2004), p. 34.

[2] Utzon was a brilliant designer and a great storyteller. His narrative for the constructability of the shells was oversimplified, but highly understandable to the lay audience. "... A factory on the site is built to produce these ribs — only one form, divided up into smaller parts, is actually necessary... the forms are movable to enable us to mass produce segments..." Utzon quoted as part of a documentary for Danish television in 1967.

[1] Jørn Utzon's "spherical solution," for the Sydney Opera House, represented in a didactic model.

referenced as a cautionary tale of collaboration between architects and engineers that has gone wrong.[3] Yet, few of the histories focus on the role of the contractor, M.R. Hornibrook, the entity responsible for developing a construction plan from the design and engineering documents. Recent scholarship by researchers Paolo Stracchi, Luciano Cardellicchio, and Paolo Tombesi, however, examines the constructability of the Sydney Opera House and the division of labor between the architect, engineer, and contractor. Their essay, titled "Not Really an Aftermath: The Role of Actual Construction in the Design Process of the Sydney Opera House Roof,"[4] details the invention of novel fabrication techniques for the precast concrete rib sections, for post-tensioning and erecting the ribs, and for developing a sequence of steps that turned the narrow site into a precast factory and staging area for the project's construction.

Stracchi, Cardellicchio, and Tombesi cite the extraordinary number of drawings produced by the architect, the engineer, and the builder as a primary driver of project development. After winning the competition, Utzon's team produced what has been referred to as the "yellow book," an extensive set of drawings describing the design. The book included 42 sheets of plans, sections, and elevations, including drawings of the spherical geometry. 2 Arup's team produced 800 sheets of engineering drawings and calculations. In preparation for the shell's construction, Hornibrook's team produced 5,200 sheets of shop drawings to document how they intended to build the structure. While it is typical for shop drawings to exceed the number of architectural drawings, this figure is used by the researchers to demonstrate the essential role the contractor played, not simply in executing the architect's design intent, but also in the intensive invention, testing, and planning necessitated by the project's unusual and unprecedented design.

It took 15 years of refinement, revision, calculation, recalculation, and testing before the project was eventually completed. The spherical solution may have been a breakthrough for the architect, but it was only one part of a series of solutions required to construct the Sydney Opera House. This particularly fraught story of the relationship between design and construction points to a persistent, though variable, gap between a design concept and a constructible building. Bridging the gap between a design and its development, engineering, coordination, and construction, however, is an essential part of the process that transforms a conceptual idea into a completed work of architecture. 4 5

The open, single-round, anonymous competition for a project like the Sydney Opera House is difficult to imagine in the current risk-averse culture of the

[3] Andrew Saint, *Architect and Engineer: A Study in Sibling Rivalry* (New Haven, CT; London: Yale University Press, 2007), pp. 371–377.

[4] Paolo Stracchi, Luciano Cardellicchio, Paolo Tombesi, "Not Really an Aftermath: The Role of Actual Construction in the Design Process of the Sydney Opera House Roof," in *Frontiers of Architectural Research 12*, (2023) pp. 242–265.

2 Utzon's spherical geometry drawings for the Sydney Opera House.

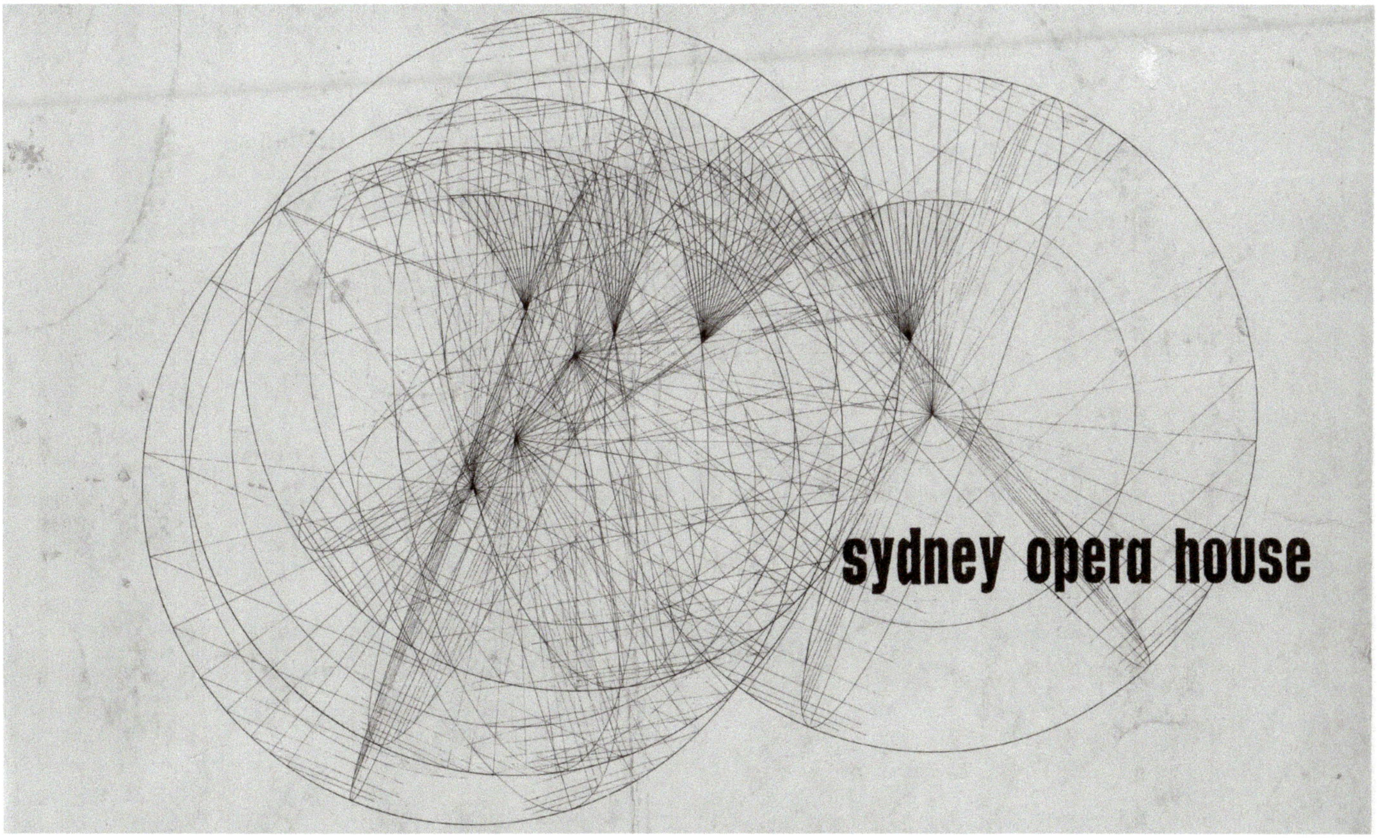

industry, which emphasizes past performance and prequalification. Yet, it highlights many of the phases, contributions, personalities, and types of expertise required to execute the iconic structure that remain relevant today. It also illustrates inevitable questions of cost, constraints, and constructability, that often accompany ambitious projects in any period. Any discussion of contemporary architecture needs to be understood against the backdrop of these forces and the different roles played by protagonists in its realization. The present state of the architecture and engineering disciplines, largely influenced by an expansive list of participants and experts, including builders, construction managers, owner's representatives, legal teams, and a host of other specialty consultants, is key to understanding both problems and possibilities in contemporary practices in design and construction.

PROCESS

The "saga" of the Sydney Opera House, as described in intricate detail by author Peter Murray, highlights many of the critical issues that continue to govern the worlds of design and construction.[5] These include matters of scope, the division of labor between disciplines, and the "instruments of service" that architects employ to represent their work. It also offers an expanded understanding of what it is that architects do within globalized networks, chains, and other contexts where building materials and products are procured. These conditions structure the processes of architectural design, from the earliest concepts to the final punch list. Contemporary design grapples with the translation from concept to constructability — the very issue that plagued the Sydney Opera House. The complex, multi-phase transformation from design to construction — encompassing representation, material processes, formal expression, ecological impact, and post-occupancy, among other critical considerations — is the primary interest of this book.

DISTANCE

Architectural design operates at a distance from its medium: the constructed building. Architects make drawings and 3D representations of buildings. Builders and specialized tradespeople construct buildings.[6] This distancing of design from implementation has also structured architectural discourse, impacting disciplinary and pedagogical frameworks as well as legal parameters for architectural practice. Surveying the state of design and construction, their connections and co-dependencies, a few key trends shaping both fields emerge, including the impact of digital tools

[5] Peter Murray, *The Saga of the Sydney Opera House*, p. vi.

[6] Robin Evans has eloquently laid out the disciplinary investments in architectural representation in his essay, "Translations from Drawing to Building," in *Translations from Drawing to Building and Other Essays* (Cambridge, MA: MIT Press, 1997), pp. 153–193.

[3] The iconic silhouette of the Sydney Opera House, as a series of nested shells or petals as show in elevation.

and workflows through building information modeling (BIM), a global network of procurement and material flows, a new consciousness about embodied energy and fair labor practices, and an expanded understanding of the built environment as a critical consideration in the age of climate crisis.

EXPANDED PRACTICE

The worlds of design and construction have evolved in parallel but along separate tracks, responding to industry forces, including globalization, financing, insurance, regulation, and relevance. Contemporary practices are expanding the limits of design with the proliferation of architecture studios and different modes of practice that include digital fabrication, computation, ecological, economic, material and forensic research, large-scale urban and landscape planning, among a multitude of new interfaces, intersections, and engagements.[7]

Despite the growing range across areas of architecture practice, designers today are more equipped than ever to collectively close the gap between the concept and construction with digital tools and workflows that nimbly produce and share new iterations, enable collaboration across platforms, and support communication and project coordination with partners and builders. Finally, many design practices around the world are doing meaningful, mission-driven work to improve lives and communities, and design discourses have significantly shifted from the strictly formal and stylistic to the strategic and purposeful potential of architecture.

GLOBAL

Contemporary design practices vary widely from scale and type to region and geography, and so designers are working across local conditions and cultural contexts, which results in the need for workflows and design documentation to communicate across time zones and linguistic barriers. The transmission of design practices across international borders, boundaries, and cultures is mirrored in material and product flows and procurement processes that span the globe. The Glass Pavilion at the Toledo Museum of Art (2006), for example, was planned to celebrate the long history of glass production in Toledo, Ohio, and the architects, Tokyo-based SANAA, employed glass as a primary material in the architecture of the museum. However, the large glass lites specified for the design were too large to be fabricated in the US. The glass was produced in Austria, curved and laminated in China, and shipped to Ohio for installation.[8] 6 The chain of procurement and fabrication demonstrates the paradoxes of the global supply chain, which, despite increased awareness of the embodied energy in

[7] I co-authored the book, *Expanded Practice,* with Meejin Yoon as an effort to think about alternative modes of practice and exceed the traditionally defined boundaries of architecture. J. Meejin Yoon and Eric Höweler, *Expanded Practice: Höweler + Yoon Architecture / My Studio*, 1st ed. (New York, NY: Princeton Architectural Press, 2009).

[8] Branko Kolarevic and Kevin R. Klinger, eds., *Manufacturing Material Effects: Rethinking Design and Making in Architecture* (New York, NY: Routledge, 2008), p. 264.

4 Study models of the structural ribs of the Sydney Opera House shells, showing the convergence toward the base of the shell.

5 Construction photo of the Sydney Opera House showing the assembly of the precast rib segments and the open trough design with cross bracing.

building materials, still relies on a complex web of procurement and fabrication to construct and deliver works of architecture.

SPANNING

The title of this book, *Design for Construction: Tectonic Imagination in Contemporary Architecture*, encompasses the processes, interdependencies, and feedback loops that guide and govern design and execution in contemporary architecture. Specifically, this book brings together the ways architectural design anticipates and incorporates methods of fabrication and construction such that the processes of designers, engineers, and builders might productively inform each other. Spanning concept and construction makes explicit the continuities between initial design conceptualizations and the final constructed outcomes. To initiate a design process with an eye toward implementation suggests a process that employs tectonic imagination.[9] The term tectonics implies an understanding of forms and forces that act on architecture and the intent to make those logics legible.[10] Tectonic imagination conceptualizes architecture within the continuum of precedent, past practices, and disciplinary knowledge and extends the material and technical possibilities of the present and future.

6 SANAA's Glass Pavilion at the Toledo Museum of Art (2006), in Toledo, Ohio.

CARDBOARD

Despite the Modern movement's emphasis on material systems like steel-reinforced concrete to produce a new formal language, many architects act with indifference to materiality. Architect Peter Eisenman famously theorized a process of abstraction and defamiliarization that he called "cardboard architecture."[11] For Eisenman, emphasizing form and plasticity was a way of thinking abstractly about form and syntax disengaged from the specifics of materials and construction. His House II (1970) project translated his theories of cardboard architecture into a built structure. 7 An aerial photograph of the house was once published in a magazine and mistaken as a model photo. The accidental confusion of a model and a building highlights one of Eisenman's goals, including the necessity of abstraction to inaugurate a new formal basis for architecture. Eisenman describes stripping away elements that articulate buildings, such as joints and flashing, "making things look homogeneous, when they really were not," was part of his agenda. "At the time I thought cardboard was as neutral as you could get it. It was pure nothing — de-material."[12]

Contemporary digital workflows reinforce abstract approaches to design by modeling geometry as lines and surfaces to which material

7 Peter Eisenman's House II (1970), in Hardwick, Vermont.

9 The phrase "tectonic imagination" was used by Italian architectural theorist Marco Frascari in a 1996 paper entitled, "Tectonics in Theory, Three Aspects of Tectonic Imagination," ACSA European Conference, Copenhagen.

10 Keneth Frampton's seminal book, *Studies in Tectonic Culture*, is essential in framing the discourse around tectonics. Kenneth Frampton, *Studies in Tectonic Culture: The Poetics of Construction in Nineteenth and Twentieth Century Architecture*, 3. Aufl. (Cambridge, MA: MIT Press, 2001).

11 "Cardboard is used to signify the result of the particular way of generating and transforming a series of primitive integer relationships into a more complex set of specific relationships which become the actual building. In this sense cardboard is used to denote the particular deployment of columns, walls, and beams as layers. It is not so much a literal recognition of the actual surfaces as cardboard and thus insubstantial but rather is meant to signify the virtual or implied layering which is produced by the particular configuration." Peter Eisenman, Michael Graves, Charles Gwathmey, *Five Architects* (New York, NY: Oxford University Press, 1975), p. 15.

12 Conversation with author July 24, 2024.

properties are applied as texture maps. Thickness and mass are also largely missing from modeling programs such as Rhino, so design workflows tend to rely on surface modeling as shorthand. Thickness, depth, and mass are added later. Digital tools and software platforms have unleashed powerful new trends in computational techniques that often result in a lag between initial explorations in representation and their eventual applications in built work. Representational tools, however, do impact what they represent. The medium of representation affects the message.

ANTICIPATION

Highly speculative projects and their representations can precede implementation and application by years or even decades. For example, Mies van der Rohe's charcoal renderings of the 1921 Friedrichstrasse skyscraper anticipated by 30 years the first all-glass curtain walls. [8] His bronze curtain wall on the Seagram Building (1958) was revolutionary when first installed; it inspired a host of imitations and became the default condition of postwar modernism. Greg Lynn's early speculations in *Folding in Architecture* (1993) [9] took years to arrive until those ideas were realized, if only partially, by Frank O. Gehry's Guggenheim Bilbao (1997). Today, digital modeling tools and specialty steel fabrication have made possible a host of complex forms and structures that would have been inconceivable a few decades ago. [10]

PEDAGOGY

Design attitudes and methods are found not only in professional practice but are also deeply embedded in the curricula of architecture schools. Questions about material and technical approaches to design are very often bound up in matters of pedagogy. When Walter Gropius reformed the curriculum at the Harvard Graduate School of Design in 1938, he famously dispensed with the teaching of architectural history. For Gropius, teaching history was an impediment to forging a "New Architecture," which sought to fuse art and social purpose with the industrial processes of the time.[13] Gropius intended to replace history with industrial systems and construction technologies as a primary source for design.

[13] Henry Cobb describes the moment at the GSD. "... the pedagogy of the modern movement, as practiced by Gropius, sought to protect the student from contamination by all of history. This erasure of the past was intended to clear the way for an entirely 'New Architecture,' liberated from the tyranny of dead styles, in which art, technology, and social purpose would be powerfully joined for the benefit of humanity." Henry N. Cobb, *Henry N. Cobb: Words & Works 1948–2018: Scenes from a Life in Architecture*, (New York, NY: The Monacelli Press, 2018), p. 21.

Today, the teaching of building technologies is but one of many aspects of design education. While integrating building technologies in the design process is a requirement of architecture curricula, the topic is often treated as a matter of compliance and building technology courses are often considered "support." How might we move the questions of building integration and building

[8] Ludwig Mies Van Der Rohe's charcoal rendering of the Friedrichstrasse skyscraper proposal (1921).

[9] Greg Lynn's production prototypes for Alessi Tea and Coffee Towers (2001).

science from the periphery toward the center of design education? With so many urgent matters confronting design and society today, the questions of what to teach and in what sequence are of critical importance. Climate change, embodied energy, social justice, and equity are all largely understood as essential to incorporate into both pedagogy and design projects. 11

REHEARSAL

In 1954, Louis Kahn wrote poetically about how representation and the ways we draw can impact reality: "If we were to train ourselves to draw as we build, from the bottom up, stopping our pencils at the joints of pouring or erecting, ornament would evolve out of our love for the perfection of construction and we would develop new methods of construction."[14] Kahn's emphasis on the sequence of building through the tactility of drawing seems relevant once more, especially given the fluid and frictionless space of the digital modeling environment.[15]

Beginning with and continuing well beyond design education, representation is a critical mode of communication that conveys design intent, whether a rendering, 3D model, construction document, or an as-built set of drawings. Architectural drawings become "instruments of service" with specific responsibility and authority.[16] Construction documents become contracts, so dimensions and notations take on legal significance. Drawings then cease to serve as representations of design intent and become receipts for what has been bid, negotiated, and agreed upon. Construction drawings created by different disciplines represent scopes of work and coordinated pathways for building systems.

Architectural drawings and specifications define the scope of the contract for construction — the what— but stop short of offering the builder specifics about means and methods — the how — of constructing a project. According to Standard Form AIA contracts, the "means and methods" of construction are expressly the responsibility of the builder, while the architect is responsible for design intent. This delineation of roles and responsibilities is designed to limit the architect's risk and empower the builder to make decisions about sequencing, trades, and processes within their area of expertise.[17]

LOBOTOMY

Architect Nader Tehrani of NADAAA describes the bifurcation of design and construction. "It is one of today's ironies that design contracts, e.g., in the USA, specify that the architect is responsible for

[14] Louis Kahn, "How to Develop New Methods of Construction," *Architectural Forum* (New York, NY: Time, November 1954), p. 157.

[15] It is worth noting that contemporary design workflows in 3D modeling better reflect the processes of construction than 2D drawing, as the 3D digital model is already spatial and volumetric. In that sense Louis Kahn's "training" to think of design as a rehearsal of construction seems more possible. What is missing in the 3D model is, of course, thickness, weight, mass and physicality.

[16] "Instruments of Service" is how the AIA defines the architect's drawings to differentiate them from products that could be purchased, versus services that are commissioned. "Instruments of Service" is also the title of *Harvard Design Magazine* Issue 52, edited by Elizabeth Christoforetti and Jacob Reidel, Harvard Graduate School of Design, Fall 2024.

10 Steel frame construction of the KAFD Metro Station, Riyadh, Saudi Arabia, designed by Zaha Hadid Architects, employs custom-rolled steel sections that would have been inconceivable before digitally controlled fabrication.

11 Raising questions about the processes of material extraction and production, the speculative practice of DESIGN EARTH engages in projects at the regional, territorial and planetary scale, including, After Oil (Das Island), 2016.

design intent, while the contractor is responsible for — and ultimately in charge of — the means and methods of construction. This legal detail sets up an adversarial relationship between these protagonists, divorcing the architect from the very techniques that help determine budget, scope, and implementation, while also distancing the contractor from the conceptual, theoretical, and organizational underpinnings of a design intent; in one simple act it neuters the architect as a builder while lobotomizing the builder as a thinker."[18] For Tehrani, as well as many other designers working today, a deep understanding of material and construction systems offers a means of reclaiming agency in a field where design is often understood as peripheral to the business of building.

With the aim of bringing the realms of design and construction into more productive dialog, this book is organized around a set of thematic chapters that elaborate on particular cases where interactions across these realms play out in specific projects. Chapter 1, "Stocktaking," begins by surveying the current state of design and construction and considers the many different forces acting on these two related but discrete fields. The present moment is characterized by a tremendous amount of construction globally alongside an increasing understanding of its significant impact on local cultures and ecosystems and a growing recognition that the built environment is a major contributor to greenhouse gases. At the same time, design techniques have made it possible to design and fabricate extremely precise, non-standard structures with the help of digital tools. And while digital processes in design and fabrication have created new possibilities, translation into built form remains limited by the economics of mass customization and the reality of erecting and assembling components of unprecedented complexity. When almost anything is possible, what we choose to design becomes increasingly important, as does how we plan to execute it.

Chapter 2, "Means and Methods," addresses the typical and atypical structure of relationships between architects and builders that have produced the divides that Tehrani refers to as a "lobotomy" for builders. And while legal structures limit the architect's role in the construction phase, creative practices are finding ways to restructure their services through alternative modes of practice. For example, design-build practices like GLUCK+ have found ways to be effective as designers *and* builders, allowing them to produce both exceptional designs and built projects.

Thinking through "Material Processes," the title of Chapter 3, architects have developed new vocabularies tied to the specifics of material systems.

17 The state of design disengaged from questions of construction is a function of what German sociologist Ulrich Beck calls "risk society." Construction is a high-cost and high-risk enterprise, where construction delays can cost money, or unsafe conditions can become hazards to health and safety. The separation of design from execution compartmentalizes responsibilities and risks between designers and builders. The delimitation of risk endorsed by the AIA also has the consequence of limiting the architect's participation during the construction phase of a project and has the effect of distancing designers from construction processes. Ulrich Beck, *Risk Society, Towards a New Modernity* (London: Sage, 1986).

18 Nader Tehrani, "Difficult Synthesis," in *Material Design: Informing Architecture through Materiality*, ed. Thomas Schröpfer (Basel: Birkhäuser, 2011), p. 34.

1 2 Facade components of the glass fiber reinforced concrete (GFRC) envelope of The Broad Museum (2015), by Diller Scofidio + Renfro.

Steel-reinforced concrete and modern architecture were historically combined to create perhaps the most significant co-development of material and design methods. Le Corbusier's Maison Dom-Ino concept for a concrete frame, distinct from its enclosure and interior partitions, illustrated the revolutionary potential of the material system. Later innovations promise to similarly transform design and construction. The contemporary interest in mass timber and engineered lumber, for example, signals a growing awareness of embodied energy and biogenic materials that has some architects calling for a new expressive language analogous to the Maison Dom-Ino for timber. 13 14 While materials and methods have historically unlocked design potential through construction systems, digital and computational tools have had a radically transformative impact on design workflows and design culture. Chapter 3 addresses questions of translation from concept to constructs and productive feedback loops between ideation and implementation. 12

Regardless of digital or analog tools, and indifferent to styles or aesthetics, buildings are constructed from thousands of individual components. The assembly of walls, slabs, studs, sheathing, windows, gaskets, and glazing is the responsibility of the builder, but what these parts are and how they are expressed is a large part of the design intent and establishing part-to-whole relationships. Chapter 4, "Part to Whole," explores different approaches to design that give expression to the realities of components and assembly. Theorists of this part-to-whole approach include historian Kenneth Frampton, who sought to resist the tendency of architecture to be apprehended as a synthetic image by emphasizing its tectonic expression.[19] Architects like Tehrani have built a body of work around the articulation of aggregation and what he calls the "ethics of mono-materials."[20]

As a counterpoint to the expressive component logic of part-to-whole design, monolithic architecture tends to subsume individual components to achieve a singular expressive form. Chapter 5, "Monoliths," revisits the 1995 exhibition *Monolithic Architecture* and explores contemporary projects that update the original framework to achieve sculptural expression despite constitutive component parts. Herzog & de Meuron's 1994 Signal Box building remains a significant benchmark, employing a single material, the copper strip, with a single detail to mask the varied programmatic elements behind an expressive screen. A host of newer projects meet the criteria of monolithic architecture by applying similar strategies with new materials and details, suggesting that the category is worth revisiting today.

Contemporary building envelopes are highly engineered, prefabricated systems that afford views

19 Kenneth Frampton, *Studies in Tectonic Culture: The Poetics of Construction in Nineteenth and Twentieth Century Architecture*, (Cambridge, MA: MIT Press, 1995), p. 2.

20 Transcript from a lecture delivered at the CCA on March 4th, 2021.

13 Mass timber frame of 11 East Lenox (2022) by Monte French Design Studio.

14 Cross laminated timber (CLT) interior of Haus Gables (2018), in Atlanta, Georgia, by Jennifer Bonner of MALL.

in and out, regulate thermal conditions such as solar heat gain, and provide the primary expression of a building. In some cases, building facades are purely cladding supported by the structure within. In others, the exterior wall is also the structural frame of the building or part of its lateral bracing system. 15 Some of these systems are designed to change their characteristics in response to exposures and shifting weather conditions. Others are integrated into the building monitoring and control system, allowing them to respond dynamically to occupants and environments. 16 Chapter 6, "From Windows to Envelopes," charts the evolution of the window as an exception to the load-bearing wall, to larger and larger openings within walls, and to the envelope independent of the structural frame.

In Chapter 7, "Force, Form, and Gesture," we turn our attention to the relationship between forms and forces and how contemporary buildings express, suppress, or amplify their structural performance. For example, the splayed geometry of the Eiffel Tower's base expresses a relationship between structural behavior and formal expression, with the tower tapering to its point as the building form corresponds to its optimized structure derived from force. 17 Contemporary buildings, like the CCTV Headquarters in Beijing, express the structural work that achieves its signature two-way cantilevered shaft by other means, namely through the density of members highlighted on its facade. 18 The relationships between force and form are rarely 1:1 and are often approximated as a structural or architectural "gesture" meant to evoke the interactions of form and force as translated into a built structure.

"Climate and Conditioning" are the subjects of Chapter 8. In addition to dramatic changes brought about by digital design tools, our understanding of the impact of the built environment on the planet and the emergence of climate consciousness has radically shifted our thinking about architecture and its construction. The built environment's significant contribution to greenhouse gas emissions and consumption of vast amounts of energy through its operations has led designers and builders to reconsider standard practices and strive to make buildings more energy efficient by utilizing renewable or recycled materials. Designers today are considering buildings in an expanded sense. Architecture is made up of myriad materials, parts, and components that have been mined, refined, extruded, assembled, transported, and installed. The emergence of an expanded awareness about the impact of buildings on the planet has transformed how we think about design, and how we build architecture today.

And then, what happens after? After the building is complete and the users move in? After

15 Diagrid exoskeleton of OMA's Seattle Central Library (2004), in Seattle, Washington.

16 Kinetic sun shading facade of the Al Bahr Towers (2012) in Abu Dhabi, UAE, designed by Aedas.

the intended uses of the building cease to be necessary and the users move out? After the effects of weather and temperature shifts have worn the building's materials? After the building is torn down, broken apart, and hauled to a landfill site? Does all the embodied energy, labor, and craft disappear into the landfill where it sits and decomposes over millennia? How might designers consider the afterlives of buildings? Contemporary approaches to adaptive reuse seek to give new purpose to old structures to retain their embodied energy. Circular construction practices catalog, process, and reuse building materials, eliminating waste and creating a closed-loop system of perpetual use. Chapter 9, "Afterlives," focuses on efforts to think of buildings as having multiple lives that can extend their present conditions and uses well into the future.

In the final chapter, titled "Construction Futures," I speculate on some of the contemporary trends that shape design and construction and where we might expect transformative innovations to reshape the discipline and the industry in the future. The disruptions caused by the COVID-19 pandemic and the supply chain issues that followed have made clear how our globalized construction industry, with many parts produced overseas, is based on assumptions of smooth logistics and the free flow of goods. Greater attention paid to procurement processes means that building materials must supply more complete documentation about their sourcing, the conditions under which they were made, and their provenance before arriving on the job site. The need to identify alternative sources, either by reuse or non-traditional material streams, will likely continue to inspire designers. 19 For example, the design and research practice HANNAH, based in Ithaca, NY, designed Ashen Cabin, a full-scale prototype built with regionally sourced lumber cut locally by a robotic arm from diseased elm trees. 20

Construction involves risks such as unanticipated conditions, labor market disruption, extreme weather events, and cost and schedule overruns; these factors will continue to affect design and construction decisions despite practices designed to minimize and manage them. The integration and coordination of information technologies within the built environment and the proliferation of machine learning and artificial intelligence will shift the way we design and construct architecture in profound ways. Installing, maintaining, troubleshooting, and operating buildings will increase in complexity. In all likelihood, labor practices, risk and its management, costs of materials, procurement, and automation will remain constant and urgent enough to transform the worlds and futures of both design and construction.

17 Gustave Eiffel's "300 Meter Tower" (1889), derives its geometry from the wind forces acting on it, translating force into form.

18 The "visually intuitive instability" of OMA's CCTV Tower (2012), in Beijing, China.

AGENCY AND IMAGINATION

This partial survey brings together a number of the forces that are profoundly reshaping the interrelated fields of design and construction. Processes informed by advances in software and digital fabrication are tending toward greater integration of design and fabrication models, while risk management maintains the culture of compartmentalization across design and construction. Meanwhile, globalized supply chains are sourcing products and materials from around the world, weaving a web of logistics that has been revealed to be vulnerable to disruption. Attitudes about climate change and social justice, and the role of the built environment in perpetuating these conditions (and its responsibility to alleviate them) are changing rapidly, demanding that we reassess the role of architecture, its fundamental definitions, and assumptions.

How, then, might we rethink the roles of design and construction to address these changing attitudes about performance, procurement, risk, reuse, embodied energy, and the laboring bodies that build our buildings? How much agency might design as both a field of study and a practice reasonably assume amidst global procurement, widespread calls to rid buildings of toxic materials, and international movements to improve working conditions in ways that would ensure our built environment is certifiably produced through fair labor practices? How might architects reclaim some agency by connecting processes of design, material specification, and the construction of architecture?

Design that incorporates material and construction logics constitutes a tectonic imagination — a way of conceptualizing design through the many parameters around implementation and execution. Developing a careful understanding of how architecture's complex matrix of processes are inevitably intertwined and codependent is a powerful way to begin to reclaim agency, reconcile the impact buildings have on lives and ecosystems, and reconnect the ways in which buildings are conceived, coordinated, contracted, and constructed.

19 The material reuse of "found" fragments of marble, reincorporated into the pavers of the Acropolis pathways (1957), by Dimitris Pikionis.

20 HANNAH's Ashen Cabin (2020), in Ithaca, New York, employs robotically sawn slabs of diseased lumber to reimagine architecture with alternative material sources.

1 The all-glass closed-cavity double facade of MIT's Schwarzman College of Computing (2024) by SOM.

Chapter 1 Stocktaking

During the construction of the Empire State Building, the deliveries of material to the job site reached a volume of 500 trucks per day, or roughly one delivery every minute during the eight-hour work day.[1] A special delivery system allowed trucks delivering cement, gravel, and sand to dump material down a chute to a basement level, where the material was sorted and routed to the floor where it was needed via one of six construction hoists or nine construction derricks. The astounding feat of construction logistics and management was achieved by the construction company of Starrett Brothers and Eken, which allowed the Empire State Building to be constructed in a staggering 13 months. Construction began on March 17, 1930, and the building was completed in just over a year, on April 11, 1931. The Empire State Building claimed the title of the world's tallest, rising 1,454 feet (443m), containing 102 floors, and housing 2.2 million square feet (209,000sm). It also claimed the title of the world's fastest. [2]

William A. Starrett, the Starrett Brothers construction company manager, described the construction of skyscrapers as "the nearest peacetime equivalent of war."[2] This analogy highlights the intensity of logistics and coordination required to erect a tall building. The speed with which the Empire State Building was completed has never been surpassed, regardless of new technologies, innovations in equipment, construction methods, and 3D models.

The Empire State Building offers a salient case study of the integration of design, engineering, and construction management. Architects Richmond Shreve and William Lamb developed their design to capitalize on the unique conditions of the moment, the real estate potential of the site, the existing technologies of building systems, and the labor market at the onset of the Great Depression. In 1929, as the building was being planned, the owner, the Empire State Realty Trust, acquired contiguous properties that allowed for a single site to occupy half of a midtown block from 33rd to 34th Streets, fronting 5th Avenue. The architects developed the building massing as a "pure setback skyscraper," with the sky exposure plane set back from the three surrounding streets, resulting in one of the clearest demonstrations of the typology and a building of unprecedented proportions and dimensions. The structure's main shaft occupies 25% of the site area, enabling a vertical extension without setback or height limit. While the scale was new at the time, the building systems were not. Steel frame buildings with masonry infill walls required typical building systems.

1 Carol Willis, *Building the Empire State* (New York, NY: W.W. Norton, 2008), p. 28.

2 William A. Starrett, *Skyscrapers and the Men Who Build Them* (New York, NY: Charles Scribner's Sons, 1928), p. 63.

[2] Designed for construction, the Empire State Building (1931) by Shreve, Lamb & Harmon, was "magnificently adapted for the speed of construction."

[3] The Empire State Building pioneered many construction innovations, such as the railroad track for material distribution.

However, the extreme height of the building, and the tonnage of steel required were unique. The difference was a difference of degree, not a difference of kind.

Calling the Empire State Building "the most colossal miscalculation of the 1920s,"[3] architectural historian Carol Willis cites a unique combination of factors that allowed the project to move forward. While the building claimed impressive records for height and speed of construction, it remained largely empty for a decade and did not turn a profit until 1950. Deemphasizing the authorial role of architects and engineers, Willis reads the history of the skyscraper as the result of economic, regulatory, and technical factors. Reviewing the correspondence between the real estate developers involved in the planning of the Empire State Building, she describes the driving factors of the project as "entirely financial, not architectural."[4] For example, the developers carefully outlined the site area, square feet, construction costs, and anticipated rents, which made up the real estate proforma, all of which preceded the hiring of an architect. Willis proposes that the skyscrapers of New York and Chicago could be viewed as emergent "vernaculars of capitalism," where building forms "evolve" as a direct response to local conditions, market forces, and regulatory constraints.[5] Willis introduces the terms "engineering height" and "economic height" to describe the limits of tall buildings. According to Willis, economic height often determines building height, while engineering height is nearly unlimited.

Construction of the Empire State Building introduced several innovative construction and management techniques that remain relevant today. The Starrett Brothers pioneered on-site techniques that ensured the smooth flow of materials and labor. For example, the building was divided into nine separate construction zones (A-H and K), each with a dedicated crane. [4] Materials were hoisted to the appropriate floors and delivered to where they were needed, utilizing a custom railroad track system for distribution. [3] During peak construction, 3,400 workers were on site, including riveters, carpenters, masons, and glaziers, among others. Feeding the workers meals required cafeterias on different levels to limit the necessity, time, and effort of going up and down the building. The Empire State Building project pioneered the development of what is known as "fast-tracking," where parts of the project are bid early, before the full drawing set is completed.

At the turn of the 20th century, American construction companies were hired to build

3 Carol Willis, *Form Follows Finance: Skyscrapers and Skylines in New York and Chicago*, 1st ed. (New York, NY: Princeton Architectural Press, 1995), p. 90.

4 Carol Willis, *Building the Empire State*, p. 17.

5 Carol Willis, *Form Follows Finance*, p. 7.

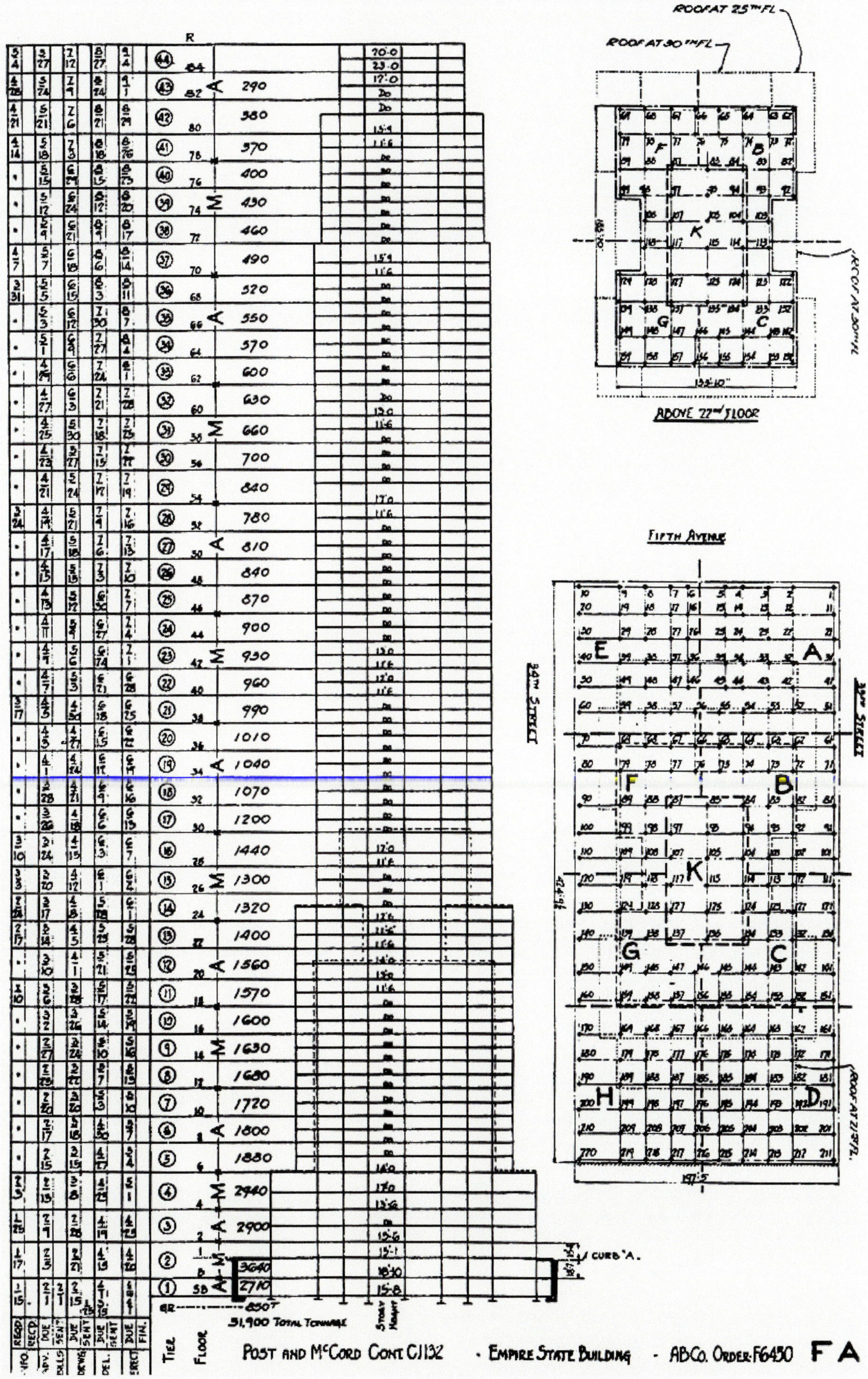

4 Structural steel erection schedule and hoisting plan diagrams for the Empire State Building (1931), produced by steel erector, Post & McCord. From *Architectural Forum* 52, 1930.

5 The historic drafting room of an architecture office as a "design site." Note the slope of the drafting tables and the adjacency to the windows.

6 The "design site" of the contemporary architecture office with flat screen displays.

structures, employing a variety of specialized trades. Structural engineer Donald Friedman, writing on the innovative processes introduced with the design and construction of the Empire State Building, notes a significant shift from physical labor to management, a condition that continues to drive design construction today.[6] With the emergence of the skyscraper and industrial processes driven by speed and efficiency, these companies employed less and less of their own labor force, preferring to hire subcontractors and shifting the emphasis from builder to construction manager.

THEN AND NOW

Taking stock of the state of design and construction, it is important to ask: What has changed with contemporary design and construction, and what has remained the same? Nearly 100 years after the completion of the Empire State Building, many of the project's innovative procurement and management processes are considered standard practice today. Despite the astounding technological innovations of the last century, moving heavy materials from one place to another, and assembling parts into a singular whole remains a costly reality for the construction industry. John Fish, Chairman and CEO of Suffolk Construction company, describes the construction industry as one of the least innovative, "second to last, just ahead of Fishing and Gaming."[7] He is quick to point out that despite seemingly entrenched construction industry practices, they have also historically been quick to adopt new technologies such as cell phones and fax machines. Contemporary job sites now employ digital surveying equipment and photo stitching technologies for site documentation.

Over the last 100 years, a dizzying number of advancements in communications, logistics, material science, and fabrication have revolutionized all aspects of design and construction. For example, 3D modeling software makes it increasingly possible to model complex structures and cloud hosting makes it much easier to coordinate between disciplines in the design process, and with trades in the construction phase. Yet, despite decades of significant technological advancement and paradigm shifts in many areas of design and construction, how a building is assembled in the field, with material and components erected and installed on building sites, today, is today, much like it was then.

6 Donald Friedman, "'A Story a Day': Engineering the Work," in Carol Willis, *Building the Empire State* (New York, NY: W.W. Norton, 2008), p. 39.

7 Remarks made by John Fish at a lecture at the Harvard Graduate School of Design, September 15, 2023.

7 The construction site of the Empire State Building (1931) with steel workers and a riveting gun. Note the "catcher" with the cone shaped container, who is waiting to catch a red-hot rivet thrown by another worker near the furnace.

8 A photo of a contemporary construction site shows many aspects that remain the same, while some things change, including the use of high strength bolts instead of rivets, and the use of personal protective equipment (PPE).

Visiting a construction site today might have a lot in common with the job site conditions of Empire State Building in the 1930s. 7 Steel beams hoisted by a construction crane deliver materials from a truck to the appropriate floor, and steel workers fasten components with great speed and agility. The sounds of a warning horn signaling an overhead crane pick, or the banging and grinding of steel assembly would be very similar. The sequence of trades on a job site would mirror those of previous times, with foundations and superstructure followed by enclosure, framing, and finishes. Steel workers would precede electricians and finish carpenters. The equipment on a contemporary construction site would reveal important differences. Tables for large sheets of drawings have slowly given way to handheld tablet computers. Digital surveying equipment, laptops with 3D information models, and handheld computers allow workers to verify site conditions with construction documents. Another visible difference would be the presence of safety equipment, safety harnesses, and temporary fall protection. The bright colors of high-visibility personal protective equipment (PPE), hard hats, vests, gloves, boots and utility glasses give contemporary job sites a distinct visual appearance that reflects an evolution in construction practices. 8

DESIGN SITES AND CONSTRUCTION SITES

The separation of the spaces of design from the spaces of construction, and the conceptual gap between the two worlds, is of central importance in this book, which seeks to find connections and translations across that divide. As described above, a comparison between the construction site of the Empire State Building in the late 1920s and today's construction sites would reveal that many industry practices have remained largely unchanged. In contrast, visiting the "design site" of an architectural practice of the 1920s and the 2020s would reveal several fundamental differences.

One hundred years ago, the architect's studio would likely include a spatial arrangement of private offices for principals and managers, and a separate drafting room of large tables with adjustable tops for hand drawing plans and details. The drafting room would likely be narrow and arranged to maximize the daylight needed to perform the tasks of hand drafting. 5 Tasks would have been specialized with specific roles for draftsmen, assistants, and errand boys. An architect's studio today is dominated by an array of upright flat-screen computer monitors. 6 Not reliant on natural daylighting, as the displays emit their

[9] Scale comparison between Empire State Building (1931) and Marina Bay Sands (2010).

own light, the architectural office today can occupy deeper floorplates. There is no longer a separation of architects and draftsmen. Two-dimensional drafting has been replaced by three-dimensional modeling, and physical outputs are done with plotters as well as 3D printers and other digital fabrication equipment.

The architects who designed the Empire State Building documented the tower with a methodology of standardization and repetition. The repetitive nature of the office tower allowed floors within segments of the tower to be documented with "representative" typical floors. Standard details for windows and stonework were drawn and documented with the understanding that they applied broadly to a multitude of conditions.

Contemporary forms of documenting architectural projects through Building Information Models (BIM) are fundamentally different than 2D drawings. BIM sheets produced today are the outputs of a singular 3D model that has been collectively "built" and layered into systems to allow for the "exporting" of 2D views and cuts. The outputs of BIM employ drawing set layout conventions but share little with documentation is produced by 2D drawing.[8] Automation will continue to impact how architecture is designed and documented in the near future.

[8] Describing contemporary architectural workflows with digital tools, architect Phillip Bernstein notes the lack of "radical innovations" in the architecture/engineering/construction (AEC) industry, noting concerns over "liability, intellectual property, or even the appropriate division of labor." Peggy Deamer and Phillip G. Bernstein, eds., *Building (in) The Future: Recasting Labor in Architecture*, (New York, NY: Princeton Architectural Press, 2010), p. 192.

MEGAPROJECTS TODAY

The construction of the Marina Bay Sands (2010) in Singapore offers a superlative example of a contemporary design and construction process that serves as a productive comparison to the Empire State Building. [9] The massive, 9-million-square-foot (836,000sm) development includes a 2,561-room hotel, casino, shopping mall, museum, and a Skypark bridging three 55-floor hotels. The project was completed in a record time of four years, following the commissioning of the developer, Las Vegas Sands, with architect Moshe Safdie.

The Marina Bay Sands project is a mixed-use megaproject comprising an ensemble of buildings that approach an urban condition. [1][2] Organized around a new subway station, the Marina Bay Sands is served by infrastructure and best understood as the implementation of a planning and infrastructure initiative. The project was conceived and planned by the URA (Urban Redevelopment Authority), the Singapore government's planning and development agency, and tendered to private developers through a developer-led competition process.

The competition brief called for an iconic structure and a new landmark for Singapore. Safdie's design for the multi-building ensemble produced

10 Construction of the splayed hotel section of the Marina Bay Sands (2010), in Singapore, by Moshe Safdie of Safdie Architects.

11 The completed Marina Bay Sands hotel (2010), with the three hotel wings supporting the cantilevering Skypark.

the signature structure with the three hotel towers surmounted by the horizontal cantilevered roof deck called the Skypark. The hotel towers each consist of a splayed and asymmetrical A-frame bar, introducing a tapered atrium condition at the base of the towers with sweeping concavities along their broad faces. 10 11 The three hotels are arrayed along a subtle curve, making each tower turn slightly, relative to the others, enhancing the sense of the sculptural composition of their forms. A glazed arcade joins the three hotel towers at the base, creating a continuous promenade and an immense interior lobby atrium. The horizontal Skypark spans the three towers and cantilevers dramatically to the north. If the architecture of casino hotels is a typology reliant on superlative dimensions and architectural spectacle, the Marina Bay Sands provides those conditions in abundance.

The four-year construction schedule of the Marina Bay Sands set its own records for unprecedented engineering and construction feats. While the construction schedule was longer than that of the Empire State Building, the ambition and technical complexity offer some useful parallels. A comparison of the two sites highlights key differences. The site of the Empire State Building occupies half of a midtown Manhattan block between 33rd and 34th streets. It has a site area of 79,000 square feet (7,340sm) and is bounded by city streets on three sides. The dense urban context meant there was no room for the staging of materials or equipment, and the logistics of deliveries were a significant challenge. The site of the Marina Bay Sands complex was a parcel of land reclaimed from the bay. Construction on the 1.6-million-square-foot site (154,000sm), or 20 times the size of the Empire State Building site, required its own set of complex logistics for material sourcing and delivery, but the geotechnical conditions introduced one of the first major challenges to the project.

The ground conditions under the three hotel towers consisted of marine clay and silt, which are unsuitable for bearing capacity and are difficult to excavate. Foundations for the towers were driven down to the bedrock. A system of circular sheet pile foundations was installed to hold back the hydrostatic pressure of the saturated soil, allowing for the excavation of the site. Additionally, portions of the project employed a "top-down" construction approach, which allowed foundations and ground slab construction to precede the excavation of basements. Once the ground floor slab was completed, construction proceeded upward and downward simultaneously, saving time on the

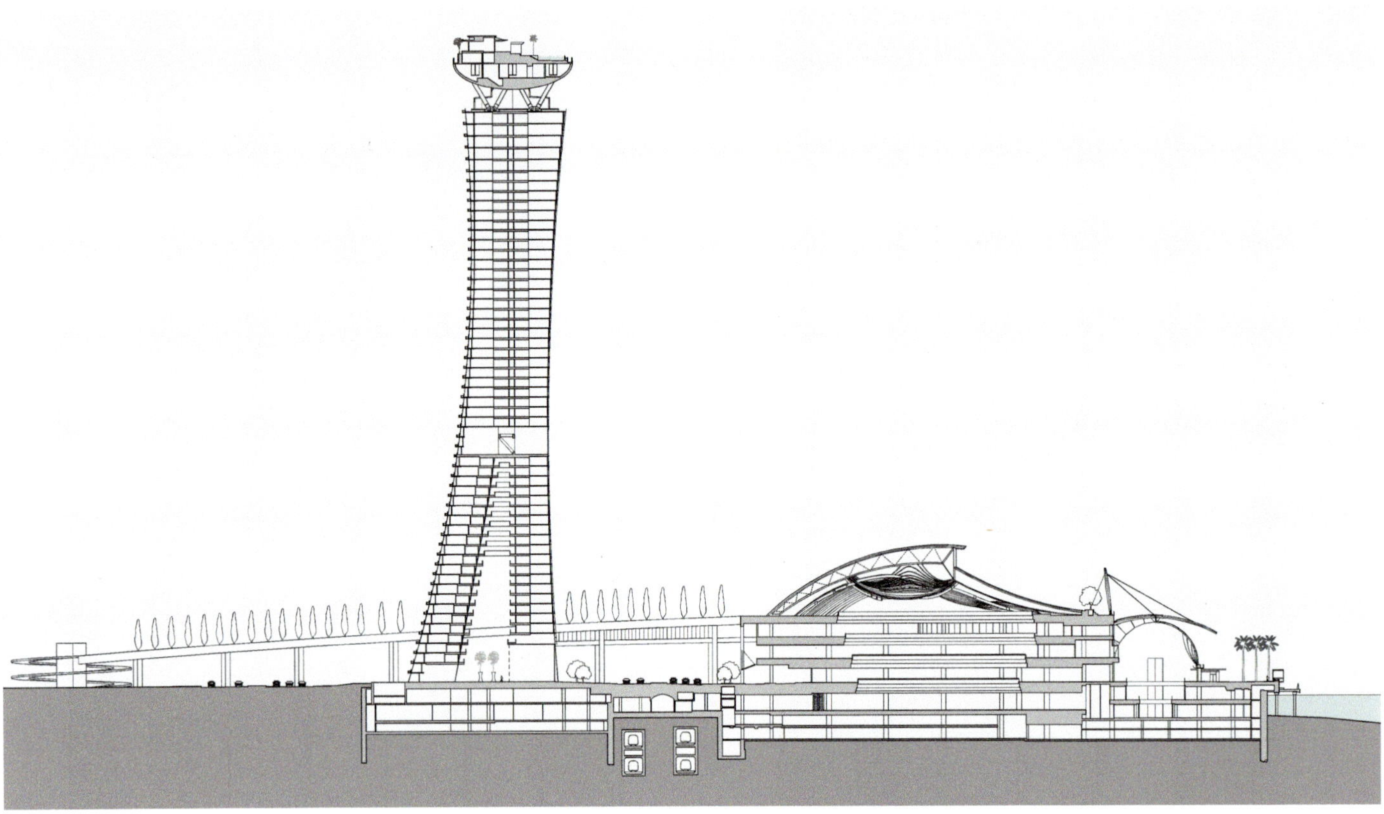

1|2 A site section through the Marina Bay Sands (2010), showing the subway line running through its center, the splayed atrium hotel, and the casino and waterfront promenade.

construction schedule.(9) Subsurface construction typically involves the most risk, given that geotechnical conditions are largely unknown. Mathew Pryor, the Sands Corporation's vice president for construction in Asia, described the foundation work as unprecedented, with 40 percent of the project below grade, miles of concrete foundation walls, and continuous pumping to keep the basement dry. The complexity of the subgrade conditions consumed an entire year of the extremely tight schedule.

Much like the Empire State Building, the accelerated construction schedule was motivated by immense financial pressure. The initial construction schedule for the Marina Bay Sands was planned to be three years. The profits derived from the casino revenue were projected at $100 million per month, making a delay of even one month extremely costly. As a result, the contractors from Ssangyong Engineering and Construction developed an accelerated construction schedule, including three shifts of around-the-clock labor and 16,500 construction workers, or nearly five times that of the Empire State Building.

Sheldon Adelson, the owner of the Sands, describes construction as analogous to "a Broadway play."(10) The sequence was carefully orchestrated so the Marina Bay Sands could meet the accelerated schedule. Unique construction management techniques used on the Marina Bay Sands included the use of top-down construction for accelerating below-grade construction, design assist contracts for technically challenging components of the design, and a global procurement strategy for materials and products.

(9) Peter Bowtell, "Touching the Impossible," in Sheldon Adelson and ORO Editions, eds., *Reaching for the Sky: The Marina Bay Sands, Singapore*, 1st ed. (Novato, CA: ORO Editions, 2013), p. 81.

(10) "Every day the people get on stage, they put on their costumes, they execute their parts — they sing, they dance, they talk. They start off at six or seven in the morning and work all day. We had three shifts a day, working twenty-four hours." Sheldon Adelson and ORO Editions, eds., *Reaching for the Sky*, p. 22.

DESIGN ASSIST

The design process for the Marina Bay Sands project was entirely concurrent with the construction phase, so rather than the typical design-bid-build, the design team developed the design with input from the construction manager throughout the process. This overlap of design and construction phases, called "design assist," meant the design team had the benefit of input from engineers as well as subcontractors who advised on constructability issues.

In proposing the asymmetrical A-frame hotel section with the curved wing, the designers could consult builders on methods of implementation, sequencing, and primary cost drivers. The hotel structure consists of a series of concrete shear walls running perpendicular to the broad faces of the hotel. In constructing the two wings independently, one vertical and one curving, the builders relied on a

1|3 The dramatic aerial assembly of the cantilevered prow of the Marina Bay Sands Skypark.

temporary propping structure to support the leaning portion until it could meet the vertical bar. Hotel Tower 1 had the most acute sloped wing, with 22 stories of inclined construction before it connected to the vertical bar with a multi-story steel link truss on levels 23 and 24.

The most challenging feature of the Marina Bay Sands was the Skypark, which spans across the top of all three towers. It compounds the complexity of three independent towers with a bridge the size of an aircraft carrier. The differential movement of the towers made the structural detailing of the bridge connections complex. The erection sequence of the prefabricated parts of the Skypark also tested the hoisting capabilities of rigging contractors. The prefabricated parts of the Skypark were built by shipbuilders in segments and brought to the site in large pieces. The parts were test-fitted on the ground before they were hoisted up to be assembled in the air at 670 feet (205m). 1|3 The cantilevered end of the deck was engineered with two massive 10-foot (3m) box beams joining at the prow. It was constructed of 14 sections with a hoisting rig explicitly designed for bridge construction, which could slide along rails to the end of the cantilever and lift a segment into position, fasten it, and then slide further out to hoist the next segment.

11 Paul Starrett, *Changing the Skyline: an Autobiography* (New York, NY: McGraw-Hill, 1938), p. 296.

DESIGN FOR CONSTRUCTION

On the Empire State Building, Paul Starrett wrote in his memoir, "Never before in the history of building had there been, and probably never again will there be an architectural design so magnificently adapted to the speed in construction."[11] In short, The Empire State Building was designed for construction. By comparison, the Marina Bay Sands was designed by Moshe Safdie through a competition process that sought to produce an iconic architecture through the design of a new landmark structure. The client for the Marina Bay Sands was the government of Singapore, which offered the development rights to the site for a fixed price, to the proponent with the most dramatic design. With the land price fixed, the differentiator was the design, producing a situation where the competition strove for spectacle while concerns for efficiency and constructability were effectively deferred.

Comparing the Empire State Building and Marina Bay Sands — two extreme structures, each pathbreaking in their time — reveals how a project can benchmark design and construction practices. Both were representative of the technological capacities of their respective eras. The Empire State Building illustrates industrial innovation as well as the

14 Frank O. Gehry's concept sketches for the Fish Pavilion (1992), in Barcelona. The cross weaving of the surface is visible in the lower right sketch.

15 The completed Fish Pavilion is one of the first architectural projects designed and constructed with the use of CATIA software.

emergence of a new building typology, the setback skyscraper. The Marina Bay Sands exemplifies and architecture of spectacle in the age of globalization, tourism, finance, and 21st-century international architectural practice. Their differences in design approaches, documentation techniques, attitudes toward standardization, material sources, and the labor force needed to erect the two structures at record-breaking speeds, speak volumes about the state of design and construction in the early decades of the 20th and 21st centuries.

SOFTWARE AND HARDWARE

Taking stock of the current state of design and construction requires us to survey industry standards and practices with respect to their time and place. What are the hallmarks, indices, and instruments that shape this particular moment? And, how are those instruments applied around the globe in different contexts and cultures? Tracking the development of design and construction practices today, few architects have shaped both sides of the equation as significantly as Frank O. Gehry. From his early experiments with his own house in Santa Monica (1978) to the billowing glass curves of the Louis Vuitton Foundation (2024) in Paris, Gehry has pushed architecture toward a new paradigm both in its design conception and in its realized construction. In design, his early adoption of computer software to achieve compound curved sculptural forms, like his iconic Fish Pavilion in Barcelona (1992), represents a sustained ambition to achieve increasingly greater complexity and curvilinearity. 14

As a document of the early days of digital design, the book *Archeology of the Digital*, catalogs a series of Gehry studies for the unbuilt Lewis Residence (designed between 1989 and 1995) in various mediums that capture a mode of working between physical models that include balsa wood and felt fabric, freehand sketches, and wireframe digital model views.[12] The catalog looks back at "the digital" as a distinct historical moment in architecture, when experimental projects employed digital tools to achieve "otherwise inaccessible results." The book's editor, Greg Lynn, argues that architects working at the threshold of the digital, including Gehry and Peter Eisenman, utilized the tools of their time to achieve unprecedented outcomes that had previously only been approximated by analog means. Lynn notes the translation of "procedural formal experiments" in Eisenman's work that began with 2D hand drafting and shifted to the computer,

12 Peter Eisenman and Greg Lynn, *Archaeology of the Digital: Project Files, Peter Eisenman, Frank Gehry, Chuck Hoberman, Shoei Yoh* (Montreal: Canadian Centre for Architecture, Berlin: Sternberg Press, 2013), pp. 145–211.

13 Peter Eisenman and Greg Lynn, *Archaeology of the Digital*, p. 12.

14 In his biography of Frank Gehry, architecture critic Paul Goldberger dedicates a chapter to Gehry's obsession with fish shapes. "The double curves of the fish form fascinated him. So did its delicate skeleton, light and graceful in itself, and the way its form suggested movement. Even the scales, which suggest that the fish is not quite so pure an object as it first appears,

[16] Construction photo of the Experience Music Project (2000), showing the makeup of the exterior wall assembly, including intermediary steel framing and metal panel cladding.

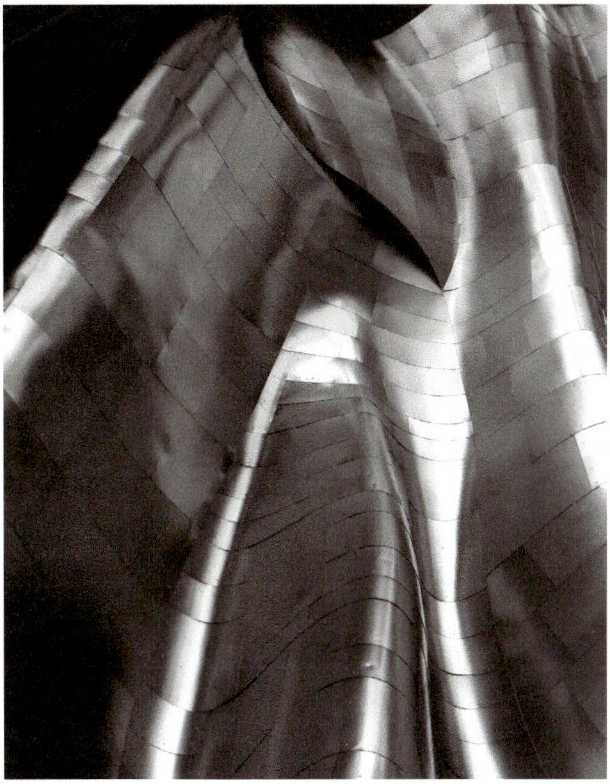

[17] The lapped panels of Gehry's Experience Music Project (2000) show the irregular tiling pattern of the cold formed panels and the exposed fasteners.

arguing that Eisenman's "rule-based" approach that preceded his use of digital tools made for a smooth transition.[13]

Lynn's account of Gehry's design methodology begins with the analog processes of surface manipulation in physical models. Gehry's use of CATIA (Computer Aided Three-dimensional Interactive Application) software, borrowed from the aerospace industry, for his unique design process pioneered modeling software capable of controlling complex and curved geometries. The Fish Pavilion on the Barcelona waterfront, is one of Gehry's many efforts to translate the supple forms of a fish into an architectural form.[14] [15] In this instance, the 3D model was used to generate the data points for constructing the pavilion's complex steel space frame. The woven stainless steel surface clad over the frame absorbs the curved geometries using the gaps between the strips for tolerance.

In developing solutions for curved building surfaces, Gehry developed metal panel cladding systems using many small units made of thin, ductile material. Gehry experimented with a variety of different metal sheet materials and systems for projects, including the Guggenheim Bilbao (1997), the Experience Music Project (EMP) (2000) in Seattle [18], which employed titanium sheets, and the Walt Disney Concert Hall (2003) in Los Angeles, which used stainless steel. In each case, the buildings consist of multiple intersecting volumes with distinctive cladding that often resemble scales. Gehry partnered with Kansas City-based metal fabricator Zahner for many of his early US projects to develop a proprietary system of intermediate steel supports between the base building structure and the metal shingled building envelope.[15] [16] [17]

The Stata Center at MIT (2004) bears many of Gehry's signature formal motifs. The 430,000-square-foot (40,000sm) building houses the departments of Electrical Engineering, Computer Science, and the Computer Science and Artificial Intelligence Laboratory (CSAIL) on MIT's main campus. In typical Gehry fashion, the building massing was initially developed as a series of stacked program blocks and evolved into an ensemble of jostling sculptural forms with different cladding materials. The exterior facades of the Stata Center employ a 3D collage technique of colliding and conjoining volumes and skins. [19] [20] The ensemble of massing and cladding allowed Gehry to mix into the composition a number of orange brick-clad volumes that gesture to the local material palette of Cambridge, enabling a kind of fragmentary contextualism. The strategy of maximum difference allowed Gehry to combine brick with

appealed to him as a reminder that such a sensuous shape can be made up of many little parts that come together, rather like his architecture." Paul Goldberger, *Building Art: The Life and Work of Frank Gehry*, First Vintage Books edition (New York, NY: Vintage Books, 2017), p. 217.

[15] Zahner worked closely with Gehry and developed full scale mockups and prototypes to test the limits of cold forming metal sheets in various configurations of lapped panels in rainscreen applications.

The dramatic curved steel ribs of the Experience Music Project (2000) under construction.

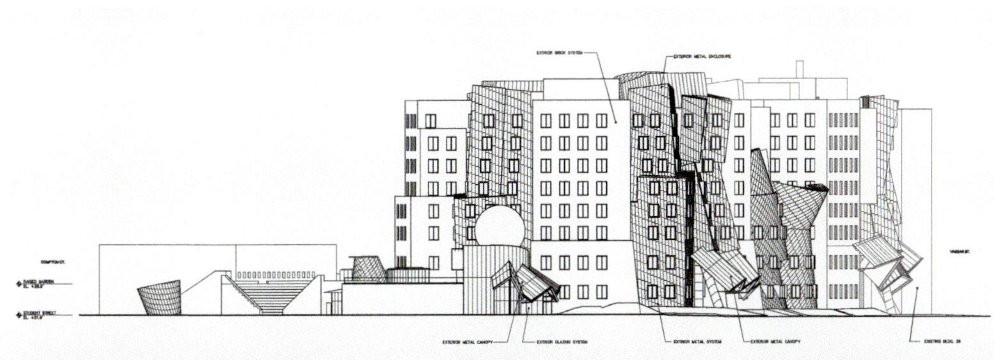

19 The "controlled chaos" of the Stata Center (2004) massing reveals the combination of normative brick-clad volumes and exceptional metal panel volumes.
20 An early Stata Center concept model shows the composition of program blocks and "accent" figures.

polished stainless steel, titanium, and bright yellow painted panels. 23

Gehry's design process includes a recursive translation between physical and digital models. The articulation of the building massing renders it a composition of chunky volumetric parts, each with its own nickname.[16] The unique expression of different component parts makes the building emphatically heterogeneous. Gehry describes the composition as a kind of "controlled chaos," which he associates with pluralism. "Democracy is pluralism, the collision of ideas. Our cities are built on a collision of thought. Look out there. There is a building by Pei, there is a bridge, there is a huge hunk in the distance. If it wasn't for democracy, it would all look like one thing. Stata represents an idea, which I think is where we are in life now."[17]

The Stata Center is an explicit case of a design process that demanded more capability from digital design tools as well as from the construction process. Translating 3D models into construction drawings and insisting that the builders also use digital models to execute the Stata Center essentially forced the contractor and all their key subcontractors to upgrade their digital capabilities. 21 22 The construction manager, Skanska, and the architect of record, Cannon, were pushed into 3D design and documentation platforms at a time when most buildings were still documented in 2D CAD drafting programs, and most builders were still working off physical paper plots. Gehry describes the technology "push" that the Stata Center produced: "We push the trades. Skanska and Cannon will go out and use CATIA from now on. They got trained here. Some of the trades are open-minded and view this as a real chance to go to the next level. There will be tremendous change in the construction industry and we are way ahead of it."[18]

The curtain wall subcontractor, Karas & Karas, was required to upgrade its digital modeling and fabrication capabilities to construct the building envelope. Dave Argus, the curtain wall foreman, said, "I don't know where I can go after this project. Where do you go after you've been to the moon? I can't see myself ever going back to another shoebox."[19]

SURVEY SAYS

One overlooked aspect of pioneering, digitally-designed architecture and its construction is the importance of digital surveying equipment at the site. Sophisticated 3D modeling software can produce formal and geometrical complexity with seemingly few constraints. Computer-controlled

16 Gehry's sculptural forms often acquire nicknames like "horse head," "giraffe," "kiva," "buddha," or "evil twins."

17 Gehry quoted in Nancy Joyce and Frank O. Gehry, *Building Stata: The Design and Construction of Frank O. Gehry's Stata Center at MIT* (Cambridge, MA: MIT Press, 2004), p. xiii.

18 Nancy Joyce and Frank O. Gehry, *Building Stata*, p. xv.

19 Nancy Joyce and Frank O. Gehry, *Building Stata*, p. 121.

2|1 A screen shot of the CATIA model showing the tiling of the metal panel units and the projecting windows.

2|2 A photo of the completed Stata Center with the staggered metal panel and projecting windows.

CNC machines and multi-axis robots can fabricate highly precise components using tool paths and cut files created by 3D software. The installation and erection of the individually fabricated parts at the construction site is the last step of the process, but it is also where all the precision in the digital world has to make the leap into the physical world of material and assembly. Construction relies on surveying techniques for erecting walls that are plumb and true. With non-standard parts, assembly and installation require advanced methods to ensure the accurate location of elements on site.

The digital surveying equipment for the Stata Center used Total Station, a system that relies on the digital 3D model and a network of distributed laser prisms to locate the exact positions of any point on the site.[20] Prisms calculate distances by measuring how long it takes for a laser to travel to the prism and back, thereby determining positions in space relative to the 3D model. Contemporary laser scanning techniques produce point clouds that can be converted to 3D Revit models, further revolutionizing architecture design and construction practices. While many aspects of construction remain stubbornly resistant to change, the proliferation of digital tools and techniques have been adopted by fabricators and builders to enable increased precision and unprecedented complexity, as well as to improve accuracy and efficiency on the job site.

20 The use of 3D surveying equipment is described in Nancy Joyce's book, *Building Stata*. Nancy Joyce and Frank O. Gehry, *Building Stata*, p. 94.

Reviewing Gehry's architecture over the last few decades, it is clear that his sustained insistence on non-planar and non-standard geometries has applied significant pressure on both the software and construction industries while increased computing power and the emergence of digital fabrication techniques has coincided with the demand for more sophisticated architectural applications. At the threshold of the digital age, design investigations — and investments — translated complex geometries from conceptual models into built realities, resulting in a profound paradigm shift in both architecture and the construction industry.

Undoubtedly, the impact of different technologies on design culture and practice at any moment in time dramatically shapes both possibility and capacity. Consider the impact of previous tools like the drafting compass, the T-square, or the mouse. Critically, Gehry's pioneering work in the early 2000s thrust design software platforms into the construction site, changing how contemporary buildings, regardless of formal language, were assembled, surveyed, and built.

2|3 Completed in 2004, the Stata Center at MIT was pioneering in its use of digital tools, including CATIA software, for both design and construction. The design team, led by Gehry Partners, used the software for the design phases, and the contractor, Skanska USA, insisted that all the major subcontractors also adopt to the same software platform.

24 One of many physical models built during the design of the Stata Center. Gehry's process still relies on the feedback loops between physical and digital models.

25 The completed Stata Center (2004) as a sprawling composition of regular and irregular building volumes.

AFTER SPECTACLE

Safdie's Marina Bay Sands megaproject, and most of Gehry's signature forms, exemplified by the Stata Center, ushered in an era that could be understood as the architecture of spectacle for an age of digitally enabled design.[21] If spectacle and expressive form were the hallmarks of the early 2000s, what are the indices and tendencies that are representative of design today? Looking west from the Stata Center along Vassar Street, MIT's two most recent buildings line the north side of Vassar Street. The new home to the MIT School of Architecture and Planning (under construction at the time of writing), and the Schwarzman College of Computing (2024), are both evidence of a contemporary approach to design and construction that might offer some insights into the current state architecture, and more broadly, cultural attitudes toward architecture at the first quarter of the new millennium.

[21] Art historian Hal Foster describes the architecture of the turn of the millennium as exemplified by the Guggenheim Bilbao, as the architecture of spectacle. Hal Foster, "Why all the Hoopla?" in the *London Review of Books*, Volume 23, No. 16–23, August 2001.

MIT's campus context, between the Charles River and the Grand Junction rail line, was historically occupied by warehouses and factories arrayed along a rail spur that cuts through Cambridge parallel to Vassar Street. The new School of Architecture and Planning is an adaptive reuse project at the site of the former Metropolitan Warehouse Building. 31 The architects, Diller Scofidio + Renfro (DS+R), have carved the original structure and inserted several glazed volumes into the structure's monumental masonry shell to bring in natural light through north-facing skylit atriums.

The new Schwarzman College of Computing, designed by Skidmore Owings and Merrill (SOM), also reacts to the solidity of the masonry context by contrast, offering an articulated all-glass box. 1 Immediately across the street from the Stata Center, the Schwarzman College of Computing aligns with its urban street wall on the north side of Vassar Street but stands in architectural contrast with its context. It is flanked by MIT's cogeneration plant to the west and the stone-clad Brain and Cognitive Sciences Building to the east. While many of the surrounding buildings employ stone and punched windows, or window walls, the178,000-square-foot (16,500sm) Schwarzman building is entirely clad in a glass curtain wall. Its compact eight-story volume fills its site with a modest setback along Vassar Street. Its orthogonal volume is articulated to create a recess along Vassar Street and divides the building into two projecting wings, each clad in a shingled glass curtain wall. The shingled glass facade is lapped in both plan and section, giving the south-facing elevation a distinctive, taut glass texture. 26

26 The shingled curtain wall facade of the Schwarzman College of Computing by SOM.

27 The overhanging curtain wall panels of the Schwarzman College of Computing with the Stata Center in the background

Observing pedestrians on Vassar Street, or the occasional student zipping by on an electric scooter, the subtle out-of-plane rotations of the glass modules overhead go largely unnoticed. Each curtain wall module projects forward of adjacent panels by about 8 inches (200mm), creating a staggered pattern of recesses, reflections, and shadows. The large, clear glass panes extend from floor to floor with no spandrel panel, and the overhanging modules extend by the full width of a panel, celebrating the large-scale modularity of the facade design. 27

The layering of the south facade, visible in the multiple reflections and planes of different degrees of opacity, reveals that it is a closed cavity double facade, the first of its kind in North America.22 Behind the sloped layer of exterior glazing is an inner layer of triple-glazed insulated glass that defines the building enclosure. The outer layer of laminated glass traps a pressurized air chamber in a cavity of varying dimensions.23 28 29 The cavity encloses an automated Venetian blind to reduce solar heat gain on the south-facing portion of the building envelope. The curtain wall manufacturer, Permasteelisa, notes that the sunshades and air pressure are controlled by the Building Management System (BMS). In the past, the building envelope and the building conditioning systems were installed by different trades and functioned independently. Today, the integration of facade systems and active shading are increasingly coordinated by the BMS.

22 The Manulife Building in South Boston, from 2009, also by SOM, also by Permasteelisa, is a ventilated cavity double facade.

23 For more on the use of air as a building and insulating material refer to Chapter 8, "Climate and Conditioning."

24 MIT buildings are numbered sequentially, and have traditionally been known by their number. Only recent buildings have names, like Schwarzman, and they essentially have two names.

25 For more on the adaptive reuse of existing buildings refer to Chapter 9, "Afterlives."

The Schwarzman College of Computing, also known as Building 45, is MIT's latest addition to the expanding campus.24 Completed In 2024, it embodies a set of academic goals and institutional values while employing state-of-the-art facade and environmental control systems. Designed to house spaces for computer science and artificial intelligence, it showcases research, as well as signifies, through its abundant use of glass, a transparency, and openness to the surrounding community and the world beyond.

PLUS MINUS

Less than a half-mile down Vassar Street, the adaptive reuse of the Met Warehouse provides an alternate strategy for contemporary design and construction. Choosing to renovate and reuse the historic masonry building introduced significant challenges.25 As a historically listed building, the warehouse facades along Vassar Street are protected, while the north facades along the railroad

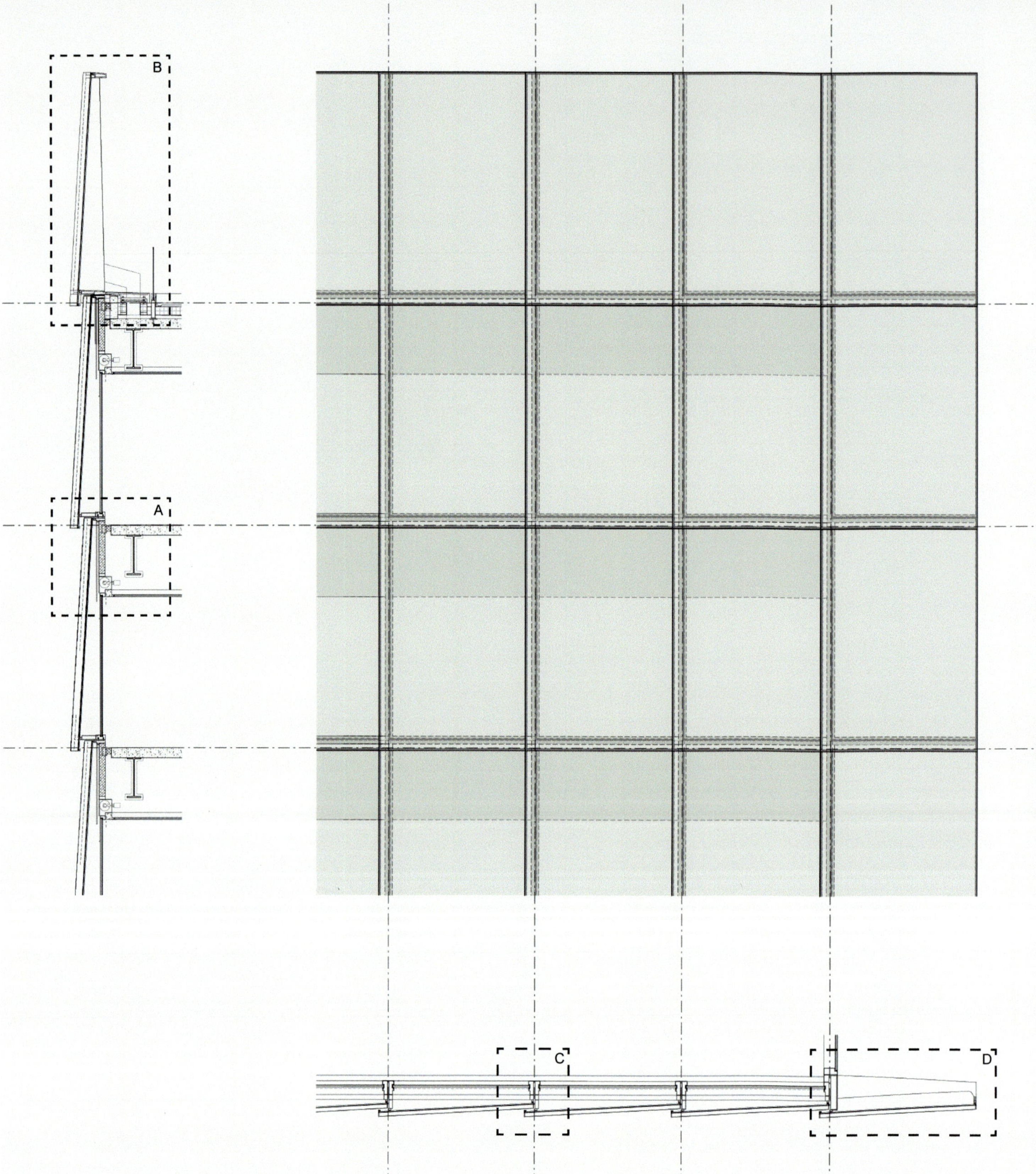

2|8 Wall type drawing of Schwarzman College of Computing, shingled curtain wall facade.

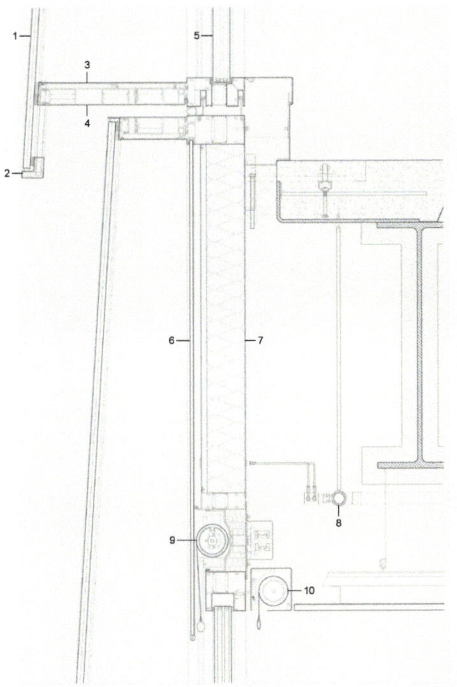

29 **Detail section of the curtain wall stack joint and shingle lap condition.**

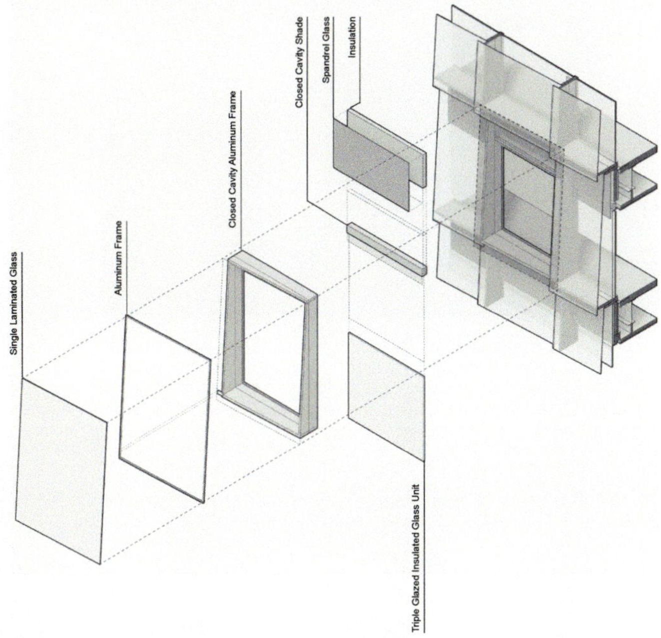

30 **Exploded axonometric diagram of closed cavity double facade.**

track are deemed less significant. The design proposal by DS+R excavates the center of the buildings to insert a series of glass-enclosed spaces that create a series of north-facing daylit atria along the 500-foot (150m) building. [32]

The original Met Warehouse consists of a series of connected brick buildings built in phases between 1894 and 1911. Designed for fireproof storage, the massive masonry walls and small windows made it challenging to adapt. DS+R's "plus minus" approach proposes a series of surgical subtractions, coupled with architectural insertions, to accommodate the 217,000 square feet (20,100sm) of studios, offices, gathering and fabrication spaces, transforming it into a "modern hub for interdisciplinary design."[26]

The new Met Warehouse will house the MIT School of Architecture and Planning, as well as the Department of Urban Studies and Planning (DUSP), The Center for Real Estate, and the MIT Morningside Academy for Design. The building will bring these departments together for the first time, and the segmented and sequential arrangement of the Met Warehouse building, punctuated by the skylit atria, creates a linear armature organized by the programmatic needs of the different academic neighborhoods.

26 See https://betterworld.mit.edu/met-warehouse/

The specific challenges for the design and construction of the new Met Warehouse are mainly due to the particularities of the existing building — the lack of robust foundations in the oldest of the original buildings required complex and expensive foundation underpinning. The thick masonry walls with small windows prompted the architects to add openings by saw-cutting the brick facades. The "intersect windows" highlight the misalignment of the original and the new windows, signaling new occupancy while maintaining historic fenestration patterns. [34] The original load-bearing masonry walls with punched windows rely on the brick jack arches to transfer loads around the opening. By adding the new windows, flush with the face of the brick, the steel frames must contend with load transfers around both the old and new openings. DS+R principal Ben Gilmartin, notes that the architecture school program utilizes the building as a teaching tool and points to the windows at a moment where the building is enlisted to raise architectural questions about material and structural modifications, as well as contemporary attitudes about embodied energy.

"The Met Warehouse project represents a commitment to preserving, conserving and adapting a landmark structure, but it is also a kind of manifesto about the future," Gilmartin states. "The idea of

31 Rendering of the Met Warehouse (2025) by Diller Scofidio + Renfro, showing the insertion of north-facing glass atriums in the masonry storage building.

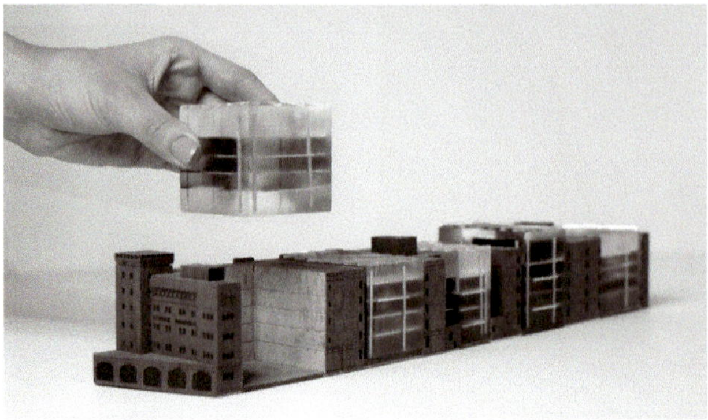

32 Physical model of the "plus minus" concept for the adaptive reuse of the historic masonry Met Warehouse building.

erasing history in favor of creating wholly new spatial, material and expressive ideas, underlying many modernist schools of architecture, is not where the future of design lies. In this space for teaching, we wanted to present a broader panorama of history through a succession of techniques that cut through layers of the past. Our approach is not one of pure reverence, but playfully reveals unexpected hybrid spaces, structural interdependencies, and intimate moments of rapport between historic and contemporary. The Met argues for the imaginative reinvention of our physical past as the new project of the future."[27]

27 Conversation with the author, July 20, 2024.

The insertion of the glass atria along the north facade punctuates the elevation with a series of glass volumes that carry daylight to the deep floor plates, but also contrasts the existing and the new. The original buildings were cellular with tight column grids and low floor-to-floor heights, whereas the new atria suspended from above, are column-free spaces with high floor-to-floor heights. The projection of the glass enclosure beyond the original building line allows the newly exposed cross sections cut through the original masonry walls to be encapsulated by the new glass enclosure, protecting it, but also framing and showcasing the wall section of the historic structure. 33

The saw-cutting strategy of cutting and drilling through walls to expose the wall thickness recalls the work of 1970s conceptual artist Gordon Matta-Clark. For example, Matta-Clark's iconic Splitting (1974) cut a vacant house through its middle, transforming an unexceptional home with a deliberate, and conspicuous intervention. In DS+R's cuts, however, the existing "host" structure is carefully reinforced to accept surgical incisions, and, ultimately, human occupancy. This detail is visible at the "intersect windows" along Vassar Street and the larger atrium insertions on the north side. 34 Cutting strategies, in this case, celebrate the misalignment of the Met Warehouse's original purpose with its current use while highlighting the significant effort to retain the historic structure and reanimate it with new functions. As a school of architecture, the transformed Met Warehouse offers didactic moments that speak to the complex conditions of the present.

Two decades separate the Stata Center from the Schwarzman and The Met. During that time the pioneering practices of 3D modeling, fabrication and coordination became common practice, allowing the two recent projects to be the beneficiaries of digital revolution provoked by the former. The different design approaches between the three offer a study of contrasts. The surgical cuts of the Met and the

33 Construction photo of the Met Warehouse building conversion from a fireproof storage facility into a new school of architecture.

34 The "intersect window" of the Met Warehouse project, where a new pattern of saw cut windows are overlaid on the historic windows of the masonry structure, resulting in a didactic detail of old and new.

taut precision of the Schwarzman are the antithesis of the Stata Center's cascading volumetric expressionism, yet both are indebted to the pioneering efforts of the older neighbor down the street.

DIGITAL WORKFLOWS

Different in scale and scope, but proximate in location, the Collier Memorial at MIT (2015) is a revealing case study in digital design workflows, fabrication, and construction. Designed by architect J. Meejin Yoon of Höweler + Yoon in the wake of the 2013 Boston Marathon bombings, the memorial occupies the site where Officer Sean Collier was killed, at the intersection of Vassar Street and Main Street, directly in front of the Stata Center. 35

The memorial consists of a solid stone, asymmetrical five-way arch supported by buttressing walls. The concept was to create a space conducive to contemplation at the busy intersection. Yoon describes the memorial as producing a "conspicuous absence," with the curved profiles of the buttress walls describing the edges of the implied central ovoid figure. 39 Constructing the initial concept of a voided center connecting five "outdoor rooms" with a vaulted stone structure, required the translation of a geometrical proposal into a material one.

The conceptual studies were modeled using a 3D modeling "Boolean difference" operation to intersect and subtract volumes and produce a hollowed-out center. Yoon's proposal to realize the memorial's vault-like form in solid stone required the collaboration of engineers and fabricators to develop a preliminary concept into a constructable design. The structural principles of arches and vaults are well understood, where a keystone-shaped stone block is supported by adjacent stones, bringing loads to the ground via compression load transfer. Yoon, a faculty member at MIT at the time, enlisted the help of colleague and structural engineer John Ochsendorf. The final form of the Collier Memorial was refined to function as a load bearing, all-compression, stone vault. Ochsendorf describes the design as five half-arches leaning toward the center, preventing each other from falling over. 40

The structural system of a compression arch supports the architectural concept of the missing figure and the symbolism of a community brought together in mutual support. Yoon notes, "The Collier Memorial is comprised of 33 geometrically unique stone blocks that fit together to produce structural stability and equilibrium." Yoon's design translated the phrase "Boston Strong," a slogan that emerged in the wake of the bombings, to turn the concept of

3|5 The Sean Collier Memorial at MIT (2015), designed by J. Meejin Yoon of Höweler + Yoon, creates a "conspicuous absence" at the center of a vaulted stone structure.

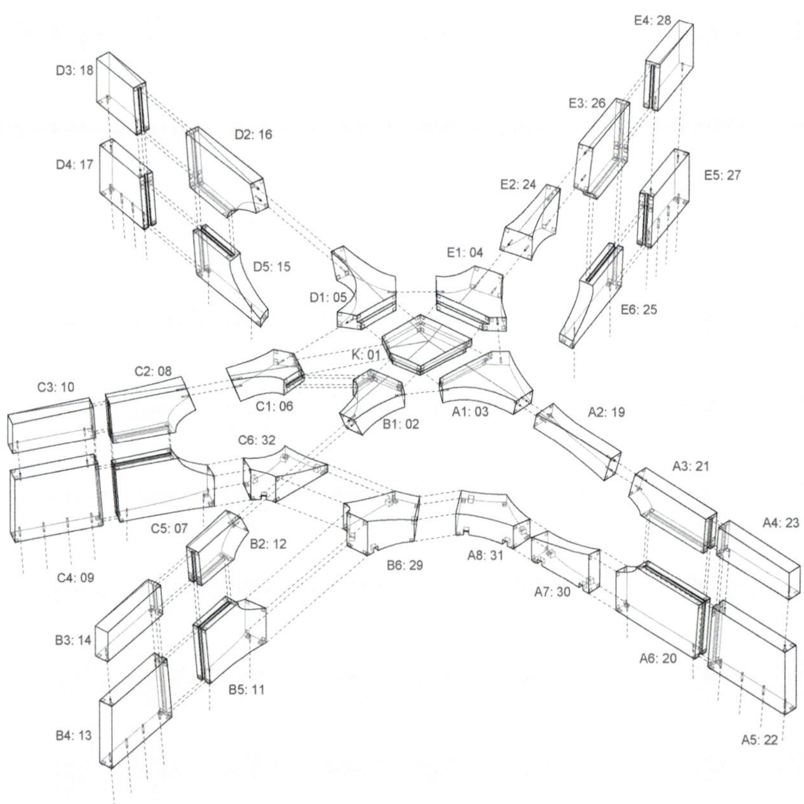

3|6 An exploded axonometric diagram of the stone blocks that make up the Sean Collier Memorial at MIT (2015).

37 CNC milling of stone blocks for Sean Collier Memorial.

38 Installation of stone blocks for Sean Collier Memorial at MIT (2015). The keystone block, labeled "K1" is visible in the lower right.

strength into a symbol of diversity and community. The arch became a symbol for a community and an institution, reeling from the trauma of domestic terrorism.

Quarra Stone was hired to provide design assist services during the design phases and give the design team input on parameters for digital stone fabrication. Stone fabricators employed an array of different types of cutting, milling, and finishing. They used computer numerical control (CNC) saws and multi-axis robots to mill the 33 unique blocks. [36] Though using computer-controlled tools, the fabrication of stone blocks still required careful calibration and quality control measures to ensure that the translation from digital model to material artifacts maintained the dimensional and geometrical precision needed to ensure precise on-site assembly.[28] [37]

The construction sequence of the Collier Memorial was developed with the construction managers, the masons, and the stone installers, who devised an erection plan that began with the five-sided keystone. [38] The installers built the Collier Memorial from the center out, from the suspended keystone to the buttresses meeting the ground. Once all the stones were in place and verified by the surveyors, the scaffolding was lowered, and the stones could transfer their loads from block to block as designed.

[28] I have written about the topics of digital fabrication, precision, and verification in "Verify in Field: Models and Other Useful Fictions," *Log 50*, Model Behavior, ed. Cynthia Davidson, 2020.

Approaching the Collier Memorial from Vassar Street, two of its stone buttresses splay out to form an embracing gesture. The shallow arch of the vault is approximately 8 feet (2.4m) from the ground, providing an intimate shelter. The missing figure at the center is the primary visual reading of the memorial, and standing under the arch provides an intimate experience of resonating sound that gives the monument a unique acoustic quality. The gray Virginia Mist granite has been flamed on all surfaces, with the exception of the Boolean faces of the ovoid, which have a high-honed finish, providing a contrast in color and texture between the rough exterior surfaces and the darker and smoother interior faces.

While a deeply meaningful public monument, the Collier Memorial also provides an illuminating case study for digital design workflows that have transformed the contemporary design process. The project raises questions about both the authority of the digital model and the agency of the architect in controlling the flow of information. It also calls up issues of risk and liability with digital workflows and the absence of shop drawings as a quality control measure, topics that we will return to in Chapter 2, "Means and Methods."

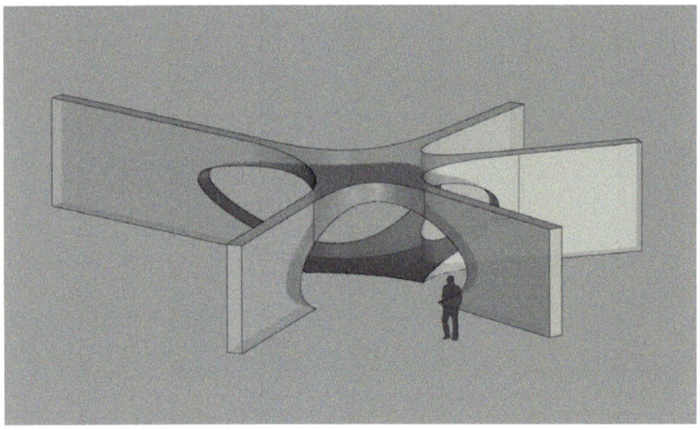

39 3D model of the Sean Collier Memorial highlighting the intersection of the hollowed out ovoid figure and the buttress walls.

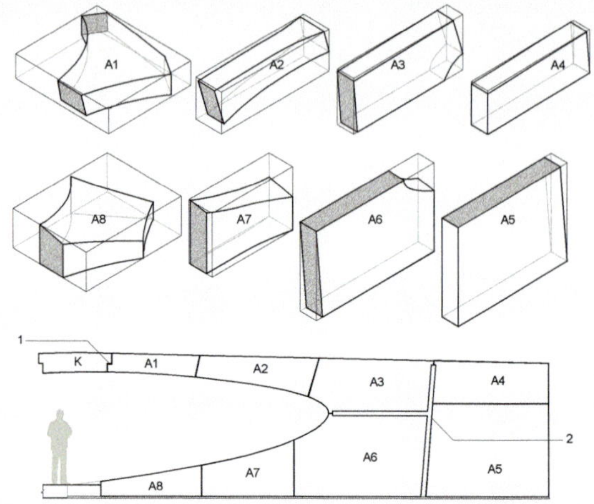

40 Axonometric drawing of stone blocks and elevation of one of five half arches that lean on each other to achieve equilibrium at the Sean Collier Memorial at MIT (2015).

Taking stock of the contemporary state of design and construction, it is clear that both processes are undergoing a transformation. As workflows evolve to turn 3D models from tools of conception into instruments of execution, we often overestimate the precision they promised. Architectural historian Francesca Hughes discusses the status of material in a moment where the digital has reset expectations for precision.[29] In her book, *The Architecture of Error*, she notes that overconfidence in the digital is complimented by an intolerance of material deviations from the digital ideal, a condition Hughes describes as the "architecture of error." For Hughes, the evolution in the instruments of measurement, design, and fabrication has produced expectations between models and material that re-categorize much of the physical world as "error." Hughes' arguments for a re-alignment of expectations and specifications for the material — with reference to the digital — are highly applicable to the design and construction processes of the Collier Memorial.

CAMPUS CROSS SECTION

The MIT campus offers a particularly rich "cross section" through architectural development over the last 150 years. Each era of MIT buildings is highly expressive of its moment of conception and construction. Reading the MIT campus buildings together highlights the sharp differences that are a function of their time and the technologies available at their moment of construction. The fact that they all share the same client provides a constant against which the differences can be read. 42

The original plan for MIT's Cambridge campus was developed by William Welles Bosworth in 1913, and serves as a useful armature and framework that accommodates growth over time. Mark Jarzombek's *Designing MIT, Bosworth's New Tech*, offers a compelling architectural history of MIT's campus plan, striking a balance between a familiar expression of neoclassical facades as designed by Beaux-Arts trained Bosworth, and a paradigm of functionalist concrete-frame structures and utilitarian planning concepts, provided by Bosworth's predecessor John Ripley Freeman.[30]

Surveying the MIT campus reveals a striking ensemble of building styles as well as construction systems. 41 The original 1915 Bosworth-designed buildings, including the Great Dome of Building 10 and the symmetrical wings that extend toward the Charles River, are reinforced concrete-frame structures dressed in neoclassical limestone

29 Francesca Hughes, *The Architecture of Error: Matter, Measure, and the Misadventures of Precision* (Cambridge, MA: The MIT Press, 2014), pp. 1–13.

30 Mark Jarzombek, *Designing MIT: Bosworth's New Tech* (Cambridge, MA: The MIT Press, 2017).

4|1 Construction photo of the Great Dome at MIT (1916), designed by William Welles Bosworth in the Neoclassical style.

4|2 Contemporary aerial photo of the MIT campus, with Baker House (1949) and Kresge Auditorium (1955) in the foreground, and the Met Warehouse visible at the corner of Massachusetts Avenue and Vassar Street in the middle left.

43 Rendering of the main group of MIT campus buildings, designed by William Welles Bosworth, with the Great Dome as the centerpiece.
44 Construction photo of the main group of MIT campus buildings (1914) revealing the utilitarian concrete frame structure under the Neoclassical facades.

facades. 43 44 Alvar Aalto's Baker House (1949) dormitory was built across Massachusetts Avenue and significantly departed from the Neoclassical style as an undulating brick facade with punched windows to the south and a set of ribbon windows facing north. Eero Saarinen added a thin shell concrete structure with the Kresge Auditorium (1955) 47 and the brick-clad MIT Chapel (1955). I.M. Pei's Green Building (1964) 45 topped out at 277 feet (84m), with the rationale that the Meteorology and Earth Sciences departments needed to mount equipment at the top of the tower to conduct research. The Green Building's concrete frame introduced the campus to a new palette of materials consisting of exposed concrete and glass, which came to be known as Brutalist.[31]

In the 1990s, MIT undertook a massive planning and building campaign under President Charles Vest, which resulted in several noteworthy buildings. In addition to the Gehry-designed Stata Center (2004), Steven Holl designed Simmons Hall dormitory (2002), Charles Correa designed the Brain and Cognitive Sciences Building (2005), and Fumihiko Maki designed the MIT Media Lab (2009) 46, as an addition IM Pei's Media Lab (1985).

More recently MIT has commissioned and constructed projects by SANAA, Michael Maltzan, NADAAA, Weiss/Manfredi, and Kieran Timberlake. The MIT campus has served as a laboratory for design and construction, showcasing design across a range of stylistic and technical paradigms. The former dean of the School of Architecture at MIT, William J. Mitchell, wrote in the introduction to his 2007 book, *Imagining MIT: Designing a Campus for the Twenty-First Century*, "When colleges and universities build, they don't just add to their inventories of floor space. They reveal — sometimes unwittingly — their prevailing values, aspirations, and preoccupations. Campuses are evolving, continuously contested representations of the communities they house."[32]

Campuses are also effective for tracking architectural evolution and transformation with a consistent institutional framework. The German word *zeitgeist* translates into the "spirit of the age." It is used to describe particular markers that seem to embody the cultural, political, and economic conditions that represent their times. Buildings are enduring markers of the passage of time. They encapsulate the ambitions and aspirations of that time. The completion of the Stata Center in 2004 marked a particular moment of optimism at MIT and in the discipline of architecture. Gehry's architecture of dynamic figures and complex

31 A history of MIT campus planning and architecture is provided by Robert Simha. O. Robert Simha, *MIT Campus Planning, 1960–2000: An Annotated Chronology* (Cambridge, MA: Massachusetts Institute of Technology, Office of the Executive Vice President, 2001), p. 31.

32 William J. Mitchell, *Imagining MIT: Designing a Campus for the Twenty-First Century* (Cambridge, MA: MIT Press, 2007), p. vi.

45 Construction photo of I.M. Pei's Green Building at MIT (1964).

46 Eero Saarinen's Kresge Auditorium and MIT Chapel (1955).
47 Fumihiko Maki's addition to the MIT Media Lab (2009). I.M. Pei's original Media Lab building is on the left.

surfaces seemed to embody that spirit. The formal expressionism and exuberance signaled the new, and innovative. Stata was state of the art, for the moment. In the early 2000s, stars aligned for the Stata Center to be designed, funded, approved, and constructed.

The cross section through various projects at MIT reveals much about the current state of design and construction as an expression of technological paradigms and cultural attitudes. The examples highlighted here raise themes that we will revisit in more detail in subsequent chapters, including the design processes and sequences, questions of legibility of material systems, and the emergence of sensibilities about energy and environment. They also represent an expanded understanding of architecture and construction as ecosystems of materials, labor, and technique that simultaneously frame how we view and shape the world through design.

1 The inflected facade of the DL1310 Building (2020), designed by Young & Ayata with Michan Architecture in Mexico City, Mexico.

Chapter 2 Means and Methods

Architects Michael Young and Kutan Ayata were not happy with the concrete mock-up. The builder for the four-story residential building in Mexico City had a produced a full-scale mock-up of the concrete bay window, but the formwork design had created awkward-looking steps in the concrete window surround. [6] The architects had designed the building with a taut massing flush with the lot line, and a series of bay windows that pushed inward instead of outward. [7] The transitions between the plane of the facade and the plane of the glass were developed as hypars, or hyperbolic paraboloids, ruled surfaces that lofted between one edge and the other. The digital 3D model created the smooth transition effortlessly, but the local builders were unsure how to construct the twisting form in concrete. The window mock-up was created with formwork built up from layers of sequential sheets of wood that had been cut and stacked to approximate the geometry of the digital model.

This technique of translating a complex geometry from a digital model into a series of sequential profiles is known by its Rhino command "contour." A selected object is intersected with an array of parallel planes at a given interval. The profiles generated by the contour command can then be cut out of planar material and stacked to form an abstract version of the digital model. This technique of laminating sequential profiles to approximate complex geometries has been typical for site models, but its use in architectural form took off with the introduction of 3D modeling software and digital fabrication tools like laser cutters and CNC milling machines.[1]

Young and Ayata met in graduate school at Princeton in the early 2000s. Designing the DL1310 Apartments, a multifamily building in Mexico City, was one of their first built commissions, and they were eager to apply their design skills to the project.[2] Observing the concrete mock-up, the architects were disappointed. The stepping of the contours — a function of the thickness of the wood formwork — was static and coarse. The smoothness they aspired to was lost in the low-resolution approximation of the ruled surface geometry. Ayata recalls that they then realized the answer might be found in the local concrete vernacular architecture and the work of Félix Candela, whose experimental concrete shells built in the 1950s and 60s gave new expression to concrete architecture. [2]

Candela's shell structures, many of which were also hypar geometries, achieved unprecedented spans and material efficiencies by combining compound curved surfaces and material systems.

[1] Laminated Object Manufacturing (LOM) is a technique of translating a 3D object into material by contouring the 3D model and cutting the profiles out of sheet material and then assembling them into a low-resolution mass.

[2] The DL1310 Apartments is a collaboration between Young & Ayata and Michan Architecture, based in Mexico.

[2] Thin-shell concrete structure Los Manantiales (1958), designed by Félix Candela, Mexico City, Mexico.

[3] The labor-intensive process of formwork and rebar for thin-shell concrete construction.

The thin-shell construction of his Los Manantiales (1958) restaurant in the Xochimilco district of Mexico City, combines four hypar saddle shapes in a radial array to produce a soaring eight-sided groin vault. [4] The structure spans 106 feet (32m) between supports, and the high point at the center is 19 feet (5.9m). The peaks and valleys of the undulating vaults produce the stiffness of a V-shaped "beam," allowing the concrete shell to be extremely thin. The leading edge of the canopy measures only 1.5 inches (4cm).

The material efficiency of the shell goes hand in hand with the labor-intensive construction of the wooden formwork. Historic photos of the construction of Los Manantiales show laborers erecting supports for narrow wooden boards that make up the ruled surfaces of the hypars. [3] Rebar was laid over the formwork, and concrete was placed over the rebar one bucket at a time. With the formwork removed, Candela's thin-shell structures of board-formed concrete appear impossibly thin and delicate, defying gravity with their sine curve geometries and smooth transitions.

For Young and Ayata, the reference to Candela's formwork technique was a breakthrough. Rather than contouring and stacking parallel sheets to approximate the ruled surface, they applied Candela's "straight line generator" to array narrow boards to define the lofted surface of the wall. The local carpenters created a new formwork, and the second mock-up produced the effect that the architects were looking for. [5] Ayata recounts the pride taken by the builders in constructing the new formwork out of wooden boards: "The imprint of the boards on the finished concrete meant that the boards couldn't be too flat, so we introduced small wedges to push the boards slightly out of plane to produce the texture of the wooden formwork."❸ He also explains the way the concrete material became the "shared medium" where both the architects' and the builders' intentions were realized.❹ [8]

The final design included 22 inverted bay windows in five different sizes to produce the "bas relief" effect of a monolithic rectilinear volume with localized inflections. Confronted with the cost of building 22 separate formworks for the windows, the designers applied another fabrication technique: the transfer mold. Local carpenters produced a unique wooden formwork for one of each window type, and then transferred the form to a reusable fiberglass mold. The relatively small scale of the window surrounds allowed the builders to reuse the mold multiple times, saving time and labor, while maintaining the integrity of the design.

❸ Conversation with the author, May 22, 2024.

❹ Conversation with the author, May 22, 2024.

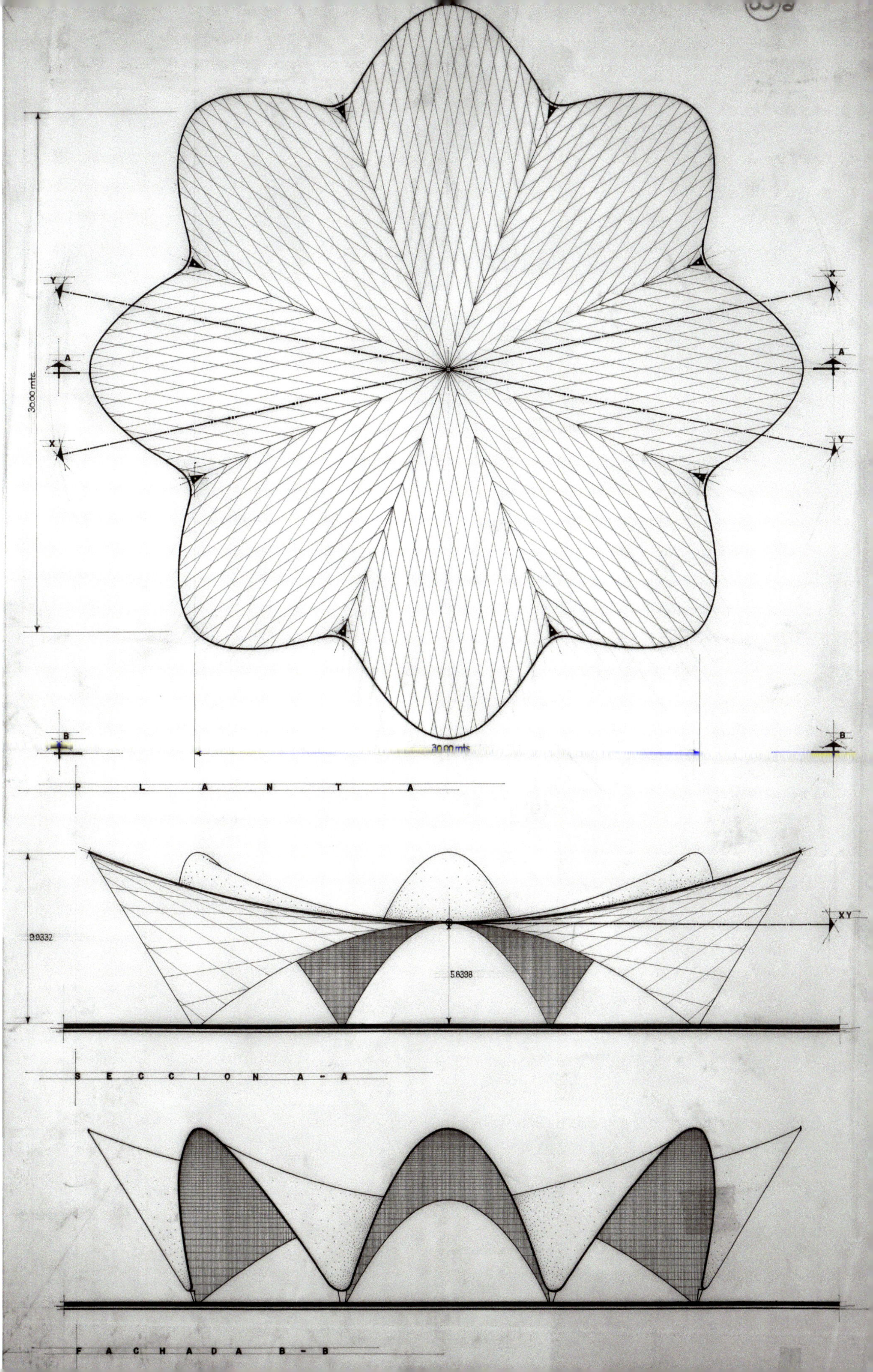

4 Plan, section, and elevation of Los Manantiales (1858) restaurant in Mexico City, Mexico.

5 Formwork for the second mockup of the window using the Félix Candela approach to the hypar geometry of the wall.

6 First mockup of the window using the "contour" approach to the hypar geometry of the wall.

The completed project includes a strikingly compact massing on its narrow lot. 1 The sweeping inflections of the inset bay windows counter the tautness of the orthogonal facades. The dark color of the glass contrasts with the light gray of the unfinished concrete, which shows the impressions of the board formwork, giving the monolithic form a fine-grain texture that clearly indexes how it was made. The DL1310 Apartments illustrate the tradeoffs between the costs of material and labor, as well as the efficiencies of reusable forms and the ingenuity of local construction practices. The project also demonstrates the non-linear translation process from conceptual digital design to physical construction. With their unique window configuration, the architects developed a site-specific approach, learning from Candela, that utilized the capacities of the local carpenters. The realization of the DL1310 Apartments project took the inertia of the original idea from concept to construction, and spanned the gap between abstract, digital geometries and material realities.

ARCHITECTURE VERB

The key to the success of the DL1310 Apartments was the architects' ability to design with implementation in mind, and to propose both material solutions and construction methods that involve various trades, including carpenters, concrete workers, and glazers. Imagining a choreography of operations in the field, incorporating the irregularities and variabilities of the wood formwork and the uncertainties of construction, the design team achieved a distinct outcome. The capacity to design within the constraints of labor, material, skill, craft, and tolerance is an example of a tectonic and material imagination.

Understanding design as a material way of thinking and as a function of a tectonic imagination, designers conceptualize architecture as a process. Architecture as a verb, as an outcome of design and labor, materials and energy, recasts architecture. No longer a static and autonomous form conceived as a pure instantiation of an idea, architecture as a verb recognizes the multiple choreographies of construction.

ORIGINS

The division of the responsibilities of an architect and those of a builder have made for an age-old divide dating back to the origins of architecture as a practice, a profession, and discipline. In the middle of the 13th century, Leon Baptista Alberti wrote:

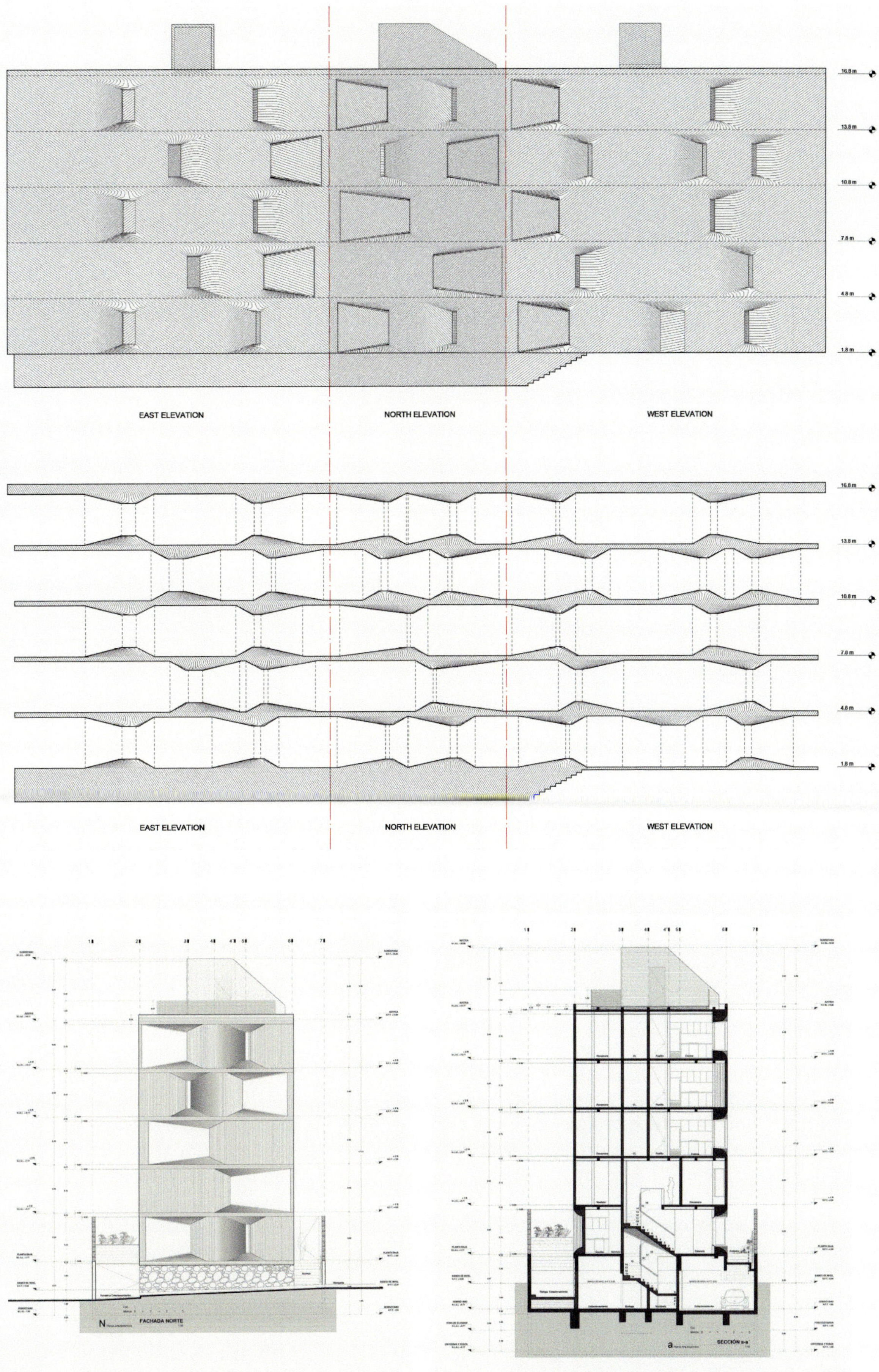

7 Section, elevation, and unrolled elevations of DL1310 (2022), in Mexico City, Mexico.

8 **Frontal view of the oblique window showing the minimized glass and the pattern of board-formed concrete.**

9 **Oblique view of the oblique window showing the full extent of rectilinear glazing and the hypar geometry of the head and sill conditions.**

"The builder is but a hammer in the hands of the architect," articulating a specific understanding and division of labor, architectural historian Mark Jarzombek in his book, *Architecture Constructed, Notes on a Discipline*, writes about how the articulation of architecture as a discipline was formed to differentiate itself as distinct from that of the builder.[5] Jarzombek shows how the word architect etymologically combines the arche-, evoking both the artful and conceptual, and -tecton, evoking the applied and the practical. The evolution of the practice of architecture from the Italian Renaissance, through the formalization of schools of architecture, such as the École de Beaux-Arts and the Polytechnique in France, to the modern understanding of architecture as a profession practiced globally with local nuances and regulating bodies, is a fascinating history of a discipline in formation, raising questions of regulation, licensure and liability as well as expertise, class, and social status.

The founding of the American Institute of Architects (AIA) and other professional bodies emerged in the late 1800s to regulate activities encompassing the design of buildings and their ornamentation, the engineering of structures and foundations, and the management of labor and construction activities. These organizations have since helped to define architecture as a profession, equipped with a specific kind of expertise, defined criteria for licensure, codes of conduct, and structures of governance.[6]

5 Mark Jarzombek, *Architecture Constructed: Notes on a Discipline* (London; New York, NY: Bloomsbury Visual Arts, 2023), p. 3.

6 For an account of the evolution of the professional bodies like the AIA, see George Barnett Johnston, *Assembling the Architect: The History and Theory of Professional Practice* (London: Bloomsbury Visual Arts, 2020), p. 2.

STANDARD FORMS

In addition to defining the rights and responsibilities of architects, as well as regulating who would have the right to call themselves an "architect," the AIA standardized practice through the drafting of contract documents meant to define norms of practice. These documents became known as the AIA Contract Documents, the "A101 Standard Form of Agreement Between an Owner and Contractor," and the "B101 Standard Form of Agreement Between an Owner and Architect." These documents have evolved over time, and have, to a large extent, codified the practice of architecture in the United States for the last 100 years.

One of the noteworthy evolutions of AIA contracts was the way the tasks of the architect were increasingly defined to limit exposure to lawsuits and mitigate risks. While architects have a professional responsibility to protect the health, safety, and welfare of the public through the regulated practice

1 0 Hypar ceiling geometry of the clerestory windows at the Daniels Building (2018), designed by NADAAA, in Toronto, Canada.

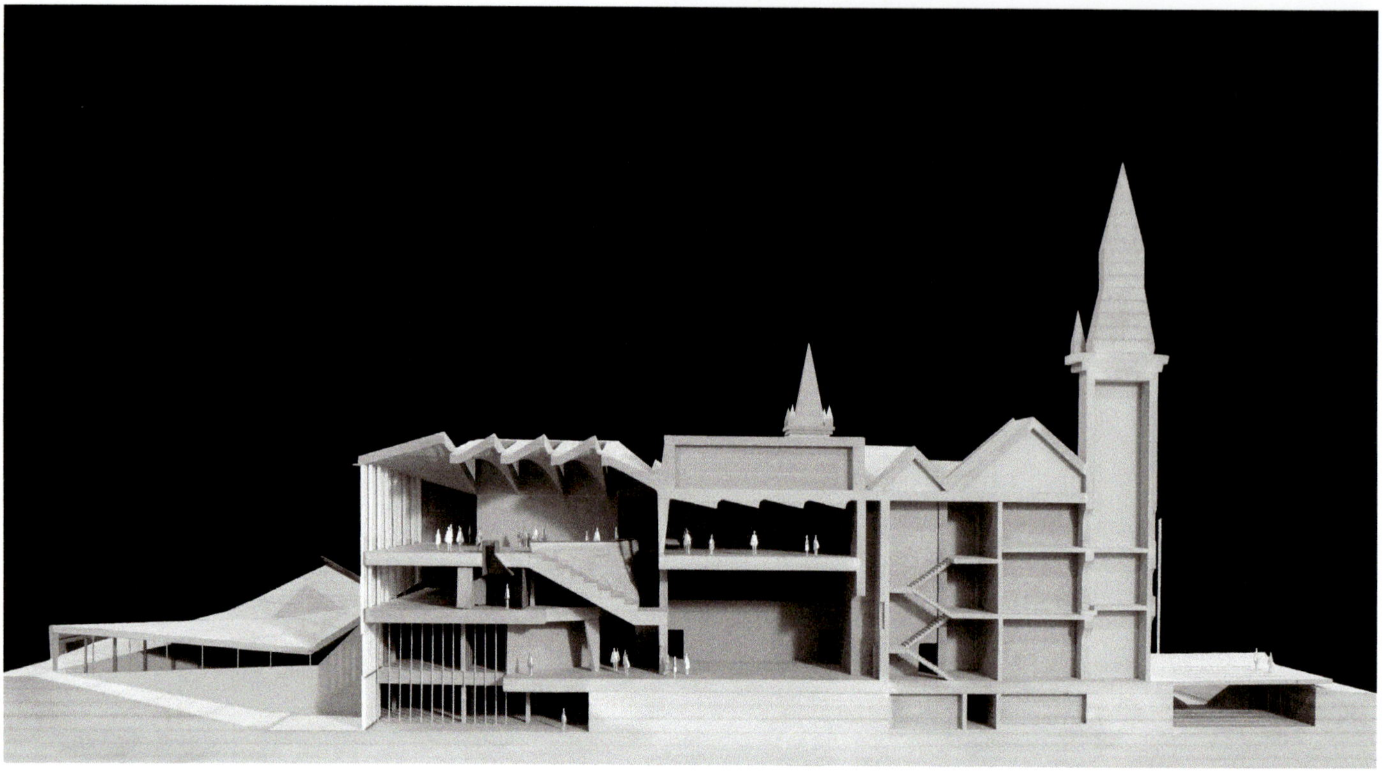

1.1 Section model of the Daniels Building showing the hypar clerestory windows and the relationship to the historic structure to which it is attached.

of architecture, the construction of buildings can be a difficult and dangerous activity. Over the years, the AIA has implemented language to define the architect's role as distinct from the role of the contractor. Where historically an architect's role during construction might have been called "site supervision," today the architect's role is described as "administering the contract." Architects on a job site are not "supervising" or "inspecting." They are making "site observations." These terms place the responsibility for design execution squarely within the scope of the contractor, while limiting the architect's risk during construction.

AIA document A201, "General Conditions of the Contract for Construction," Article 4.2.2 Administration of the Contract states: "The Architect will not have control over, charge of, or responsibility for, the construction means, methods, techniques, sequence or procedures, or for the safety precautions and programs in connection with the Work, since these are solely the Contractor's right and responsibilities under the Contract Documents, except as provided in Section 3.3.1."[7]

[7] AIA Contract Document, A201-2017, General Conditions of the Contract for Construction.

The architect, as author of the contract documents, is responsible for the "design intent" as reflected in the drawings. The architect also acts as the owner's representative during construction by visiting the site at "intervals appropriate to the stage of construction," and keeping the owner "reasonably informed" about the progress and quality of work. The language of the AIA Contract Documents significantly limits the architect's role during construction, while ceding control of, and responsibility for, the execution of the project to the contractor.

DIGITAL PRESSURES

One of the paradoxes of contemporary design and construction is the simultaneous pressures exerted by, on the one hand, the use of digital tools that tend toward continuity from the architect's 3D models and the contractor's digital fabrication files, and on the other, the legal frameworks that aim to partition risk by ceding the architect's control over the means and methods of construction. Architects produce design intent while also defining many of the key parts of building, including interior and exterior geometries, materials, modules, and levels of finish. Much of this information in the BIM model can be used directly to produce area take-offs for estimating, or even by fabricators for producing cut files for ductwork fabrication, for example. The BIM model is such an information-rich data set that it is understandable

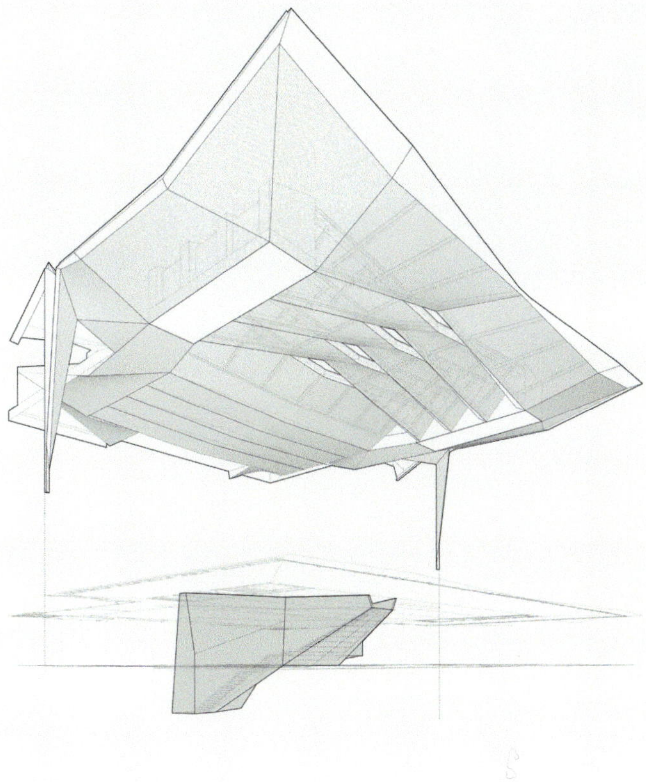

12 Perspective of the underside of the hypar ceiling soffit of the Daniels Building.

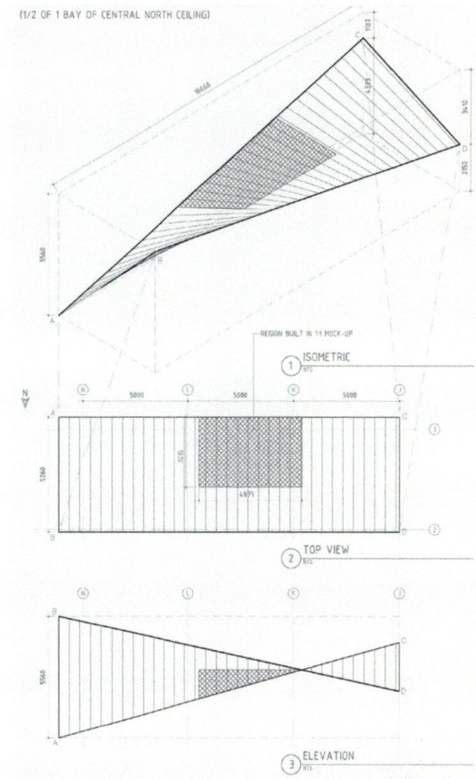

13 Drawings of the ceiling mockup of the Daniels Building soffit.

why all parties seek to make use of the information embedded within it, and why the BIM model is replacing the 2D sheets of drawings.

PROOF OF CONCEPT

Few architects are as articulate about the interactions between design and construction as Nader Tehrani of NADAAA. Tehrani's work often takes as its motivation the deliberate articulation of material systems and is characterized by expressive material units, design processes of aggregation, and part-to-whole relationships. Tehrani has written on the ethics of mono-material systems and the notion of a "catalytic detail" as a local condition that can proliferate to become an articulation of a whole building.[8] He has also expressed concerns over the "dismembering" of architecture under the current separation of design intent from construction implementation as structured by the AIA's General Conditions.

Tehrani refers to early projects, such as the Mantra restaurant hookah den, where he and his team took on design and installation roles to execute projects that were deemed too expensive or not constructible for the given budget. He also describes close collaborations with specialty fabricators, like millworkers, to deliver unique solutions involving their design processes based on fabrication parameters like bed sizes and assembly logics, to enable certain efficiencies and economies between design and construction.

8 Transcript from a lecture delivered at the CCA on March 4, 2021.

9 Transcript from a lecture delivered at the CCA on March 4th, 2021.

During the design process for the Daniels Building at the University of Toronto (2018), Tehrani designed the open studios on the upper level of the addition with north-facing clerestory windows inspired by the concrete architecture of Candela. 11 12 13 In Tehrani's words, the twisting skylights combined a "structural strategy in dialog with a daylighting strategy and a hydrological strategy."[9] Initially, the proposed concrete hypar roof system was deemed too expensive by the contractor, and Tehrani was encouraged to substitute a conventional flat roof with discrete skylights that would save approximately a million dollars.

Unwilling to accept the "not constructible for the budget" argument, Tehrani embarked on a proof-of-concept study by building a ruled surface ceiling in the basement of his own office to demonstrate the economical constructability of the light-gauge framing as an alternative approach. The team built a 16 feet x 10 feet (4.9m x 3m) section of the ceiling using 2x4 metal studs and two layers of quarter inch "high flex" gypsum wall board. 14 15 The

14 Light gauge metal framing (LGMF) of the full-scale mockup of the hypar surface of the Daniels Building ceiling, constructed in the basement of the NADAAA office.

15 Drywall finish of the full-scale mockup of the hypar surface of the Daniels Building ceiling, constructed in the basement of the NADAAA office.

entire process was documented and posted on the firm's blog. According to one post, "the mock-up convinced the construction team and reduced estimated costs by more than 50%."[10] The project eventually adopted Tehrani's approach to preserve the unique bowtie soffit and clerestory lites, giving the architecture studios their signature ceiling. 10

The design and construction of the DL1310 Apartments (2020) in Mexico City and the Daniels Building (2018) in Toronto both reference Candela as a precedent yet required adaptive approaches to design and implementation that responded to local conditions in order to be realized. Both projects began with formal ambitions for specific facade or daylighting conditions. Both relied on software-enabled abstraction to conceive of unique geometric proposals. The plasticity of abstraction yielded formal solutions that required translation into materials for implementation. For DL1310, the translation to wood formwork and fiberglass, before becoming a concrete wall, required reconceptualization of the construction processes. For the Daniels School, the evolution of form and geometry into a steel-frame with drywall finishes required multiple phases of testing and negotiation before implementation. 16 17 In both projects, a hands-on approach to implementation allowed for unprecedented solutions and the unique materialization of architectural concepts. The next two projects began as questions of material, making design and design processes one and the same.

10 NADAAA blog post, https://www.nadaaa.com/blog/dfald-hyperbolic-paraboloid-ceiling-mock-up/

DESIGNING PROCESSES

As much as a material, concrete could also be understood as a process. Concrete has been favored by architects not only for its durability but also its ability to transfer and transform the temporary geometry of the formwork surface onto the permanent form and volume of a solid, which led architects to experiment with new processes of forming and casting.

The Bruder Klaus Field Chapel (2007) is a small, one-room structure designed by Peter Zumthor for the small town of Mechernich outside of Bonn. Zumthor was asked by a local farmer and his wife to design the small chapel to honor Nicholas von Flüe, the patron saint of Switzerland, also known as Bruder Klaus. Zumthor conceived of the chapel as a simple conical volume hollowed out of a rammed earth concrete structure. He proposed to make the formwork for the chapel out of wooden logs arranged as a conical bundle. Local farmers built the wall, one 20-inch (50cm) layer, or lift, at a time, and at a rate of one layer per month for two years. Once cured,

16 Steel framing of the hypar surface of the Daniels Building ceiling soffit.
17 Drywall finish of the hypar surface of the Daniels Building ceiling soffit.

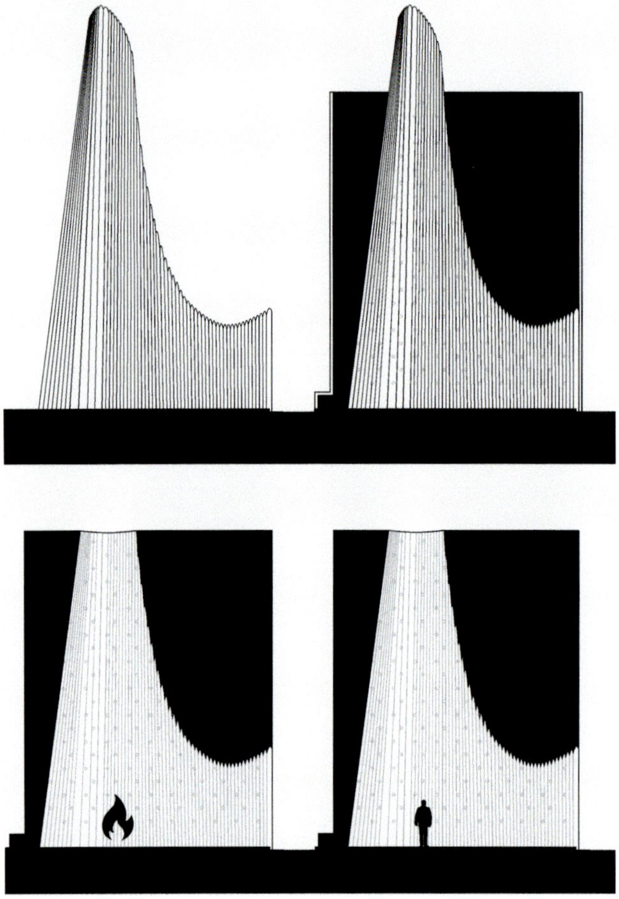

18 Construction sequence diagrams of the Bruder Klaus Field Chapel (2007), designed by Peter Zumthor, in Mechernich, Germany.

19 The rammed earth concrete exterior of the Bruder Klaus Field Chapel masks the charred conical interior space within.

the wooden formwork was removed from within by burning the logs, leaving a charred, hollowed-out interior in the shape of the bundle. 18 The rough texture of the walls bears the ribbed, concave imprint of the logs, and the smell of burnt wood pervades the intimate chapel. A teardrop-shaped oculus brings light in from above, while pinpricks of light, channeled by clear glass spheres inserted into the aggregate are scattered across the rough walls. The vertical exterior of the chapel contrasts with the conical interior volume within, making the spatial experience of the interior a discovery. Zumthor designed the form and configuration of the chapel, but more importantly he designed the methods of its making. The farmers' labor in constructing the chapel, step by step, was an integral part of the design and the choreography of construction.

In a similar act of material design process, the husband-and-wife team of Ensamble, Débora Mesa and Antón García-Abril, constructed the Truffle House (2010) on a waterfront site at Costa da Morte in Northern Spain. 21 The architects describe the process: "To build it, we made a hole in the ground, piling up on its perimeter the topsoil removed, and we obtained a retaining dike without mechanical consistency. Then, we materialized the air building a volume with hay bales and flooded the space between the earth and the built air to solidify it. The poured mass concrete wrapped the air and protected itself with the ground. Time passed and we removed the earth, discovering an amorphous mass. The earth and the concrete exchanged their properties. The land provided the concrete with its texture and color, its form and its essence, and concrete gave the earth its strength and internal structure. But what we had created was not yet architecture, we had fabricated a stone."[01]

The architects designed not a house, but a process. They used the excavated ground and a stack of hay bales as formwork to produce the unique shape of a one-room home. They used a large-diameter quarry saw to expose the front and back faces of the volume. To remove the hay bales, they were assisted by a cow, named Paulina, who ate her way through the hay over a 12-month period. 20 The house was completed by a large glass window mounted to its waterside, and a prefabricated door mounted to its landside. The interior, painted white, has the qualities of a grotto, with ledges to accommodate a sleeping platform and a hearth for a fireplace. From the inside, the exterior glass is not visible, giving the illusion that the seaside face is open-air.

Both Zumthor's Bruder Klaus chapel and Ensamble's Truffle House are construction

01 Architect's website. https://www.ensamble.info/thetruffle

20 Construction sequence diagrams of the Truffle House (2010), designed by Ensamble Studio, in Costa da Morte, Spain.

21 The exposed section cut face of the Truffle House hints at the unusual process of its making.

choreographies as much as they are architectural designs. In each case, the plan for implementation involved unusual processes and non-traditional labor sources. The North American separation of "means and methods," from "design intent," are necessarily transgressed in these projects, allowing the architects to speculate on a continuum of design and implementation that would be impossible following the AIA's General Conditions.

CREATIVE PROCUREMENT

Innovative design processes can also extend to the procurement of specific materials or material systems. When architects Jing Liu and Florian Idenburg of SO - IL designed the Kukje Gallery (2012) in Seoul, they imagined a composition of program volumes draped by a taut translucent veil, which they described as an "architectural cloud."[12] 22 23 In researching available materials for such an approach, they proposed to clad the gallery in a tailored sheet of stainless steel chain mail. Several varieties of coiled scrims exist as products, but the concept required a sheet that would be unbiased in its orientation. Only a true screen of welded interlocking rings would achieve the desired optical and material effect.

12 The architects describe the Kukje Gallery as "crafting clouds," in *SO - IL, Solid Objectives: Order, Edge, Aura* (Zürich: Lars Müller Publishers, 2017), p. 291.

Idenburg recalls searching on the e-commerce site Alibaba.com for a company that could make the chain mail sheet. They located a fabricator, the Anping County Risheng Wire Mesh Factory in Hebei, China, who could fabricate a 12-inch x 12-inch (300mm x 300mm) stainless steel "wire cloth" sheet for $149. In a leap of faith, the architects provided a credit card number and placed an order. Several weeks later, the sample arrived in the mail. To their surprise, the quality of fabrication was good, and they were reassured that the custom veil for the gallery facade was an achievable goal.

The architects then developed the design as a central gallery and a number of projecting volumes, including an entrance vestibule, a cylindrical elevator space, a mechanical room, a curved walkway, all draped by the stainless steel veil. Idenburg and Liu enlisted the support of facade consultants, Front Inc., to assist in the design of the custom stainless steel skin and glazing systems. 28 They were able to calculate the forces in the mesh, understand its unique stretching behavior and anticipate the sag in the chainmail due to self-weight.

The next steps in the design development of the project included visiting the factory to discuss the procurement of the mesh. Idenburg and Liu traveled to a remote town in Hebei China to visit the factory.

22 The draped veil of SO - IL's Kukje Gallery (2012) in Seoul, South Korea.

23 Physical model of the Kukje Gallery's primary volume and draped protrusions.

According to Idenburg, they arrived to find a small shop house with one person in the back welding stainless steel rings. 26 When asked about the capacity of Risheng Wire Mesh Factory to produce the required quantities of chain mail, the solitary welder pointed out that all the storefronts in the town were capable of welding stainless steel rings. Idenburg and Liu proceeded to enlist all the welders in town to produce the necessary quantity of the stainless steel wire cloth.

In an improbable scenario, the combined efforts of the multiple welding shops in the remote town, the stainless steel mesh facade of 510,000 rings was produced and procured for the Kukje Gallery. The design team built a full-scale mock-up to observe its behavior and developed the details to incorporate their findings. The mesh is designed to span from a hemmed detail at the ground to a support rail at the roof. The top rail is equipped with tension springs to maintain a specific level of tautness in the mesh and prevent sagging. 24 27

The effect of the stainless steel veil is what the architects call an "atmospheric surface" that deliberately blurs the edges of the building, transforming the concrete volume into something mysterious and alluring. Liu and Idenburg describe the effect: "In its finished state it forms a visual haze that diffuses the building's solidity into something indeterminate, developing a strangely simultaneous quality of form and formlessness."13

13 *SO - IL, Solid Objectives*, p. 296.

PROJECT DELIVERY

These alternative modes of design and construction involving self-performed work, atypical means, and unconventional methods, challenge normative modes of practice and project delivery. Within the North American context, conventional project methods include design-bid-build and design-build, typically separating the process of design and building into discrete tasks, and specialized, professional roles and responsibilities.

In a typical project, the owner hires an architect and a contractor separately. During the design phase, the architect hires a team of consultants, including the engineering disciplines, to produce a set of contract documents. In the bidding phase, the owner uses the completed set of contract drawings to hire a contractor for an agreed upon sum. In the construction phase, the contractor then hires a team of subcontractors, including the various trades, to perform the work. The architect's role during construction is to represent the owner and administer the contract. The architect has no contractual

24 The spring loaded "hem" detail of the chain mail veil at the roof of the Kukje Gallery (2012).
25 Close up photo of the chain mail "weld cloth" of the Kukje Gallery (2012) facade.

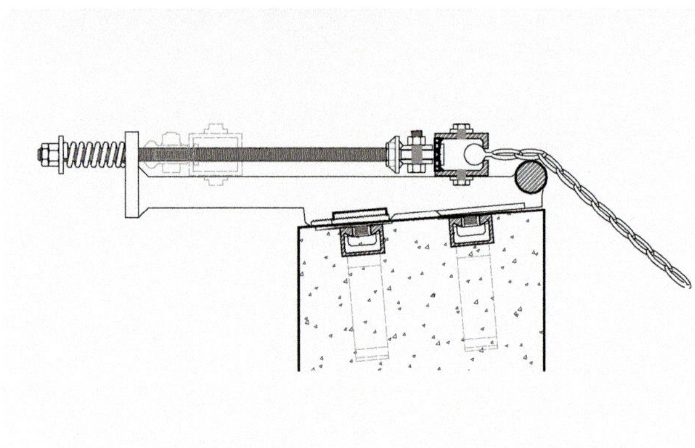

26 Labor intensive fabrication of individually welded stainless steel rings to form the chain mail screen facade.

27 Detail section of tension spring at the "hem" detail of the chain mail veil at the roof of the Kukje Gallery (2012).

28 Installation photo of the chain mail screen showing the catenary curvature under its own self weight.

relationship to the contractor, and thus interacts with the contractor and the subcontractors through the owner's contract. The contractual separation of architect from builder is inherent to the design-bid-build model of project delivery and a function of risk mitigation.

The design-bid-build model assumes that the design is completed before the contractor is invited to bid on the drawings. The architect's drawings become part of the contract for construction, meaning that they are legal documents that the contractor commits to adhere to throughout construction. Changes to the drawings after the contract is awarded require a formal process of contract modification: the change order. There is little room for collaboration or shared problem solving in this model, as the design and its documentation take on significant legal and financial dimensions. This model insulates the architect from construction and its associated risks. It also severely limits the architect's ability to interact with builders in a meaningful way.

DESIGN-BUILD

An alternative to design-bid-build is the design-build model, where the designer and the contractor are the same entity. From an owner's point of view, the idea of hiring one entity to handle both seems appealing. Indeed, many builders also offer in-house, design services allowing an owner the benefit of one-stop shopping. In North America, contractor-led design-build is common, especially in the residential market. 29

Architect Peter Gluck of GLUCK+ is a pioneer of the architect-led design-build model. For Gluck, the ability to "follow through" on the design from its conception to its construction makes design-build an obvious choice. Gluck notes, "Nobody knows your design as intimately as you do," and "nobody is going to build your design as carefully as you would."[14] According to Gluck, handing a contractor a roll of drawings with the bid documents, is inherently unfair to both parties. The contractor is at a disadvantage, as they have a few weeks to price something that has likely taken the design team more than a year to assemble. A lump sum price effectively commits the contractor to multiple years or work, taking responsibility for all the contents of the drawing set, and the specifications. The architect, tasked with administering the construction contract is cast in an adversarial role, ensuring that the documents are adhered to, while defending the owner against change orders and delays.

[14] Peter Gluck comments during a lecture at the Graduate School of Design, February 3, 2021.

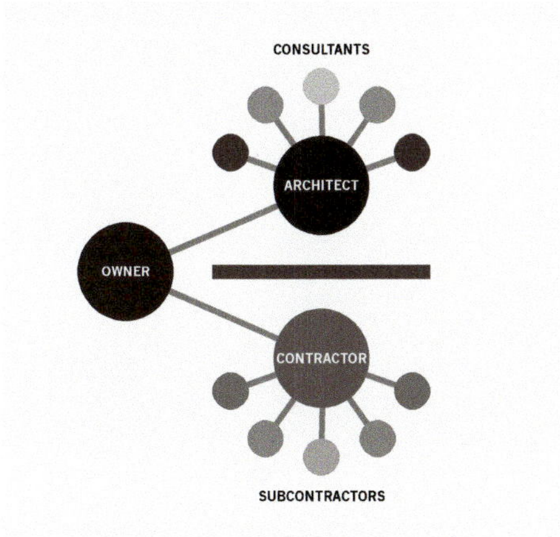

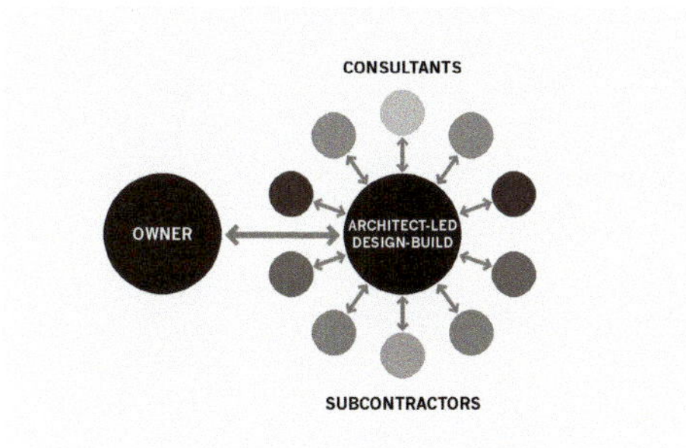

29 Diagram of traditional design-bid-build project delivery with typical separation of the architect from the contractor. Diagram by GLUCK+

30 Diagram of Architect-led design-build with the architect the center of both the consultants and the subcontractors. Diagram by GLUCK+

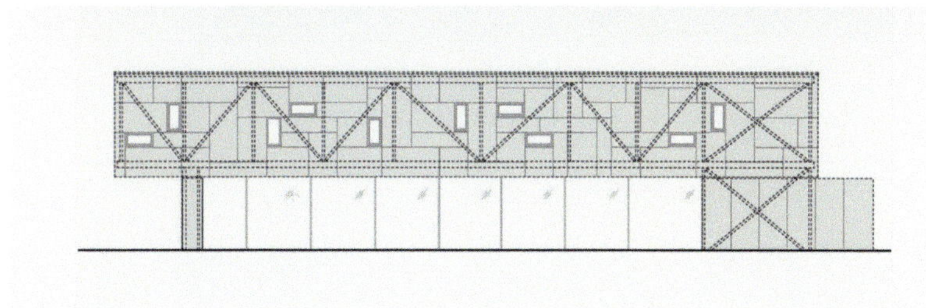

31 The Floating Box House (2006), Austin, TX, designed, and built, by Peter Gluck acting as architect and contractor.

32 Elevation drawing of the Floating Box House showing the full-story steel truss to achieve the long span house design.

Recognizing the pitfalls and failures of traditional models of project delivery, Gluck formed a construction company to self-perform construction on his own design projects. In the design-build scenario, the owner has a single source of responsibility with the architect-led design-build team. 30 As a licensed contractor, he offers clients the assurance that the design will be faithfully executed by a knowledgeable team of builders who are already invested in the successful outcome of the project. Gluck argues, "Architecture has often been likened to frozen music. If so, the construction of a building is the equivalent of a music performance, which is in fact the only thing that makes it real. To realize a design, in my view, the architect ought to be not only the composer but also the conductor, the more so because with a building there is only one performance."[15]

Since architecture is the art of a single performance, Gluck chooses to build it himself, ensuring a high-quality end product. Limiting uncertainty and eliminating the padding that often comes with the risks of bidding produces significant cost savings. Acting as designer and builder also increases expertise among his staff. Experience on the job site empowers designers and changes their design process. Architects familiar with construction have a clearer understanding of issues that arise as a project is built. Gluck+ also packages their drawings differently. Rather than organize their sets by discipline — A for architectural, S for structural, M for mechanical, and so on — the set is divided by trade. A unique set of drawings is produced for the excavation and foundation subcontractor, and another set for the steel subcontractor, and so on. Drawing with the trades in mind changes how information is organized and how architecture is represented.

GLUCK+ has designed and built most of their projects to date, resulting in thoughtfully designed and constructed projects as well as highly innovative construction techniques. By leading and integrating construction and design, Gluck has developed new approaches to below-grade construction and excavation, as demonstrated in projects such as their Floating Box House in Fort Worth, Texas. 31 32 They have also had success with prefabricated modular construction for affordable housing urban infill sites like The Stack in Manhattan.

With construction comes risk, but Gluck would rather be in control of the risk rather than subjected to its effects. Processes involving problem solving and attending to details is what he describes as the overlap between "thinking and making." He points out that by building and realizing projects,

15 Peter Gluck & Partners, Architects, *The Modern Impulse*, ed. Oscar Riera Ojeda and Peter L. Gluck (Philadelphia, PA: ORO Ed, 2008), p. 23.

33 Construction photo of the steel frame erection of the Floating Box House. Peter Gluck acted as the architect and the contractor, allowing him to both design and build the house.

34 Exterior view of the Floating Box House (2006), showing the dramatic cantilevered upper volume and the subterranean floor below.

he is limiting risk. For him, being on-site makes things safer. "All knowledge learned translates into confidence." The experience of constructing the buildings that he has designed allows him to translate experience into knowledge, and, as he describes it, "knowledge into wisdom."[16]

WHAT MORE?

Michael Murphy and Alan Ricks were students at the Harvard Graduate School of Design in 2006 when they attended a lecture by Dr. Paul Farmer of Partners In Health (PIH). Farmer spoke about a project to build a hospital in Rwanda. After the lecture, Murphy spoke to Farmer, inquiring about the architect, and imagining a potential internship. "Architect? Why would I ever use an architect? I just drew the last hospital on a napkin!" was Farmer's response. Murphy and Ricks volunteered to work on the hospital and flew to Rwanda and spent the summer working on a hospital in Butaro, a remote region near the border with Uganda. Their summer internship led to a leave of absence from their graduate studies, and they formed MASS Design Group as a non-profit 501(c)3 while still students.[17]

MASS Design, which stands for Model of Architecture Serving Society, is dedicated to working with communities and stakeholders to use architecture as a force for good. Their early work with hospitals in Rwanda and Haiti gained the group visibility, which exposed them to new opportunities to work in an atypical form of practice. Their model involved moving to the sites they were working on and partnering with local institutions to find the best-suited design solutions. Murphy and Ricks also assessed resources available in terms of local materials and building crafts, be it compressed stabilized earth blocks, or wooden planks sawn by hand. The MASS Design model of alternative practice also included an extreme resourcefulness in terms of not only design, but also execution.

MASS Design today is a 200-person organization with offices in Boston, Kigali, Poughkeepsie, and Santa Fe. They employ architects, landscape architects, furniture designers, writers, researchers, and filmmakers. They also have a construction division based in Kigali, Rwanda, called MASS.Build. The construction team includes 160 full-time salaried staff, and 2,100 contract employees.

The Ellen DeGeneres Campus of the Dian Fossey Gorilla Fund (2022) is a 48,500-square-foot (4,500sm) ensemble of buildings in Kinigi, Rwanda, where MASS Design acted as architect, landscape architect, furniture designer, exhibition designer,

16 Transcribed from a lecture through the Architectural League. https://www.youtube.com/watch?v=CGzZ-d4y11KY

17 The story of the forming of MASS Design is recounted in the book *Justice is Beauty*, Michael Murphy, and Alan Ricks, eds., *Justice Is Beauty: MASS Design Group* (New York, NY: The Monacelli Press, 2019).

35 The Ellen DeGeneres Campus of the Dian Fossey Gorilla Fund (2022), designed by MASS Design Group, is integrated into its mountainous site in the Kinigi, Rwanda.

construction manager, and contractor. 35 Murphy notes that taking on more scope allowed the team flexibility in both design and execution. They also realized that in the remote part of Rwanda there was a limited pool of skilled labor to draw from, so they developed a program to train local Rwandans in various trades including welding, carpentry, and stone masonry.

The multi-building campus occupies a 12-acre parcel adjacent to Volcanoes National Park. The plan includes several organically shaped, freestanding buildings integrated into the contours of a gentle hill. The buildings are planted with green roofs that help them blend into the carefully landscaped site. MASS Design describes their approach as "purpose built," meaning that the design team worked with the Fossey Fund to develop a program that met their needs based on their fund's mission. The campus buildings house a research center, a conservation gallery, an education center, as well as housing for researchers and visiting students.

The three main buildings are designed to integrate with the contours of the hilltop site. Their footprints are rounded to match the terraced landscape, their extended slab edges provide shaded outdoor spaces. The buildings take advantage of passive design strategies like self-shading, daylighting, and natural ventilation. The design team took advantage of local materials like the volcanic stone for the exterior walls and regionally sourced pine for the ceilings and soffits.

The stone-clad walls of the Fossey campus are an interesting case study of the evolution of detailing and craft, as well as resources and labor. 38 The volcanic basalt stone is an abundant locally-available material, but is not typically used in vernacular construction. The architects from MASS Design had used the stone as a cladding material on their first Rwandan project, the Butaro District Hospital in 2011. The irregular-shaped polygonal stones are packed together to form the face of the masonry wall. The technique of cutting and fitting the volumetric stone was developed with local craftspeople. 36 The stone walls of the Fossey buildings employ a similar stone-setting technique to Butaro, but also reveal a detailing refinement. The slightly rounded corners of the new buildings showcase the ability of the stonemasons to form volumetric walls. Similarly, the wall transitions to the clerestory windows also have a filleted corner profile that softens the transitions, but also highlights the dimension of the stone material. 37 Unlike veneer stone applications, where the thinness of the stone is evident when it turns the corner, the irregular and volumetric stone walls of the

36 The process of cutting and assembling the local volcanic stone into the irregular pattern of the masonry walls involved details that evolved over time, with input from the local builders.

37 The puzzle-piecing of irregular stones allows for a high utilization of available material and a degree of improvisation from the local builders.

Fossey buildings reveal the processes of forming, fitting, and assembling the unique stone setting methods that were jointly designed.

The campus was the first project constructed by MASS.Build allowing MASS Design to test principles of resource allocation as well as to maximize the local social and economic impact of the project. MASS partnered with the German Corporation for International Cooperation (GIZ) and the Integrated Polytechnic Regional College Musanze (IPRC) to provide training in carpentry, plumbing, electrical, masonry, and landscaping.

The 2,400 people employed by the project contributed to the local economy during the construction of the project, but the capacity-building with the training of the local labor force has proven to be a deeper, longer-term impact of the project. 41 Murphy argues that the architectural project supports the Fossey Fund's mission to catalyze conservation and species protection while simultaneously leaving a lasting legacy of skills and training for the local community. MASS Design describes the process of working together with local craftspeople and stakeholders as a process of "accompaniment."18 Rather than design that dictates what is to be done, the MASS Design approach offers a partnership with those who will build the structures. By working together with craftspeople and co-authoring the details, the process ensures that the design is tailored to the available materials and capabilities, but also enlists the craftspeople in determining the outcome, which builds skill, but also a sense of ownership and stewardship. By accompanying the local builders in the processes of conception and construction, MASS Design provides support and guidance through a project of joint development that leads to positive, long-term impacts. 40

Describing the mission of MASS Design, Murphy asks, "What more can architecture do?" In designing and building a research center in Rwanda and training the labor force with skills in fabrication and construction, Murphy conceptualizes both the "footprint" and the "handprint" of the project. Environmental footprints are often understood as the negative impacts of a project on the natural environment. The human handprint of the project, however, acknowledges the people that made the project, and the project's social handprint is representative of how it has improved conditions in the local community. Murphy describes their four principles as "Hire locally, source regionally, train where you can, and invest in dignity."19 MASS Design's ambition is to minimize the project's environmental footprint, while maximizing its social handprint.

18 Dr. Paul Farmer describes "accompaniment" as journeying together. "I'll share your fate for a while, and by 'awhile' I don't mean 'a little while.' Accompaniment is much more often about sticking with a task until it's deemed completed, not by the accompagnatuer, but by the person being accompanied." MASS Design Group, Michael Murphy, and Alan Ricks, eds., *Justice Is Beauty*, p. 38.

19 See https://www.ted.com/talks/michael_murphy_architecture_that_s_built_to_heal?language=en

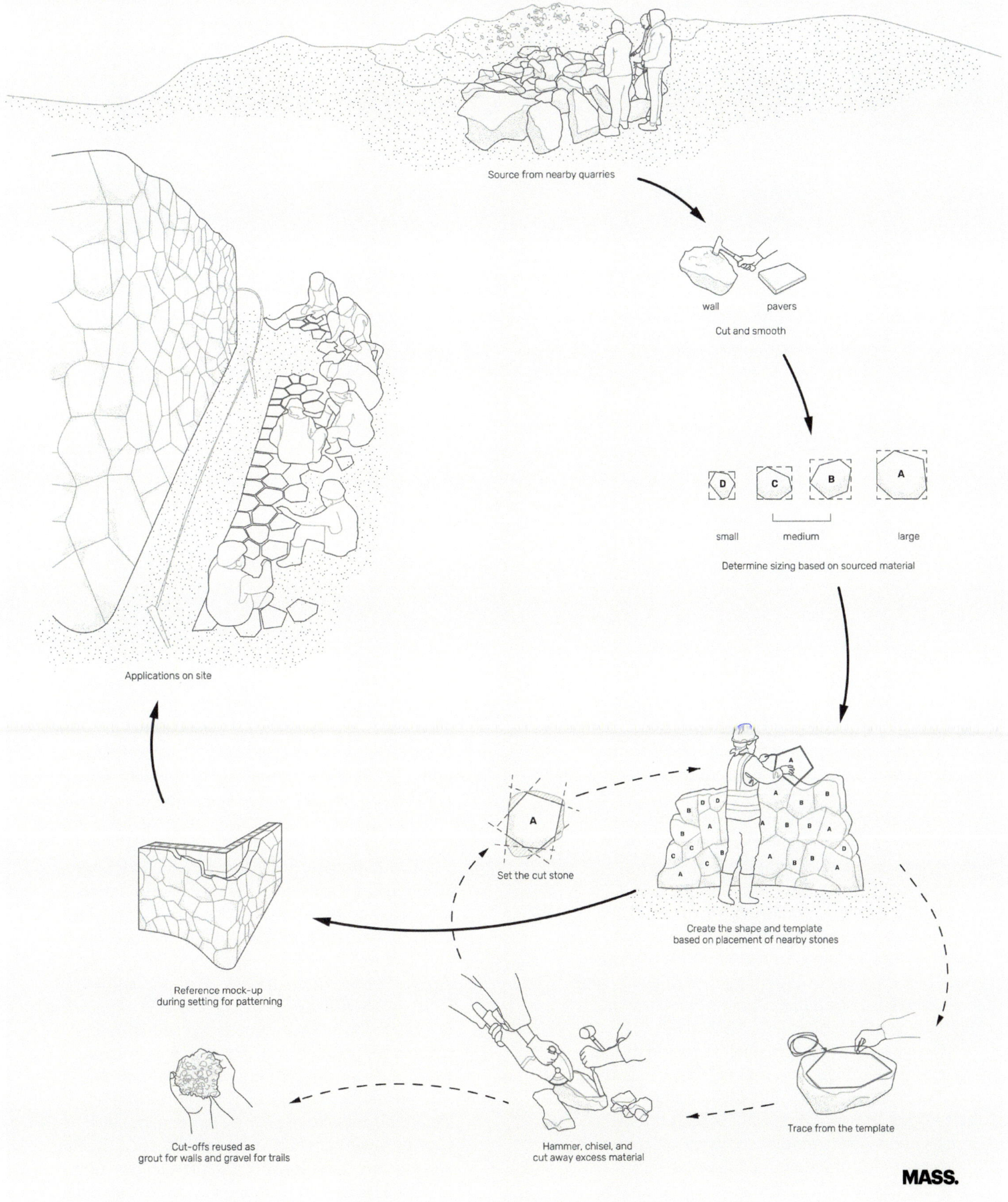

38 **Workflow diagrams of the masonry walls and pavers of the Ellen DeGeneres Campus (2012).**

39 Following the principle of "handprint," the MASS Design team saw the project as an opportunity to build capacity and skills among the local community, in addition to building the building. A mason in training, cutting stone.

40 A mason sets irregular stone blocks in a unique pattern as part of the exterior wall of the Ellen DeGeneres Campus (2022).

RISK SOCIETY

Conventional architectural practice in North America draws a sharp line between design and implementation. The architect's role, as defined by AIA contracts and insurance policies, is concerned primarily with the design intent — the *what* — but has no responsibility for the specifics of the *how*. Allocating the means and methods by which a building is constructed to the builder narrows the architect's scope to solely design intent. The separation of intent from implementation, and its legal rationale, reflects the evolution of architectural practice in the context of construction litigation and the emergence of what sociologist and theorist Ulrich Beck calls "risk society."[20]

[20] Ulrich Beck, *Risk Society: Towards a New Modernity*, Theory, Culture & Society (London; Newbury Park, CA: Sage Publications, 1992).

Even as risk mitigation pushes design further away from execution and implementation, digital workflows and BIM modeling press design toward greater integration with fabrication and construction. As the medium of communication of design intent, architectural drawings and digital models are acknowledged as instruments of service to the architect. Yet, contemporary developments in design software, BIM, project delivery, and digital fabrication have blurred the boundaries between representation and execution with significant implications for risk and responsibility. Digital documents, where the 3D model is the deliverable, transfer design intent from concept to construction with little need for drawings. The instrumentality of digital design offers opportunities for greater architectural agency while simultaneously requiring new mechanisms for risk management and quality assurance.

Contemporary processes by which architecture is conceived, modeled, and mocked up illustrate the messy process of translation from concept to construction. How should architecture leverage the capacities of computational tools for fabrication against the backdrop of a risk-averse construction industry and a litigious society? How can design be tailored to the situational context of construction even as we practice in an increasingly globalized world with uneven distributions of technology, skill, and labor? And how can design and its expanded capacities address larger questions such as environmental responsibility, social justice, and fair labor practices? If architecture is to have greater agency in addressing the many challenges encompassed by the built environment, then questions of material procurement and reuse, embodied energy, and building trades and craft cannot be separated from design.

41 MASS Design acted as the architect, landscape architect, furniture designer, exhibition designer, construction manager and contractor for the Ellen DeGeneres Campus of the Dian Fossey Gorilla Fund (2022), which employed 2,400 workers to build the campus in Kinigi, Rwanda.

While the means and methods of construction are, in the North American context, the responsibility of the builder, a more holistic approach would extend from concept through to construction. Questions of sequencing, constructability, and tolerance are integral to a comprehensive design process. Limiting liability has distanced designers from the processes of construction, ceding authority over materials and methods, the very medium of architecture. The contingencies of construction necessitate the development of techniques for absorbing and incorporating contingency and call for new strategies that allow verification, integration, and tolerance to become drivers for design.[21] By maintaining a comprehensive view of the design and construction processes, architects can better address questions facing the built environment today. Questions calling for designers to better understand the what and how — and critically, the *why* of our choices.

21 The notational convention of verify in field (VIF) is one such acknowledgement of the limits of architectural representation, and an indication that more information is needed. For more on VIF and contemporary design workflows that incorporate elements of feedback and verification, see Eric Höweler and J. Meejin Yoon, *Verify in Field: Projects and Conversations Höweler + Yoon* (Zürich: Park Books, 2021).

1 The Association of the Northwest German Textile (2018), designed by Behet Bondzio Lin Architekten, uses custom bricks with varying degrees of projection to produce the fabric-like effect.

Chapter 3 Material Processes

The gardens behind the pavilions at the University of Virginia's Academical Village are enclosed by an undulating brick wall in an S-curve pattern. The serpentine walls, as they are known, are only one brick wide, but the curvature provides the stability for them to stand up to 8 feet (2.4m) tall.[1] [3] A straight brick wall that high would be unstable and would necessitate intermittent buttresses. Thomas Jefferson, the designer of the wall, understood the principles of structural stability and material efficiency. His annotated drawing shows his calculations for the number of bricks needed for the curvilinear wall as compared to a buttressed straight wall.[2] [4]

The Academical Village, as envisioned by Jefferson, would house an entire community, including students, faculty, servants, laborers, and enslaved peoples. Students residing in the single rooms facing the Lawn would attend classes in the Pavilions, which served as teaching spaces, but also as residences for the faculty and their families. The daily life of a student included a short walk to "the hotels" behind the Pavilions, several times a day for meals. The serpentine walls that enclosed the work yards behind the Pavilions also separated the daily lives of the students from the various activities in the yards. [5]

Peering over the serpentine walls today, we see lush gardens with mature trees, plantings, and intimate settings. The gardens, however, are not original to Jefferson's Academical Village, and were planted in the 1940s to evoke landscape styles of both the French and English garden traditions. The original use of the walled yards was much more utilitarian, as they were work yards where the enslaved labored to grow crops, raise and slaughter livestock, wash laundry, and chop firewood. Architectural historians Louis P. Nelson and Benjamin Ford describe daily life in the yard: "While the Academical Village was a functioning university, it was simultaneously a plantation landscape, with a whole world of activity behind the fences and walls that dominated everyday life for the majority of the enslaved who lived and worked at the university."[3] Jefferson used the natural site topography and constructed brick walls to segregate the students from the world of the enslaved labor.

The bricks that make up the serpentine walls were sourced from local brickyards. The brick walls laid in the undulating design allowed Jefferson to enclose the largest area with the fewest bricks. Jefferson had experience producing bricks for his own home at Monticello, where records show that one hundred thousand bricks were produced from

1 The serpentine walls were originally eight feet (2.4m) tall and were modified in the 1960s to their current five-foot (1.5m) height.

2 Desmond Guinness and Julius Trousdale Sadler, *Mr. Jefferson, Architect*, A Studio book (New York, NY: Viking Press, 1973), p. 143.

3 Maurie Dee McInnis and Louis P. Nelson, eds., *Educated in Tyranny: Slavery at Thomas Jefferson's University* (Charlottesville, VA: University of Virginia Press, 2019), p. 75.

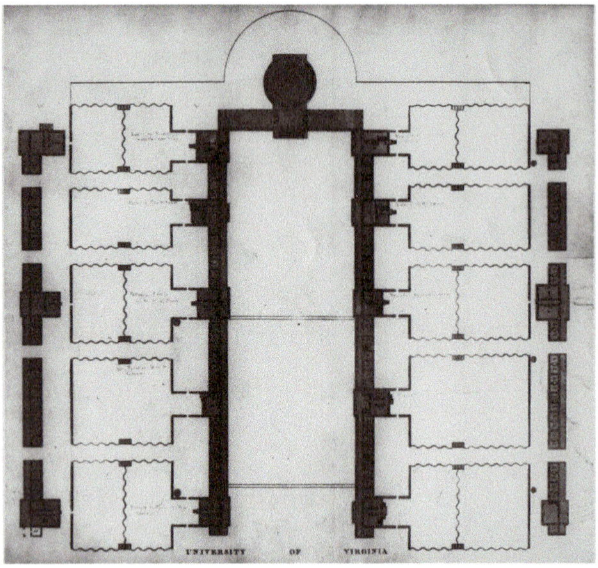

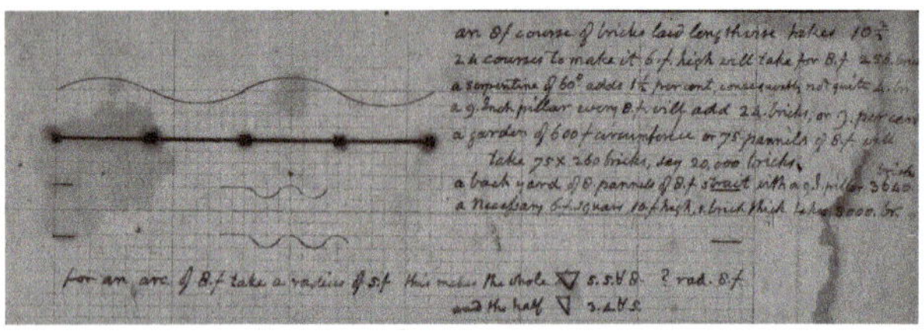

2 The plan of the Academical Village shows the arrangement of buildings defining the Lawn, with the Rotunda as the focal point. The spaces behind the pavilions are enclosed with the undulating serpentine walls.

3 An annotated sketch by Thomas Jefferson shows his calculation of the number of bricks needed to construct the serpentine wall versus the planar wall with intermittent buttresses.

4 Imprints of fingers in the bricks at Monticello reveal hidden labor in even the most common of materials: Brick.

1773–74.[4] The task of digging the clay, carrying the water, mixing, forming, and firing bricks fell to enslaved laborers, many of them boys. Some of the bricks incorporated into Jefferson's Monticello show the imprints of the hands of the brickmakers. Brickmakers regularly pressed the brick with their fingers to test the pliability of the brick before firing. The presence of the finger marks on the bricks is a poignant reminder of the hidden labor that pervades architecture. 4 If a fingerprint is an indicator of the individual, then the imprints of fingers are profoundly meaningful as a record of the labor of enslaved Africans whose identities remain largely unknown.[5]

Materials like brick are inextricably bound to the particularities of the places where they are made, through the flows that link the brick to the brickyard, and the orange, iron-rich clay from which it was made. The bricks in Charlottesville tell powerful stories of the hands that shaped them and the labor and skill that assembled them into the serpentine walls surrounding the yards, the proud facades of the pavilions and the white-washed columns that front iconic Lawn. Materials convey stories — stories of their origins and their applications, of the craftsmanship of their builders, and the intentions of their designers.

A world apart, in a northern suburb of Shanghai, an industrial shed has been repurposed as an architecture studio and workshop. 6 Approaching from the street the visitor encounters a concrete block wall that appears to be rippling like a fabric. The single cell concrete masonry units (CMUs) are rotated so their open faces are oriented horizontally, allowing the wall to read as open or closed, depending on the rotation of the block within the wall. This wall is made up of thousands of individually laid and rotated units that form a gradient pattern of open and closed blocks. Along a single course, the CMUs rotate 180 degrees with 10 different angles forming a gradient of opacity, producing the visual effect of a pixelated undulation. 7

The designer of the concrete block wall, architect Philip Yuan, a professor and researcher at Tongji University, describes the wall as fabric-like, calling it the Silk Wall. He is also founder of the architecture studio Archi-Union, which is housed behind the Silk Wall. A project description on the firm's website states: "The facade of the factory building is based on real materials and construction performance. Using the cheapest hollow concrete blocks, we try to express the texture of silk from this plain basic material, respond to the site where a large number of cotton textiles were once produced, and

4 Palmer C. Sweet, "History of Brick in Charlottesville and Albemarle County," *Virginia Minerals*, Vol 44, No 3, August 1998, p. 17.

5 In the process of designing the UVA Memorial to Enslaved Laborers, a group of UVA's President's Commission on Slavery and the University (PCSU) compiled a list of all known names of the enslaved based on diary entries, letters, records, and property logs. The number of the enslaved that worked on the UVA grounds is estimated to be 4,000.

5 The serpentine walls surrounding the yards of the Academical Village.

6 The undulating effects of the Silk Wall (2010) project designed by Philip Yuan of Archi-Union.

create a new architectural soul. The reading of the grayscale in the texture of silk becomes the medium for translating silk into the way of wall masonry. Through gradual changes and a certain regular rotation angle, the wall presents a soft and wrinkled effect like fabric."[6]

The sequential rotation of the CMU blocks affects its visual porosity as well as the shadows it casts. The "grayscale" in the project description refers to the image of an undulating fabric as it is translated from a rasterized image with variations in tone, to an angle of rotation of the physical block. Yuan describes his design methodology as parametric, meaning that attributes such as degrees of rotation are manipulated algorithmically, based on variable inputs, or parameters.

The Silk Wall, and workshop space behind it, was built in 2010, when parametric software was just becoming widely available. Architects interested in the expressive potential of computational scripting as a means of form-finding were testing translations between digital form making and material implementation. Projects like Silk Wall were early examples of algorithmically determined forms and organizations "outputted" as material assemblies. As with many pioneering projects, the ambition of parametric design and its physical translation required certain improvisational steps in their implementation.[7] The Silk Wall project was meticulously constructed by Chinese masons who set individual CMU blocks by hand with the help of wooden templates. The wall has the appearance of a 3D computer model, or a set of physical components determined by a multi-axis robot, but the reality is that it is assembled by human masons to give the appearance of digital design and robotic assembly.

In the wake of Silk Wall, Archi-Union has been researching digital fabrication techniques from robotic welding to large format 3D printing to robotic laid bricks in prefabricated facade panels. Silk Wall was an early test for more advanced parametric brick patterning and robotic bricklaying. Yuan is busy at work on the next generation of brick screen walls which are assembled with an epoxy adhesive instead of the traditional mortar mix. Yuan calls this process, "cyborg fabrication,"[8] mixing robotic and human labor to assemble full scale wall components with the variation and precision that is hard to achieve with human masons.

Two hundred years separate Yuan and Jefferson's undulating walls. The S-curve patterns of the serpentine walls have a wavelength of about 20 feet (6m) while Yuan's undulating assemblies

[6] Project description appears on the Archi-Union web site. See http://www.archi-union.com/col.jsp?id=158, accessed on 6/6/2023.

[7] A parallel could be drawn with Gerrit Rietveld's iconic Schröderhuis in Utrecht, built in 1924. Rietveld aspired to an abstract composition of colored planes and rolled steel sections, but was constructed of available materials, including brick walls and timber beams with two yellow painted I-beams to signal industrial materials and systems.

[8] See https://www.dezeen.com/2018/11/23/archi-union-venue-b-conference-centre-cyborg-fabrication-shanghai/, accessed 6/6/2023.

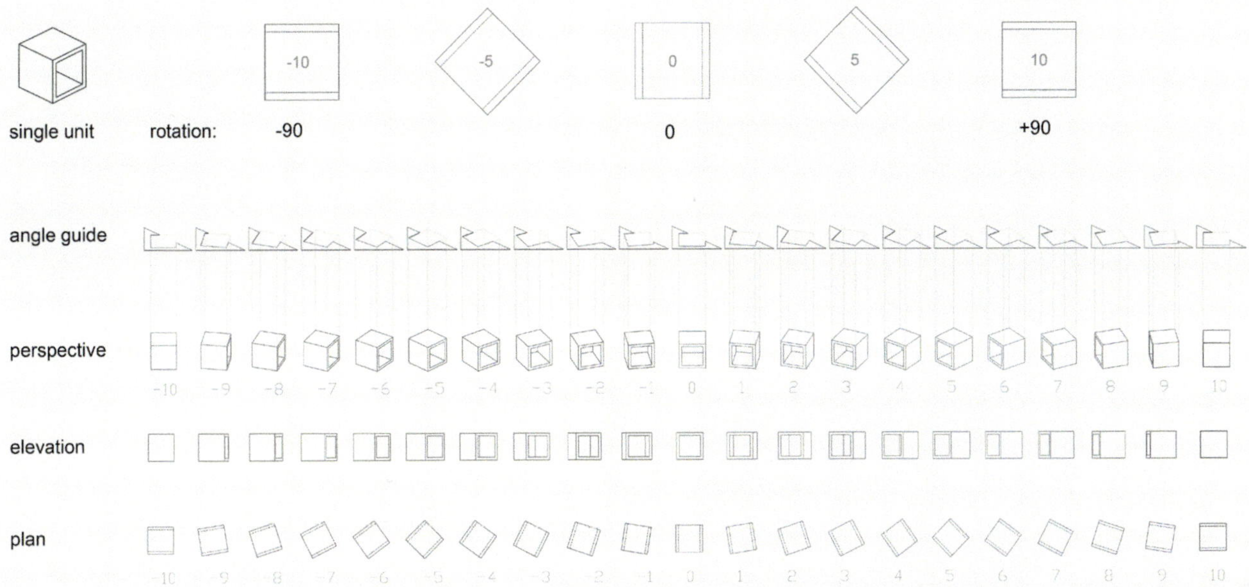

7 Block rotation diagrams for the Silk Wall with 10 angles of rotation. Note the "angle guide" template for assisting the mason in setting blocks. Instructions for installation, such as these assuming human labor. In the near future, robotic block setting will change the way design is documented, and instructions are communicated.

are each rotated individually along the same plane. Jefferson's curvilinearity was a function of structural stability and material economy, while Yuan's formal manipulations are motivated by the production of geometric effects and testing the limits of new technologies.

The comparison of the two wave pattern brick walls raises questions about the nature of material assemblies, their relation to craft and labor, and the use of digital tools in conception and construction. We might ask: What is a brick when it is no longer set by a brick mason? Is it possible to separate material from their modes and methods of installation? [9]

From ancient times and across cultures, the brick has typically been scaled to the human body. Its dimensions are determined by the physiology of the mason who would hold the brick in one hand while wielding a trowel in the other. The body of the mason is inscribed in the wall, just as the arcs of a Jackson Pollock painting are related to the bodily gestures of the artist. At Monticello, the presence of finger marks in the bricks are poignant traces of the enslaved brickmakers that directly connect the building material to the labor that assembled it. In thinking about architectural materials today, we cannot look at a brick wall without also seeing the raw materials that went into it, the labor that produced the brick, the transportation that carried the brick from its source to its site, and the skill that was employed to lay the brick in its configuration as architecture. This expanded understanding of materials includes the trades, the skills, and the craftsmanship that has been practiced as a mode of knowledge and transmitted from tradesmen to apprentices.

ASK THE BRICK

On June 6, 1972, Louis Kahn, speaking at the International Design Conference in Aspen, engaged in a fictional dialog with a brick, asking, "What do you say brick?" [10] To which the brick responded, "I like an arch." Kahn's lecture touched on a range of topics, including the sense of wonder, wellbeing, ecology and the act of becoming, meaning, and the immeasurable. "Design demands that one understands the order. When you are dealing, or designing in brick, you must ask the brick what it wants. Or what it can do." [11] Kahn goes on to describe concrete as molten stone that is "marvelously capable." For Kahn, it is the capability of materials and systems that make design a manifestation of the nature of things. This exchange, between the architect and the brick, reinforced

9 Some material and building systems are defined by a single word that is both a noun and a verb. For example "render" is a cementitious exterior stucco that is applied by "rendering," the verb.

10 https://www.youtube.com/watch?v=ZaUtcKqdL5E

11 https://www.youtube.com/watch?v=ZaUtcKqdL5E

8 Louis Kahn's query to the brick, and the brick's, response: "I prefer an arch," suggests a preference or a propensity. Kahn himself experimented with brick arches and concrete tension ties, like this arch at the Indian Institute of Management (1974) in Ahmedabad, India.

9 The "ethics of brick," as manifested by Sigurd Lewerentz's baptistry in the Sankt Petri Kyrka (1966), Klippan, Sweden. Lewerentz' rule that no bricks be cut produced a number of different expressions, including the variable mortar joint and the ragged edge.

a long tradition of correspondence between a particular material and its application. From Leon Baptista Alberti to Gottfried Semper to Frank Lloyd Wright, architects have advocated for working in the "nature of materials," and finding an "honest" expression of a material's essential qualities.

The parable of the animate brick, expressing a preference for an arch, also implies a correctness as to how it should be used. Kahn's order of materials suggests that materials have essential qualities, and that there is a right and wrong way to use them. But do materials possess inherent qualities that presupposes specific uses? Is there an implicit ethic of materials? Of brickwork?

Bricks comes with rules determined by their proportional logic that allows them to aggregate. Brick proportions vary significantly with geography and culture. A standard North American brick is proportioned so that its length (L) is nominally equal to two times its depth (D) and three times its height (H). This logic allows for rotating bricks in a Flemish bond to tie across the two wythes of depth and maintain the module within the course. It also allows three bricks in a Soldier course to align with one brick laid horizontally. The proportioning system of L = 2D = 3H (6:2:3) also considers the nominal dimension of the mortar joint.

12 Eleanor Beaumont, "Revisit: Lewerentz's St Peter's," *Architectural Review*, August 2018. See https://www.architectural-review.com/buildings/revisit-lewerentzs-st-peters, accessed 6/10/2023.

MATERIAL METHODS

The Swedish architect Sigurd Lewerentz built several of the most remarkable brick buildings of the 20th century. Lewerentz developed a rule-based approach to the use of brick, insisted that no brick be cut, and that each be used in its given proportions. This rule-based constraint resulted in innovative formal techniques derived from the logic of the material system. The variability of the mortar joint was exploited by the masons to comply with his rule set, but also to allow for variation and expression. In some areas, the mortar joints thicken to accommodate an irregular dimension. In others, they jostle and reveal an irregular, seemingly unfinished, edge. 9

At the Sankt Petri Kyrka (Saint Peter's Church) (1966) in Klippan, Sweden, Lewerentz used brick for the walls as well as the vaulted roof and the floors. His mono-material approach produced a unique geometry for the vaulted ceilings. The monolithic walls reveal their thickness in the doors and windows, and the glass window is pushed to the exterior with a unique glass clip as a result of the brick detailing. The floor pitches toward the altar and the pattern of bricks converges and folds in what has been described as, "colliding and splintering across the floor like the threads of a tapestry."[12]

10 In designing the Skycourts Building (2011) in Chengdu, China, Höweler + Yoon specified a local grey brick typical to the region, and corten steel for the window surrounds.

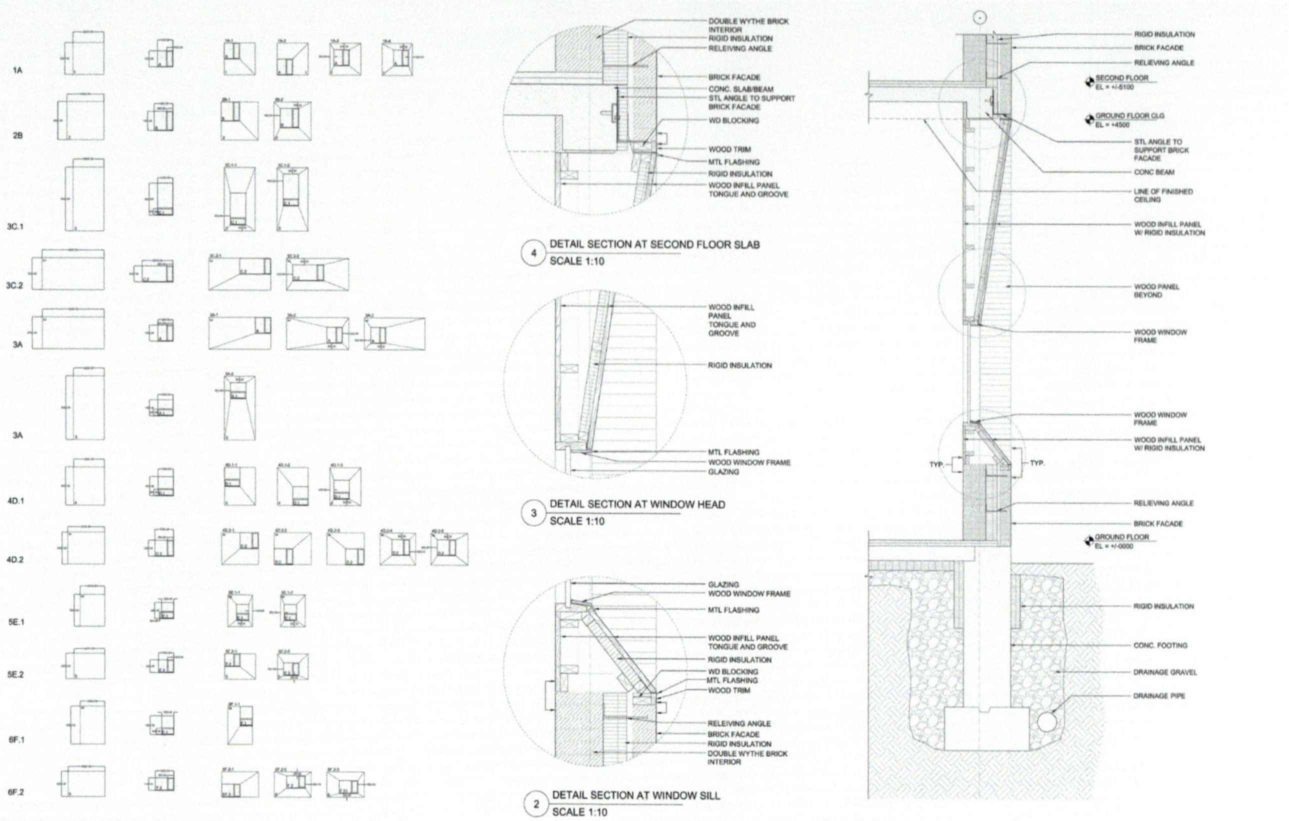

11 The facade drawings for Skycourts (2011), include an array of window types and a typical corten steel chamfered window surround.

Lewerentz's brickwork is most stark in the altar, baptismal font, and reading alcoves. The altar is a massive brick platform that seems to erupt from the floor. The baptismal font is a cut in the floor, expressed with an irregular edge, 9 as if unfinished, and the paired seating alcove is also a showcase of modular bricks arranged to accommodate the ergonomics of the seated figure. Here, the mortar joints absorb sloped geometries to reconcile the orthogonal masonry units and the proportions of the body.

Lewerentz worked closely with the masons on site to convey his mono-material concept. Even though he was in his 80s when the building was being constructed, the architect spent significant time on site with the builders. 13 Many of Lewerentz's essential details were drawn at 1:1 scale, and in some cases full-scale plywood templates assisted in the setting of specific bricks. Lewerentz's work begs the question: How is design intent conveyed from the conceptual to the constructed? How should the architect interact with the builder and the tradesperson? For Lewerentz, the obsessive detailing and material logic finds its inherent rules in the logics of construction and in the techniques of the tradespeople.

13 Carlo Scarpa is another architect who developed a design process that involved on-site detail design work in collaboration with the building trades.

BRICKLAYER'S LOGIC

The Skycourts building by Höweler + Yoon is designed around a rule set for the brick facades. 10 The project, completed in 2011, is sited in Chengdu, China. The client made a specific request that the building relate to its cultural context and that it convey "Chinese feeling." As architects based in Boston, Massachusetts, we worked to organize the building around a set of courtyards in keeping with a local typology. The *siheyuan* building refers to the courtyard, or the framing of the sky. The facades of Skycourts consist of large opaque areas punctuated by windows in chamfered steel surrounds. For the exterior walls, we selected a local gray brick for the walls and weathering corten steel for the window surrounds. 11

The concept for the brick detailing is a simple, rule-based design intended to give the building a unique appearance, and something that could be easily understood by the brick masons in China. The rule for Skycourts: All bricks to maintain the cardinal directions, as if each brick had a built-in compass. The building footprint consists of orthogonal and diagonal geometries derived from the master plan of the site. Walls running in the cardinal directions would be flush brick in a running bond pattern. Walls

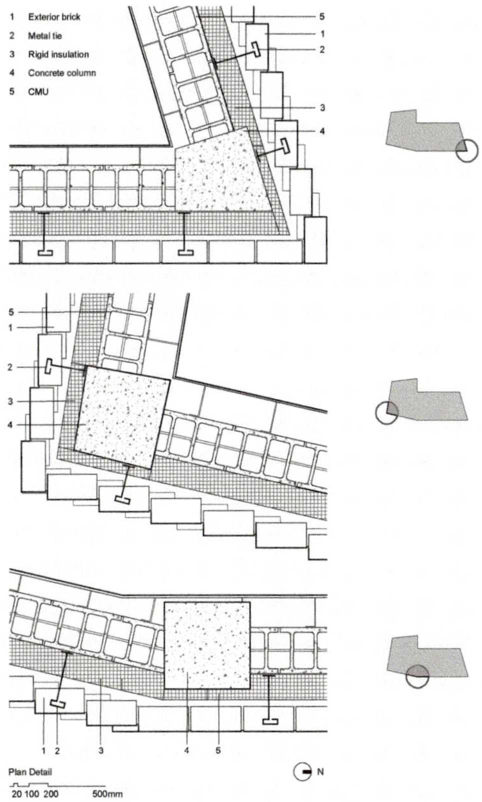

1 2 The brick logic of Skycourts is to keep the bricks oriented to cardinal directions, regardless of the local geometry of the wall. The result is a combination of flush and staggered brick textures.

1 3 The corner condition of Skycourts, where an oblique wall turns to another oblique wall, is registered as a sawtooth brick pattern on both faces. This condition is drawn in the middle detail to the left.

oriented diagonally would be constructed with a staggered pattern, as bricks would maintain their global orientation despite their local relationship to the plane of the wall. The result is a building that has regular textured faces and smooth faces in a subtle display of orientation. The outer corners are moments when the brick patterns transition from smooth to textured, or from textured to textured. 1 3 From the outset, we developed the design with cross-cultural transmission in mind. Detail drawings illustrate the brick logic as they turn the corner. 1 2 Its brick-specific rule set was designed to span the gap between intent and implementation.

Architectural design is a means of operating on the world at a distance. In many cases, builders are geographically, and culturally remote and architectural drawings must convey information about the design intent in the clearest possible way. The Skycourts brick pattern design was conceived of as a communicable logic that would be unequivocally understood. Edwin Lutyens is said to have observed, "A drawing is a letter to a builder."[14] If so, what kind of letter is it? And what tone does it take? Is the tone imperative? What is assumed about the reader? Does it leave room for interpretation? What is left unsaid? And what is read between the lines of the architectural drawing?

14 This remark is referenced by Adrian Forty in his introduction to the book, Stories from Architecture by Philippa Lewis, *Stories from Architecture: Behind the Lines at Drawing Matter* (Cambridge, MA: The MIT Press, 2021), p. 9.

INTERACTION OF MATERIALS

Tucked away and around a corner of the lobby in the two-story Graduate Center at Harvard Law School, a beige brick wall is marked with the shadows of deep recesses within the brickwork. 1 4 The wall's zipper-like patterns are made by missing bricks within a Flemish bond pattern. The brick mural, titled *America*, is the work of German artist Josef Albers who collaborated with the building's architect, Walter Gropius.

Like many artists at the Bauhaus in Dessau, Anni and Josef Albers immigrated to the US after the Nazis closed the school in 1933. Josef Albers became the head of a new art school, Black Mountain College, in North Carolina, where both taught until 1949. During their time, Josef and Anni developed an experimental curriculum in design ranging from furniture, glass, textiles, and jewelry. In 1950, their friend and colleague Walter Gropius, then Chair at the Harvard Graduate School of Design, invited Josef to design the mural for the new Harvard Graduate Center designed by Gropius' firm, The Architects Collaborative (TAC).

Albers designed *America* to be made of the same material as the building, a matte beige brick. Rather than treat the mural as something applied

14 Josef Albers' *America* mural (1950) at the Graduate Center at Harvard Law School, consists of a deliberate subtraction of bricks from the brick wall of the building. The gaps between the bricks and the shadows they cast constitute the artwork for Albers.

to the building, Albers re-ordered the elements of the existing building. The Graduate Center typically employs a Flemish bond brick pattern with a flush mortar profile. Albers' design selectively removed bricks and introduced new cut bricks to create a shadow relief between the two layers. The new cut bricks are precisely proportioned to create square brick faces and square voids. The voids between bricks are accentuated by the shadows and take on a figural property like many of the Anni Albers' textile and graphic compositions. The shadow gaps seem to weave and stitch the courses of brick together. Albers described his mural as an "organic unity," with the architecture, as it is made of the same materials as the building and is premised on subtraction rather than addition. His careful composition works with the proportions inherent in the aggregation logic of the standard module and manages to make a solid brick wall appear porous and textile-like.[15]

[15] American author Christopher Benfey, who is Anni Albers' great nephew, traces the influence of Pre-Columbian graphic patterns and masonry details from Mitla and Monte Albán. Benfey argues that Albers' relational theory of color, making the perception of color dependent on its adjacent colors, should also be applied to materials. Benfey's account of the Albers' transatlantic travel, pedagogy and influence is woven into his book, *Red Brick, Black Mountain, White Clay*, where he journeys across America in search of a sense of craft and place. Christopher E. G. Benfey, *Red Brick, Black Mountain, White Clay: Reflections on Art, Family, and Survival*, [1st U.S. pbk. ed.]. (New York, NY: Penguin Books, 2013).

Josef Albers' relational theory of color, which he published in the book, *Interaction of Color*, illustrates how the perception of one color is highly dependent on its adjacent colors.[16] For example, A tan square appears lighter against a light blue background than it does on an orange background. Albers stated, "There is no such thing as the object 'in itself.'"[17] How we perceive a thing is a function of its context. This applies to color squares as well as materials. Albers' *America* highlights the interaction of materials with light and shadow, figures and fields.

The bricks used by Albers at Harvard, and Jefferson for the UVA serpentine walls, came from local brickyards that produced a brick with a distinctive mix of colors that has become an institutional signature brick. The deep brown, almost purple, color of Lewerentz's church Saint Peter's Church in Klippan is sourced in Helsingborg and the minerals in the clay are responsible for the rich shades of color. The pale gray tones of the Skycourts bricks are particular to the region. Just as wine connoisseurs describe the *terroir* of a wine — how a region's climate, soil, and terrain affect the taste of wine — there is also a *terroir* to brick that tells of the mineral makeup of the earth from which it was made, the heat with which it was fired, and the skill of the brick-masons that laid it.

BRICK EFFECTS

The colors of the brick and the mortar are significant decisions in the design of a building. The geometry of the joint is also significant. Frank Lloyd Wright is famous for insisting on the horizontal articulation

15 The V. C. Morris Gift Shop (1948), designed by Frank Lloyd Wright, showcases the versatility of brick, creating a multilayered arch set within a horizontally articulated field, and even transitioning to a brick screen at the left edge.

of a brick wall by specifying a raked mortar joint for the horizontal joints to catch the shadows, and a flush mortar joint on the verticals to de-emphasize the verticals. Wright's V. C. Morris Gift Shop (1948) in San Francisco utilizes a pale yellow, horizontally-proportioned Roman brick to emphasize the horizontality of the large blank portion of the facade. 15 Wright specified a pigmented mortar for the verticals to blend the horizontals into seemingly continuous strips. The stepping between rings of the arches and the strong horizontal articulation gives the facade a strong graphic presence. Wright locates a vertical slot window at the transition of the left-hand edge of the projecting volume and expresses the three-dimensional quality of the brick by removing every other brick, turning the brick into a screen. How a brick wall turns a corner reveals the thickness of the module and, for Wright, confirms the modularity of the brick from which it is made.

Robert Venturi's Fire Station Number 4 (1967), in Columbus, Indiana, uses two different colors and finishes, an orange matte brick and a white glazed brick, to create a super-graphic effect. 16 The single-story fire house is a modest building housing an apparatus bay on the right and a dormitory on the left. A fire hose drying tower marks the center of the facade and creates a vertical presence for the station with the graphic sign at the top, identifying it as "Fire Station 4." Venturi treats the entire facade of the building as signage, using the graphic treatment of the two-tone brick to render the building as a billboard. A false parapet on the lower left-hand side helps it to align with the higher parapet on the right. Superimposed on an orange background, Venturi overlaid a white figure of glazed brick that seems to land arbitrarily in the middle zone but stops short of the edges. The white glazed brick area stops awkwardly at the quarter point of the large square window, where an expansion joint marks the transition to the orange brick.

Venturi's frontal brick color transitions seem arbitrary, but his corner transition at the edge of the hose tower reveals Venturi's graphic, a-tectonic sensibility. The brick outside corner is mitered, making the white glazed portion appear paper thin, as it does not wrap the corner. 17 For Venturi, who saw the fire station as a civic building in need of a signifying facade, the thin layer of white glazing on the brick achieves a modest monumentality and highlights the flatness of the surface. The central bay of Venturi's firehouse facade, unlike Wright's Morris Gift Shop, is entirely flush. For Venturi, the brick facade is a primarily a signifying surface, and secondarily a tectonic assembly.

16
Josef Albers, *Interaction of Color*, 50th anniversary edition; 4th edition, (New Haven, CT: Yale University Press, 2013).

17
Christopher E. G. Benfey, *Red Brick, Black Mountain, White Clay*, p. 132.

16 Robert Venturi's Fire Station Number 4 (1968), in Columbus, Indiana, uses two colors of brick to give a modest building a civic presence.
17 The white glazed brick is mitered at the corner, making the centered white square read flatly – like a 2D graphic.

18 The Vassar Street Residence Hall (2021) at MIT, designed by Michael Maltzan, seems to reference Venturi, and turns the corner without giving the brick a thickness, and making it read like a texture map.

Michael Maltzan's Vassar Street Residence Hall at MIT in Cambridge could be understood to be in dialogue with Venturi's Fire Station Number 4, with its orange and white composition and graphic brick treatment. The building massing is pushed north toward the adjacent railroad track, allowing the south face to push in and out to form a backdrop building with four short wings that extend to Vassar Street. The entire building is clad in a highly uniform brick, with areas of red, white, and blue bricks. The transitions between the brick color fields create the appearance of a patchwork, sometimes transitioning at a corner, and sometimes aligning with a window opening. The juxtaposition of red and white brick evokes Venturi's "misregistration" of glazed brick at Fire Station 4.

Maltzan's attitude about the graphic treatment of brick is also evident at the corners. Maltzan employs a thin brick product, which consists of a one-and-a half-inch (40mm) brick profile that is snapped into a metal profile tray designed to receive the brick. The metal tray is supported over a cavity so the whole system acts as a rainscreen. Maltzan transitions between the three colors without revealing the dimension of the brick. Transitions in the face of the building are coplanar, while transitions on corners are mitered, and tight. Like Venturi's Fire Station 4, the outside corner transitions are mitered rendering the brick as a graphic treatment, or like a texture map in a rendering. 18

While these three projects employ bricks as a material, they also establish a rule set of standard modules and techniques of aggregation. If the standard module changes, then the rules change, and different design opportunities emerge. The following three projects illustrate examples of the potential of systems employing custom masonry units and their aggregative possibilities.

185 Grand Street in New York City, is in Little Italy, a neighborhood known for its pre-war building stock of tenement buildings with robust brick facades. The architect Morris Adjmi created a new version of a punched window brick facade by producing a new custom brick with a unique "domed" profile. 19 Each pair of stacked bricks is articulated with two spherical projections that look like the raised marking of a braille text. Adjmi uses the "braille bricks" to overlay a punched window pattern that is deliberately misaligned on the facade. The braille brick graphic on the north elevation includes 10 implied windows overlaid on the façade with eight real punched windows. 20 The misalignment of the arched windows of an ornamented facade and the actual windows allows Adjmi to reference

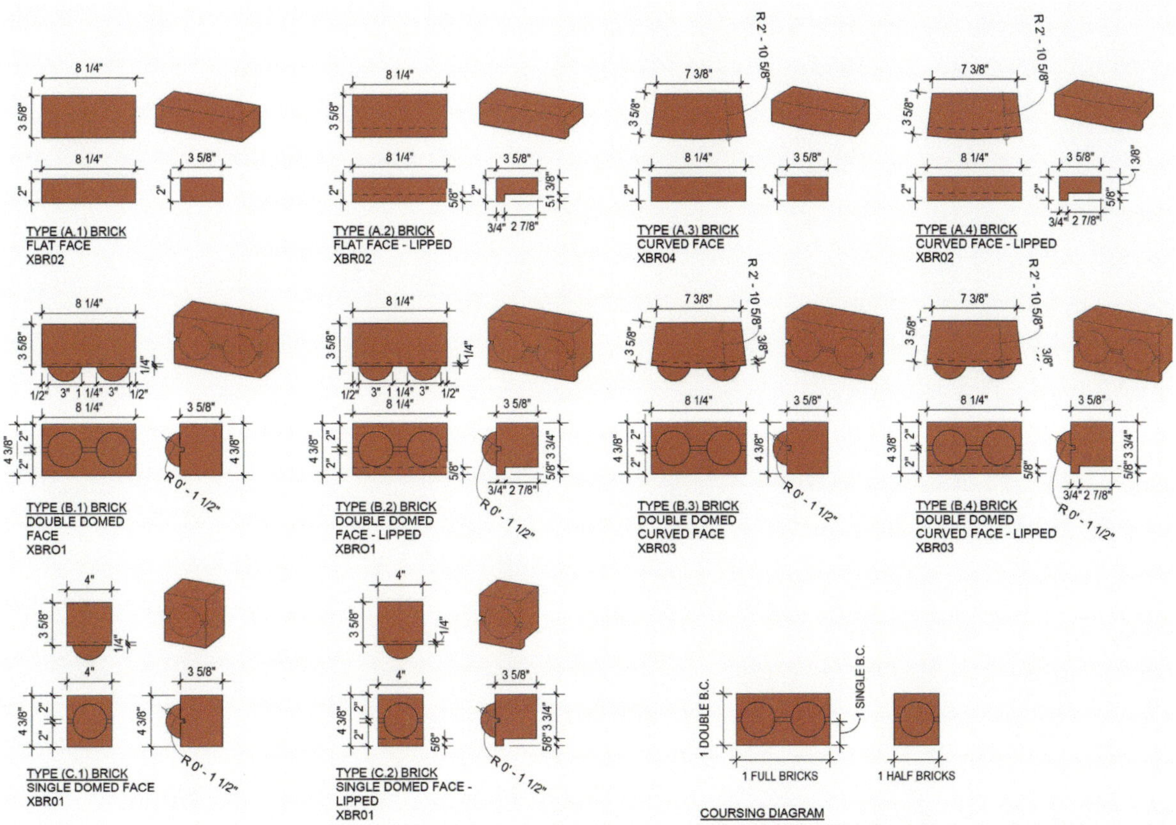

19 Morris Adjmi designed several custom bricks with a domed profile for 185 Grand Street, in New York City.

20 The facade at 185 Grand Street showcases the raised brick textures of the custom bricks to overlay another window pattern with arched windows and a deeper cornice line. Adjmi's architecture uses the full expression of the custom brick modules to create a subtle gesture to the historic context of the Nolita neighborhood.

For the Las Americas housing project (2021), in Leon, Mexico, SO – IL developed a frame and infill logic for the facades, where the infill is custom masonry unit with a saw-toothed profile and a vertical proportion.

2
1

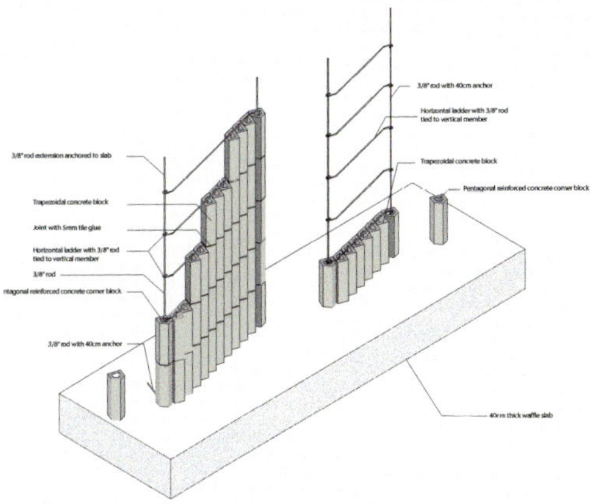

22 The custom CMU blocks stack five-high to infill between the two concrete slabs.
23 The single cell CMU blocks are reinforced with rebar and grout.

24 The play of shadows on the saw-toothed CMU walls highlights the unique geometry of the walls.

the historical context through the decorative brick pattern while maintaining a modern architectural substrate underneath. Adjmi's clever use of customized brick modules allows him to create a highly expressive and varied facade with a limited number of unique brick types.

Las Americas (2021), SO - IL's social housing project in Leon, Mexico, creates a high-density prototype for multifamily residential building as an alternative to the single-family, repetitive homes in the sprawling suburbs. 21 SO - IL employs a figure-eight multifamily courtyard typology of four, five, and six stories complete with filleted corners and bowed facades. The entire building is constructed as a cast-in-place concrete structure with a custom cement block walls between slabs. Floor-to-ceiling windows and entrance doors infill between wall segments.

The large quantities of CMU blocks needed for the project convinced the client to produce a custom block. Where traditional two-cell CMU blocks are rectangular, in a 1:1:2 proportion, SO - IL's custom block creates a single cell block in a five-sided extruded figure that is 20-inches (50cm) tall. 23 Five courses of the tall blocks infill from the floor slabs to the underside of the slab above, while the sawtooth pattern gives the whole project a fine vertical grain. 24 The CMU walls are set back from the edge of the slab, allowing the slab overhangs to provide shade and cast dramatic patterns on the facade. SO - IL's custom CMU block plays within the rules of cellular blocks but finds its own unique expression through changing one variable: the standard module.

18 The image of the undulating wall was inspired by a stone sculpture of Beethoven by Max Klinger, where the marble is so skillfully carved that it looks like the soft folds of a fabric sheet. In Klinger's sculpture the stone masquerades as fabric. In the Behet Bondzio Lin building, the brick becomes a billboard for a building-scale image of stone masquerading as fabric.

IMAGE EFFECTS

Producing custom brick modules when a project has a sufficient economy of scale allows designers to expand the expressive range of masonry construction. The brick facade of the Association of the Northwest German Textile and Garment Industry in Münster, Germany, demonstrates what is possible with seven custom brick modules. 1 25 The architects, Behet Bondzio Lin Architekten based in Münster and Leipzig, designed a new palette of brick modules with the long face rotating outward in two and a half degree increments. 26 27 The masons set 74,000 bricks in a specific sequence, creating the gradient pattern of light and shadow cast by the bricks and producing a graphic image. The process involved the translation of a raster image into a seven tone pattern. The brick pattern on the facade creates a material mural of an image of fabric, referencing the textile industry that it represents. 18

25 The Association of the Northwest German Textile (2018), designed by Behet Bondzio Lin, creates a pattern that produces a low-resolution image in the play of shadows cast by the projecting bricks.

26 Expanding the palette of typical rectilinear bricks, the architects developed six additional bricks with oblique projecting faces.

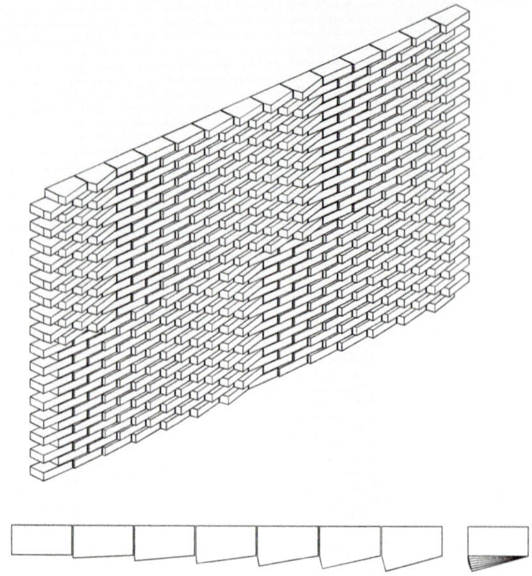

27 Custom brick units illustrated in plan show the varying degrees of rotation out of plane, and an axonometric representation of the custom bricks laid to produce an image effect.

MATERIAL POTENTIALS

It has been more than 50 years since Louis Kahn asked the brick what it wanted to be. In Kahn's "order of materials" there is an assumed material predisposition toward certain uses. The past 50 years, however, have also shown designers pushing materials beyond their traditional applications. Rather than focus on the essence of a brick and what it would "naturally" become, designers are exploiting the potential of brick to be something that it has never been before. This shift from inherent qualities to innovative potential allows a brick to be an arch, but also a vault, a billboard, and an image. Ask a brick what it wants today, and it might say, "I'd like to dance."

Through the discussion of brick and its imagined desires we can develop a sense of an expanded materiality, focused on potential. This expanded sense of materials includes the sites from where materials come, and the labor and skill that transform them from something generic into something more specific. It includes both upstream flows and downstream consequences. Thinking expansively about materials leads us to ask: Where did it come from? Who made it? And, what happens to it when it is obsolete?

Historically, the introduction of new materials has acted as a catalyst for new forms of architectural expression. The driving narrative of the Modern Movement was to find an architectural syntax appropriate to the age and derived from the expressive possibilities of new materials. This is a good moment to ask: What is the state of materials today? It seems that architecture has never had so many choices, and that we have never known as much about these materials. How best to apply this knowledge?

EXPANDED MATERIALITY

Materials for architecture are typically the result of elaborate processes and procedures. The basic brick has been excavated, mixed, molded, fired, batched, packaged, and transported before it arrives in the hand of the mason. The 2x4 available in your local lumber yard, has also been planted, maintained, logged, processed, sawn, kiln-dried, tagged, batched, and shipped before landing at the job site. Materials are products. When they are assembled into cavity brick walls, platform framing, and rainscreen facades they become systems. And throughout these extensive processes, these materials require labor and energy.

28 The Anandaloy Center (2019), in Rudrapur, Bangladesh, was designed by Anna Heringer using traditional earthen construction, like cob, which combines clay, earth, sand, and straw.

29 In the construction of the Anandaloy Center, local craftspeople were enlisted and traditional techniques were incorporated.

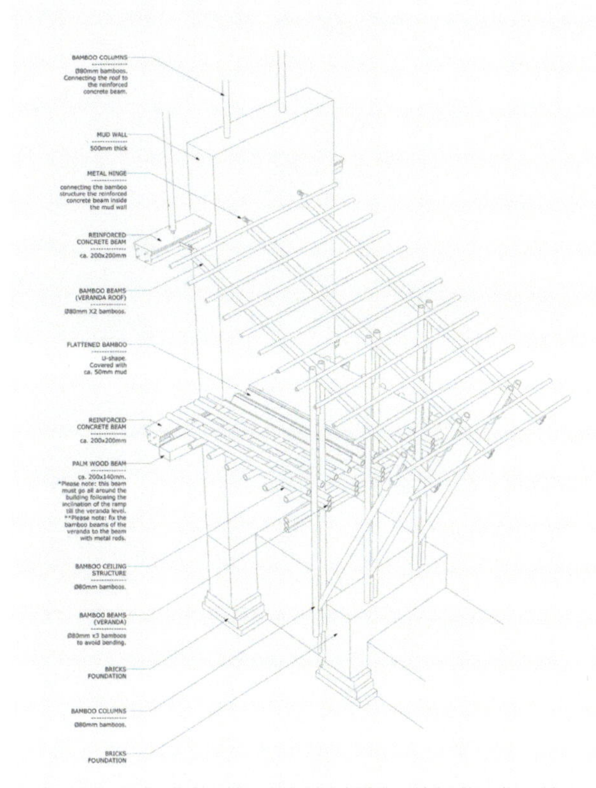

30 Anandaloy Center employs both traditional and modern techniques. Rammed earth walls of cob and bundled bamboo members are used to create walls and roofs. Heringer's emphasis on low-energy and local material results in a construction that can be executed by local craftspeople.

Low-energy materials and systems like earthen construction are increasingly seen as an alternative to high-embodied-energy materials like steel and concrete. Earthen construction is also uniquely able to express the "genius loci," or site specificity, through its unique color and texture as well as reinforce local building traditions and indigenous knowledge. Rammed earth also has a high thermal mass due to its necessary thickness, so it is well-suited for hot climates with large diurnal temperature swings.

German architect Anna Heringer is a proponent for sustainable, site-specific architecture and has been developing techniques for building with local materials and vernacular techniques. Her Modern Education and Training Institute (METI) Handmade School (2006) and her Anandaloy Center (2019) both in Rudrapur, Bangladesh, employ thick earthen walls of cob — a mix of clay, earth, sand, and straw. The cob mix is produced locally with the participation of local builders, and cows are used to mix the cob in a large pit adjacent to the site. 28 29

Heringer first developed her design for the METI School in Rudrapur as a student project and was able to advocate for its construction by arguing that her design was an alternative to more energy-intensive techniques using bricks and CMUs. The METI School was constructed by using local materials, techniques, and labor. The lower level consists of three classrooms enclosed by thick earthen walls. The upper level is constructed of bamboo framing and covered with a corrugated metal sheet roof. Bamboo is an abundant local material, and the craft of lashing and framing bamboo members is familiar to local builders.

For her Anandaloy Center, Heringer applied lessons from the METI School to a different site with similar needs. 30 The center houses facilities for people with disabilities as well as a small studio to produce fair-trade textiles. The exterior rammed-earth ramp demonstrates how the incorporation of accessibility into traditional structures is achievable. The two-story earth construction also pushes the material to showcase its wider applicability.

Heringer's designs hybridize local labor practices and techniques with innovative modifications. The foundations are improved by employing bricks and an impermeable underfloor membrane to protect from moisture. The earthen walls are much thicker than is typically constructed, improving stability and durability, and they are punctuated with more doors and windows, which improves cross ventilation. The generous roof overhangs protect the walls from damage from heavy rains.

31 The Ricola Kräuterzentrum (2014), in Laufen, Switzerland, was designed by Herzog & de Meuron as a prefabricated rammed earth building with a hybrid structure. Large earthen modules were brought to site and stacked in a running bond pattern.

Heringer's work in the region brought international attention to local and traditional building practices and offers an alternative design approach to importing construction techniques that are often more energy-intensive and maladapted to the local climate and construction cultures.[19] She describes architecture as a tool to improve lives, both through the programs it houses, and the techniques and skills used to build the structures.

19 The METI School project won the Aga Khan Award in 2007 and was featured in the 2010 MoMA exhibition, *Small Scale Big Change, New Architectures of Social Engagement*, curated by Andres Lepik.

In 2014, Herzog & de Meuron used rammed earth for a new industrial building for the Swiss herbal products company Ricola in Laufen, Switzerland. After having designed several buildings for Ricola, the Basel-based architects opted for an ancient technique for the new facility, the Kräuterzentrum. The choice of rammed earth is consistent with the firm's ethos of material experimentation. As a technique, rammed earth typically consists of thick walls installed on-site with wooden formwork, which retains the mix of earth, clay, sand, silt, water, and an optional binding material. Sequential layers of mix are installed between the formwork and compacted by hand or mechanical means. The material cost of rammed earth is very inexpensive, but it is labor intensive.

For the Kräuterzentrum production facility, Herzog & de Meuron used a mix of local earth, loam, and marl (an unconsolidated soil composed of clay and lime) with a volcanic truss lime mortar, but they also specified a prefabrication process where the wall was constructed in segments and craned into place. 31, 32 This modular construction technique minimized on-site construction and ensured higher quality control. Special care was taken to grout the joints with a compatible mix to mask the joint lines. The new Ricola warehouse exceeds the size of traditional rammed earth construction by using prefabrication methods, but also because the walls are relieved of their structural requirements as there is a concrete frame within the building footprint that supports the roof. 33 The rammed earth forms a very thick, self-supporting envelope that signals a strong connection to place and an increasing interest in low-embodied-energy material systems.

DESIGNED ARCHAEOLOGIES

The Stone Garden (2020) apartment building in Beirut, by Paris-based architect Lina Ghotmeh, employs the materials and methods of earthen architecture to produce a sculptural building that highlights its connection to place. Ghotmeh describes memories of the architecture and landscape of Beirut, marked by conflict and war,

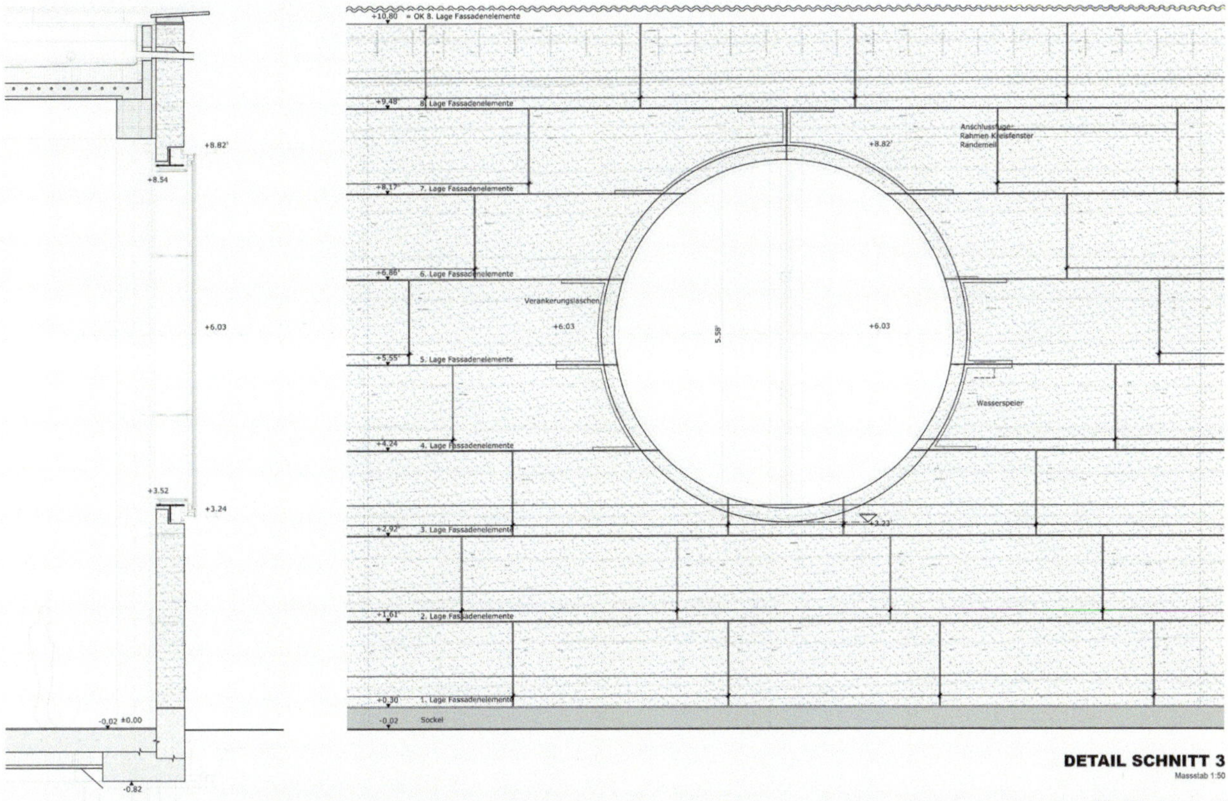

3|2 The large format module of prefabricated rammed earth appears like a large masonry wall.

3|3 The Kräuterzentrum employs the rammed earth as a building envelope with high thermal mass, but the structure of the warehouse is a concrete frame, seen behind the wall in this construction photo.

3 4 Lina Ghotmeh's Stone Garden housing project (2020), in Beirut, Lebanon, was conceived as a highly site-specific structure. The richly textured surfaces recall the strata of the earth beneath it.

35 Ghotmeh's design process involves conceiving of material and texture, references and allusions, but also tools and techniques. For the Stone Garden Ghotmeh designed a long "comb" with which to rake the facade.

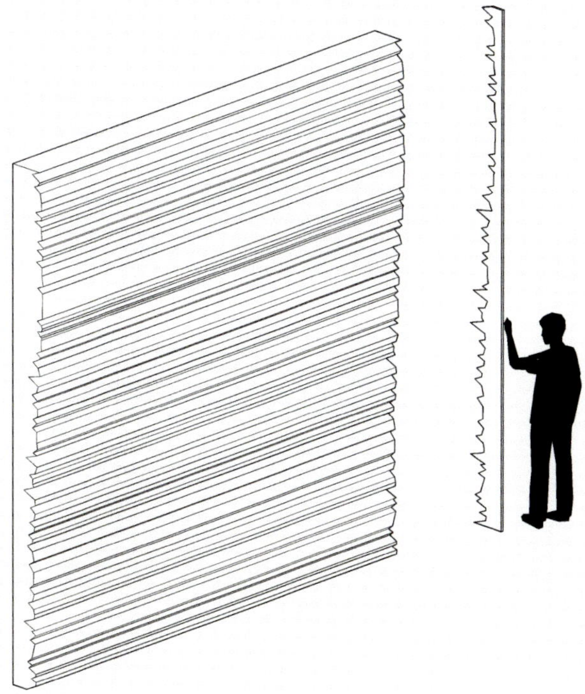

36 The custom tool, with its unique profile, produces the unique texture of the facade, making it both place—and process—specific.

where buildings bear traces of violence on their scarred and pock marked facades. The Stone Garden building rises 13 stories from an irregular site and a context of non-descript buildings. 34 The facades are punctuated by deeply recessed square windows of varying sizes. Small windows on the south elevation correspond to bathrooms and kitchens, while much larger windows on the north facade contain recessed terraces.

Ghotmeh describes her approach as an "archeology of the future," creating work that expresses continuity with the past even as it imagines a different future. Working with local craftspeople Ghotmeh developed a technique for patterning the exterior faces with a horizontal ribbed texture that evokes a stratified section of the site and the marred facades of its context. Ghotmeh describes the building as "emerging from the ground, as a built form of life and death, presence and absence, evanescence and timelessness, beauty and rawness."[20] The exterior walls of Stone Garden are finished with a custom profiled "comb" that is used to "rake" the building with a fine texture. 35 36 The two-inch-deep (50mm) grooves wrap the building's exterior facades, but do not turn into the window returns, making the textured skin appear like an earthy crust on the perimeter.

GAP IN THE FOREST

Underscoring the link between a building and its site, and between materials and their sources, architectural historians Robert Atkins and Hope Bagenal note that, "for every considerable building there is somewhere a quarry or a brickfield or a gap in a forest."[21] The first classrooms built at Cornell University shortly after its founding in 1865 were housed in three stone buildings along the western edge of what is today the Arts Quadrangle. Morrill Hall, McGraw Hall, and White Hall, known collectively as Old Stone Row, share an architectural vocabulary and material palette of Ithaca bluestone. Morrill Hall was the first building built, based on a Second Empire design by Harlow Wilcox and Cyrus Porter. The facades include rusticated stone and narrow vertical arched windows and voussoirs. The stone for all three buildings was quarried from the adjacent slope that is today known as Libe Slope (Library Slope). Standing in front of McGraw Hall, the bowl-shaped depression in the side of the hill is the visible outline of the original quarry, making a direct connection between the subtracted ground and the additive building.

A similar gap exists in the limestone quarries outside Bloomington, Indiana, where the stone for

[20] Project description from Ghotmeh's web site. https://www.linaghotmeh.com/en/stone-garden-el-khoury-foundation.html

[21] Robert Atkinson, and Hope Bagenal, *Theory and Elements of Architecture* (New York, NY: Robert M McBride & Company, 1926), p. 30.

The striking facade of 15 Clerkenwell Close (2017), designed by Amin Taha of Groupwork, challenges our material sensibilities and expectations. His work charts a different sensibility of material, processes, energy, and labor.

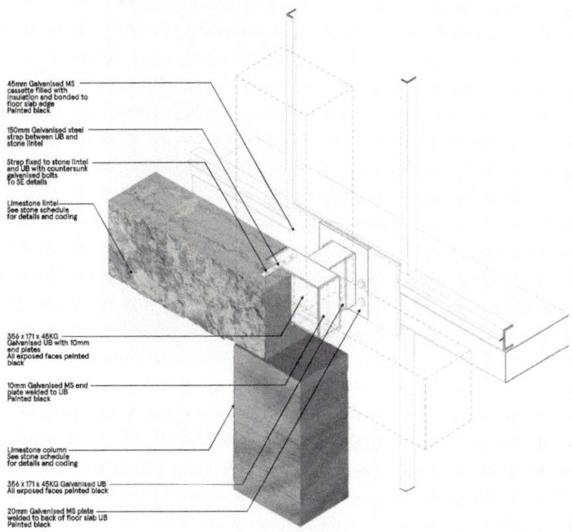

38 The connection detail, at 15 Clerkenwell Close, between the "as is" stone piers, and the custom built and finished elements like the glazing and structural slab systems.

39 The sublime qualities of a stone quarry include the immense scale of quarry—the scale of landscape—and the scale of human intervention, the tiers, or "benches" of sawn blocks. A quarry is simultaneously natural and manmade, familiar yet strange.

the Empire State Building was quarried. The Empire Quarry provided 18,630 tons of stone for the facades of the Empire State Building. Visiting the quarry today, the rectilinear void with steep walls and tiered benches, it looks like a negative imprint that the famous skyscraper would fit back into, seemingly connecting the finished building in New York back to its material origins in the stratified ground.

The experience of visiting a stone quarry can be astonishing, if not sublime, as the man-made geometries of quarried "benches" (the layers of material extracted) merge with the irregular geometries of the mountain. 39 The techniques for breaking the stone free from the ground have gone largely unchanged for hundreds of years. The Romans extracted blocks of Carrara marble by drilling sequential holes into the stone to create a plane of weakness. Wood wedges were inserted into the gaps to pry the block apart. Cables running in loops were used to saw blocks of stone apart. All of these processes of breaking and sawing the stone leave traces on the blocks.

London-based architect Amin Taha, of Groupwork, appreciates these traces of making in the irregular faces and quarry marks on blocks of stone. For his design of a six-story building in a historic neighborhood in London, Taha opted to use dimensional unfinished stone. The building, located at 15 Clerkenwell Close, employs a stone frame of columns and lintels as its primary facades. 37 38 A floor-to-ceiling glass facade sits behind the stone frame. The striking aspect of the stone elements is its "unfinished" appearance, which is sometimes rough and sometimes smooth. Appreciating the cleft and drilled faces of the stone blocks in the quarry, Taha specified the beams and columns to retain the "undressed" face of the rough block, even as other faces are sawn straight and smooth. The members on the facade each have one rough face and three flat faces. Adding to the "found" nature of the stone, Taha has chosen to orient the rough faces outward for most of the elements, but occasionally rotates a column or a beam to show the smooth face. The seemingly random arrangement of rough and smooth members in a casual pattern gives the impression that they are salvaged parts of a previously ruined structure.

The association with ruins and salvage is not accidental. Taha's use of rough-faced stone in large dimensions is a deliberate reference to the presence of a ruined 11th-century Norman abbey on the same site. The unfinished stone highlights the artifice of the surrounding context with its "pretense to history" and authenticity, when its stone facades are only

40 The Final Wooden House (2006), designed by Sou Fujimoto, uses large format wood blocks to create a unique spatial configuration that is seemingly both additive and subtractive.

"skin deep." The facades of 15 Clerkenwell Close provoked the local council who felt that the building was not appropriate to the historic context and called for its demolition on the basis that it departed from the originally submitted permit drawings. The demolition order has since been lifted.

For Taha, the cleft face of the stone recalls the process of splitting it from its bed in the quarry. It has an authenticity in the traces left by its fabrication and processing. It is also the real structure of the building, as the stone columns carry the loads at the perimeter of the building. Taha worked with structural engineers Webb Yates to use the stone piers structurally as an exoskeleton.

Taha's use of dimensional stone members is motivated by the desire to better account for embodied energy in building materials. For Taha, the stone piers have a bearing capacity akin to a site-cast concrete column but are produced with a fraction of the carbon. Taha's studio employs custom life cycle assessment (LCA) calculators to track and account for all the embodied energy of all materials employed in the construction of a new building. The "undressed" stone piers are the aesthetic outcome of an applied and explicit material ethic, informed by material properties and material processes.

The rough stone frame of 15 Clerkenwell Close reveals the processes that materials undergo in their transformation from raw materials to building products. It is precisely its unfinished-ness that drew the ire of the local authorities, and the unfinishedness that highlights the processing that most building materials undergo. This is true for both stone and wood, where materials are extracted or harvested, sawn and standardized, sorted and graded.

MONO MATERIAL

The Final Wooden House by Sou Fujimoto is an experiment in a modular and mono-material approach to a wooden house. 40 Closer to a log cabin than a balloon frame house, Fujimoto's house appears to be an eroded cube made up of large-scale wooden blocks. Each block has a square cross section of 14" x 14" (350 x 350mm). The large format blocks are used as structure and enclosure, as well as furniture and internal stairs. The interior volume of the house seems to be eroded from the solid like a three dimensional puzzle. Sections cut through the house suggest that it is about 50 percent solid and 50 percent void. 43 Despite the insistence on a singular modular dimension for all parts of the house, it is

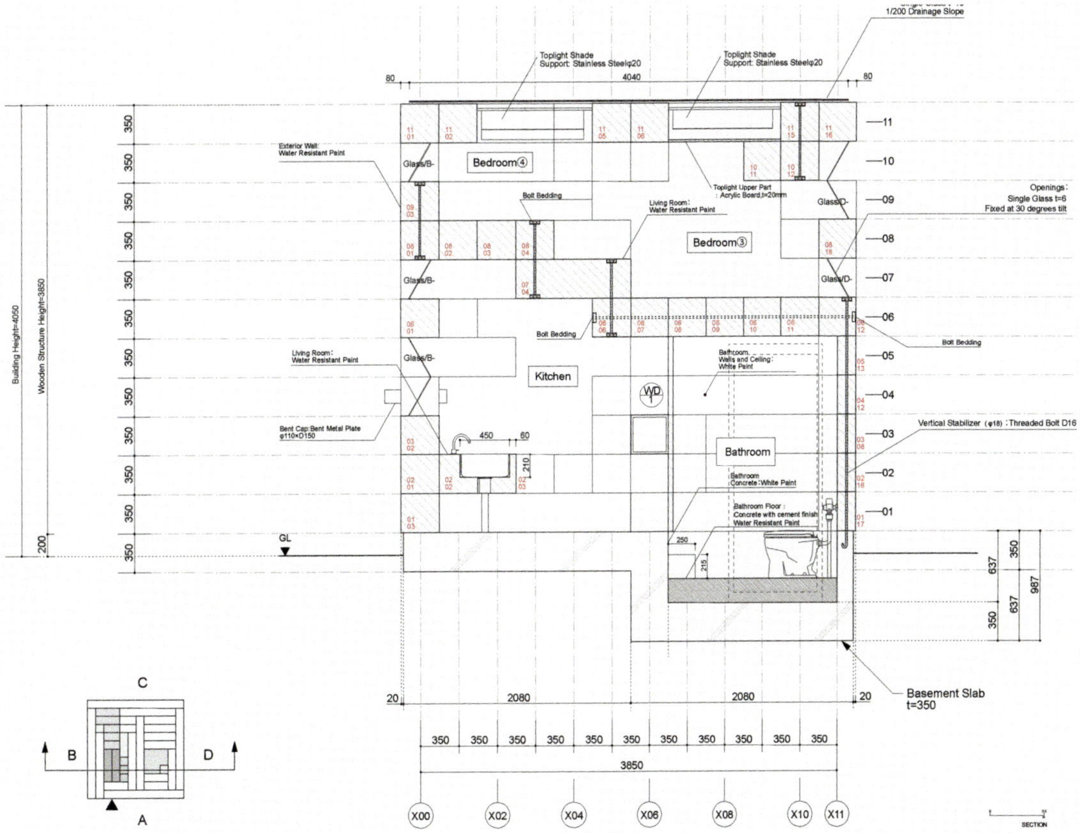

4.1 The section of the Final Wooden House reveals the effort it takes to construct the house like the conceptual model. Tension ties shown in the drawing allow the blocks to span and corbel like the model, while also providing a degree of freedom in the case of expansion and compression of the wood material.

4.2 The stepped interior geometry of the blocks lends itself to ergonomic proportions.

4.3 Physical model of the Final Wooden House.

4|4 The Aspen Art Museum (2014), designed by Shigeru Ban, showcases several types of wood used in contemporary construction, including mass timber trusses and the woven wooden boards seen on the facade.

4|5 The rooftop cafe at the Aspen Art Museum sits under an S-curved mass timber space frame truss.

4|6 The fabrication of the wooden truss involved CNC milling of large panels of laminated veneer lumber (LVL) with notches for interlocking joints.

remarkable how well it accommodates the human body. 4|2 The 14-inch (350mm) module lends itself to the ergonomic dimensions of bench height with one module, and counter height at two. The interior "landscape" of the wooden blocks forms a pixelated, cave-like interior that also acts as a stairway to the roof.

Fujimoto's house stacks self-similar blocks of dimensional lumber held together by fasteners and tension ties. 4|1 As with masonry construction, which relies on stacking units, the amount of overhanging blocks, or corbelling, is limited. Fujimoto ties the blocks together with tension ties, consisting of threaded rods that run through multiple blocks and are secured with an anchor plate and closed with a wooden plug. The detail at the other end is enhanced with a spring connection, allowing the structure to absorb some material expansion due to moisture, or absorb lateral seismic loads.

While Fujimoto's experimental house employs solid timber members which rely on mature trees, contemporary engineered lumber products utilize younger trees of smaller cross section which are laminated and glued as wood fiber into composites. Glue-laminated, or glulam, wood has been used for decades, but it is currently experiencing a surge in demand as mass timber becomes more desirable by architects interested in how the lower embodied energy of wood construction and changes in building codes allow for wider applicability in mass timber structures.

Shigeru Ban has devoted his career to experimenting with wood and paper in architecture as a means for making architecture do more with less. From his earliest experiments with materials, Ban has sought innovation by embracing reuse as a strategy, as much as a necessity. He has sought out partners and engineers like Frei Otto and Hermann Blumer to realize projects that push the limits of lightweight structures. His projects with paper tube construction for refugee shelters sought new uses for waste materials, and his paper tube Japanese pavilion for the 2000 Expo in Hanover demonstrates his use of recyclable materials arranged in a weaving-patterned shell structure.

For the Aspen Art Museum, Ban designed a permanent structure with a combination of laminated timber frame roof and a woven wooden facade system. 4|4 The base building structure is a more conventional steel and concrete frame, and the wood and paper elements serve as visual accents throughout the museum. The top floor of the museum includes an outdoor sculpture court

47 The Trumpf Canteen (2009), designed by Barkow Leibinger provides a light-filled, social space for factory employees at the center of the industrial campus. Conceived of as an organic structure, analogous to a leaf, the roof is subdivided into cells that are constructed of glulam timber.

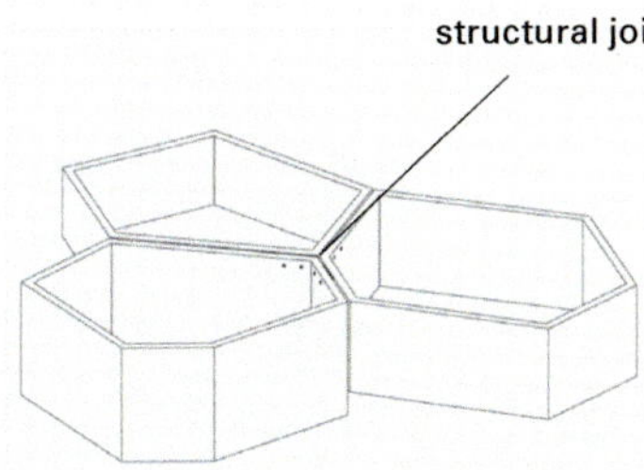

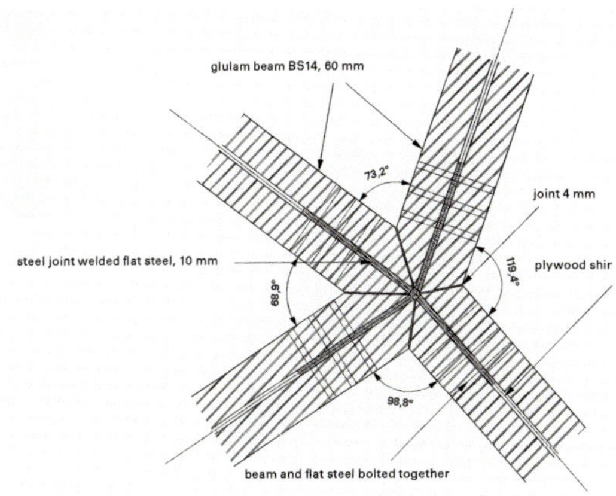

48 The glulam cells of the Trumpf Canteen pack together to create the infill surfaces between the primary steel structure.

49 The irregular geometry of the cells meets at intersections, where steel plate joints tie the individual cells together.

and cafe, expressed as a mass timber wooden spaceframe supported by steel posts. 45 The webbed members are designed as a sine curve, running diagonally between nodes of an orthogonal grid at the bottom chord of the truss. The architects specified spruce laminated veneer lumber (LVL) for the cords, which were optimized for length, and birch cross laminated lumber (CLT) that was used for the web members and benefits from the alternating orientation of wood plies, create greater stiffness and at the same time, minimize member sizes. The Aspen Art Museum's trusses were prefabricated to the largest transportable dimension and then connected in the field with long stainless steel screws. 46

The top floor of the museum benefits from the natural daylight around the perimeter as well as skylights at the center of the plan. The effect of daylight and passing through the woven spaceframe truss celebrates the warmth and lightness of the wood. The building facades employ a woven wooden screen that unifies them and provides solar shading benefits. The facade members are proportioned like planks to allow them to bend in and out of plane to achieve the weaving effect. The wooden screen is supported by a steel tube frame behind the screen and braced back to the building with outrigger struts. Approaching the museum from across the street, the woven pattern recalls a crafted woven object and produces an interesting effect of scale indeterminacy. The screen masks the various volumetric and material conditions beyond view yet adds a familiar reinvented texture to the neighborhood.

Berlin-based Barkow Leibinger has pioneered a material- and fabrication-driven design approach that is marked by their deep research into both material possibilities and applied methods, which they then use to inform specific projects. Their Trumpf Canteen project (2008), for example, demonstrates the use of CNC cutting to produce an irregular wood cell system capable of infilling between deeper, long-span steel members. 47 Working with structural engineer Werner Sobek and wood fabricators Holzbau Amann, the architects developed a hybrid system of glulam panels to form cells, which are bolted together with steel spindle-plates to connect them to irregular angles. 49 The cells are one of three types: skylights, artificial lighting, or acoustic treatment. The long spans between steel columns and the abundant daylight make the space feel light and airy, while the wood-coffered ceiling captivates with its irregular cell pattern that recalls the microscopic cell structure of a leaf.

50 The model of the CLT House (2024), designed by nArchitects, explores the use of rough sawn boards as a cladding material to contrast with the engineered lumber of the house structure.

51 The "live edge" siding of the CLT House celebrates the original state of wood material complete with the irregular profiles, knots, and "imperfections."

LIVE EDGE

The CLT House (2024), designed by Mimi Hoang and Eric Bunge, of nArchitects, is located outside Rhinebeck, New York, and employs a range of different wood types, serving as an essay on the expanded materiality of wood. 51 The walls and slabs are built of CLT panels that serve as structure and interior finish. The exterior finish of the house is made of large, rough-sawn cedar planks in a lapped configuration. For cladding, they use a full board with an irregular edge, which is called waney board, or live edge siding. Between the two layers, the wall assembly includes a vapor barrier, insulation, furring, and an air cavity. The structure, built of mass timber construction, attests to an enthusiasm for prefabrication and renewable materials. The solid CLT walls and slabs speak to contemporary engineered lumber systems that source fiber from small-caliper trees and produce large format blank panels through the lamination of layers. The CLT walls and soffits give the house a distinct warmth. The lap siding on the exterior retains the natural profile, or "live edge" of the tree, recalling the rough edges of vernacular architecture that so often reflects and reveals the natural resources used for construction.

22 In The Book of Tea, by Kokuzo Okakura, the "worship of the Imperfect" is understood as central to the philosophies of the tea ceremony. A description of the tearoom is included in Arthur Drexler, *The Architecture of Japan* (New York, NY: The Museum of Modern Art, 1955), p. 211.

IRREGULAR IMPERFECT

The traditional Japanese tea house is often regarded as embodying the principles of Japanese architecture and its unique aesthetic sensibilities. Traditional tea houses are modest structures, designed to host the tea ceremony in an intimate setting. The tea house typically consists of a single unfurnished room with tatami mats and an in-floor hearth. A recessed alcove, called a *tokonoma*, is typically adjacent to the tearoom, and sometimes includes a raised floor and built-in shelf to display a carefully composed flower arrangement. The tea house employs traditional Japanese carpentry, which consists of precisely sawn, carefully assembled interlocking wooden joints. Exposed roof rafters and screen walls showcase the skill of Japanese carpentry. The framing of the tokonoma also typically includes at least one column that is left "unfinished," retaining its original geometry, including leaving its bark intact. A column (*tokobashira*) and a lintel (*otoshigake*) might include a naturally curved tree branch, or a forking member that acts as a column and a beam. 52 The presence of a "natural" tree element within the tea house is consistent with the philosophy of the tea ceremony, which includes the worship of the imperfect, and the transient.22

5|2 The Murin-an Tea House in Kyoto, Japan exemplifies the refined aesthetics of Japanese carpentry and craftsmanship. The presence of an unfinished wood element, the *tokobashira*, in the composition introduces an element that refers back to natural states of things.

5|3 The tree trunk columns in the ground floor galleries of OMA's Kunsthal (1992) seem intended to provoke other associations, from the pop cultural and surrealist "chance encounter" to the disciplinary "primitive hut" of Marc-Antoine Laugier.

The Japanese concept of *wabi-sabi* is an aesthetic principle that prioritizes the incorporation of an authentic imperfection within a highly intentional aesthetic system. *Wabi* can refer to the functionality of objects and their imperfections. *Sabi* can refer to the natural process of weathering and aging that highlights the impermanence of things. The presence of an unfinished post or beam within a precisely assembled structure recalls the forest from which the wood has been harvested — sometimes from the site itself — processed, and assembled to form the tea house. The crooked tree branch is precisely imperfect to remind us of the natural sources of materials as well as the careful labor of carpenters that transformed the irregular tree branch into a work of architecture.

Similarly, the use of unfinished stone and wood gestures toward material sources and processes that transform raw, natural elements into materials, products, and then into buildings. At a basic level, a finished building is the sum of all the materials that are assembled in one place to form the concert house or the office building. It is also an accumulation of the labor and embodied energy that it takes to harvest, saw, finish, assemble, polish, mine, process, extrude, transport, and assemble those parts to form the finished building. Instances and traces of the unfinished within buildings serve as reminders of this sense of the expanded materiality of architecture.

UNFINISHED TECTONICS

The Kunsthal, designed by OMA, caused a sensation when it opened in 1992. The low, boxy building, partially perched on the Westzeedijk in Rotterdam's Museumpark, appears at once familiar and strange. Its hovering steel roof, exposed I-beam, colonnade, and ribbon windows evoke familiar motifs of 20th century architecture, yet the Kunsthal produces something altogether new as a result of its self-conscious referentiality. Approaching the museum through the Museumpark, it appears from a distance to be a modernist box of stone and glass. Upon closer inspection, the building is not at all as simple as it appears. Rather, it is a carefully designed orchestration, of citations, materials, and motifs that are familiar. An oblique ramp runs diagonally through the middle of the building, connecting to the upper level of the Westzeedijk. From the ramp, a sloped-floor auditorium becomes visible on one side and glimpses of gallery spaces appear on the other. A colonnaded loggia defines the southern edge of the building, facing the busy roadway. An orange-

54 The gold painted "wood grain" pattern on the interior of OMA's Sala Suggia in the Casa de Musica (2005) combines the "high" and "low" brow aesthetics and associations of the opulent and the ordinary.

painted, wide-flange beam casually overhangs the building on the east side, recalling the projecting beams of traditional Dutch town houses. The "slipped" beam is a hint that the structure of the Kunsthal is not as straightforward as it might appear.

The columns that make up the colonnade are irregularly spaced, and each one is unique: a square concrete column, a white-painted steel cruciform column, and a white- painted castellated column that seems like a readymade object purchased from a catalog. The collection continues inside the museum with a series of inclined concrete columns that lean to remain perpendicular to the oblique ramp. Within the lower-level galleries, columns clad in natural tree trunks span from the concrete floor to the black-painted ceiling soffit, creating the uncanny quality of an indoor forest. 53

Structural engineer Cecil Balmond describes the syncopation of structure as an expression of the "informal."23 The tree trunk columns in the galleries recall the trees in the Museumpark beyond, blurring the boundaries between the interior and the exterior, the natural and the manmade. Within the concrete box of the Kunsthal, the trees have a surreal effect, a reference perhaps to Salvador Dalí's "Paranoid Critical Method," which Koolhaas popularized in his manifesto, *Delirious New York*. The assortment of column types suggests that these are citations; tree trunk columns might refer to Laugier's "primitive hut," while cruciform columns recall those of Mies van der Rohe's Barcelona Pavilion, or the Neue Nationalgalerie in Berlin.

23 Cecil Balmond, Jannuzzi Smith, and Christian Brensing, *Informal* (Munich; New York, NY: Prestel, 2007), p. 62.

Another OMA project, the Casa da Música (2005) in Porto, Portugal, offers a similarly surreal material palette on the interior of the Sala Suggia concert hall. The hall is a classic "shoe-box" proportion, for acoustic reasons, but the interior finishes are rich with color and texture. The side walls and ceiling are clad with standard plywood sheets, decorated with gold-painted wood grain patterns. The gold figure is an exaggerated image of the wood grain scaled up to almost appear like a graphic hatch pattern from a CAD drawing. 54 The spare plywood shoebox and the sumptuous pipe organ may typically be incongruous but are unified by the application of white and gold painted patterns. Other spaces within the Casa da Música also offer rich material finishes. In a nod to local craft traditions, the VIP lounge is clad in floor-to-ceiling *azulejo* ceramic tiles. Other spaces are clad with the green-colored, high-relief absorptive surfaces seen in recording studios, or button-tufted upholstered ceilings. Recognizing the closed nature of many performance venues,

55 Venturi, Scott Brown's Institute for Scientific Information (1979) in Philadelphia is widely regarded as a "decorated shed," with its straightforward boxy form and playful use of color and pattern to elevate the ordinary and the everyday.

the architects elected to amplify the differences between interior and exterior spaces while offering glimpses inside to the wider public. From the exterior, the building is like a geological form, hard and faceted. The broad faces of white concrete stand in sharp contrast to the buildings in the surrounding context.

The presence of unfinished wood in the Kunsthal tree columns, or the amplified signifiers of wood like those found in the Casa da Música's gold-painted plywood panels, raises the questions of artifice and authenticity typically conveyed through the materiality of architecture. Traditional uses of stone and wood provided the appearance of luxury and wealth. For example, the Church of San Marco in Venice displays an opulent array of figured marbles that speak to the wealth acquired by Venetian merchants. Mies' large format marble slabs in the Barcelona Pavilion showcase the richness of the natural stone and draw on its associations of opulence and ornament. Koolhaas' selection of materials for the Casa da Música refers to both local crafts and global procurement patterns. The "azulejo" wall finishes gesture toward local tradition, while aluminum plate floors speak to a global web of production, fabrication, and transportation. The super-graphic gold pattern on the plywood sheets highlights the tension between "high" and "low," evoking Pop Art signifiers of wealth and value. There is a Warholian readymade aspect to the silkscreened gold grain in the concert hall that resonates with the presumed elitism of a cultural institution in a working-class neighborhood. Koolhaas' materiality operates on multiple registers to produce surrealist, scale-ambiguous effects and provoke popular associations, all the while maintaining a continuous dialog with disciplinary questions of ornament and artifice.

24 The extreme abstraction of the De Stijl movement reducing art and architecture to primary colors and orthogonal compositions in space, is essentially a non-material movement, replacing material with pigments.

POSTMODERN PALETTES

While 20th-century modern architecture made claims about structural integrity and the so-called honesty of materials, several pioneering projects of the modern movement focused heavily on the image of mass production and industrial materials more than a true material expression. For example, the abstract white planes of Le Corbusier's Villa Savoye, or Gerrit Rietveld's Schröder House were, in fact, masonry walls with painted stucco finishes. 24 Architect Peter Eisenman's early houses celebrated the non-materiality of "cardboard architecture" as a technique of abstraction that would draw attention to formal and geometrical relationships. Famously,

56 The Best store in Richmond Virginia, known as the Peeling Building (1971), is the work of James Wines of SITE. The brick wall is curled on the ends to expose a layer of white painted brick behind.

57 To construct the Peeling Building, Wines detailed the curved wall as a double wythe brick wall with wire reinforcement between courses.

the group of architects affiliated with Eisenman, the "Five Architects," who produced a number of abstract white-painted modern houses, were also dubbed "The Whites."[25]

As a reaction to the "purist" and "abstract' work of "The Whites," Robert A. M. Stern organized a response in which he presented the work of "The Grays," including Romaldo Giurgola, Allan Greenberg, Charles Moore, Jaquelin T. Robertson and himself. The work of "The Grays," came to be known as the "postmodern" architects, who in addition to recovering ornament and western classical motifs, also brought back an expanded architectural palette, and a sense of color, materiality, and a heavy dose of irony.

When the Institute for Scientific Information (ISI) building opened in Philadelphia, Pennsylvania, in 1979, it was quickly heralded as a demonstration of the constructed concept of the "decorated shed." 55 Designed by Venturi, Scott Brown & Associates, the straightforward office building, clad in a brown, CMU block, fronting Market Street incorporated a colorful pattern of colored CMU blocks to animate what is otherwise an unremarkable elevation. Venturi is said to have derived the pattern from a Victorian pattern book. The use of colored CMU blocks to create a graphic pattern on an otherwise generic building is illustrative of his discourse on everyday architecture and his critique of the "less-is-boring" abstraction of modernism. The ribbon windows of the four-story building are separated by flush vertical metal panel piers which are also articulated with color. The use of painted metal panels and colored blocks reinforces the center of the facade as a symmetrical pattern. Early interpretations of the facade likened the pattern to an early computer punch card. A more contemporary reading of the facade might be that of pixels from a low-resolution graphic such as an 8-bit or "de-rez" graphic.

25 The book, *Five Architects* featured a white cover, black and white photos, and houses that were consistently painted white.

The matter-of-factness of Venturi's decorated shed was a critique of the claims made by prominent late-modernist architects. His reference to Victorian patternmaking displayed on an otherwise banal box of a building showcased Venturi's affinity for integrating architectural ornament, often described as something non-structural, non-essential, or non-functional. It is additive, or surplus. Venturi's ISI facade, however, is not additive. It is made up of the existing masonry components of the building. In specifying the CMU block colors, Venturi's facade introduces a pattern that is integral to the masonry, without changing the form of the building. Venturi's postmodernist approach to masonry units as pixels

58 James Stirling and Michael Wilford's, Social Science Center (WZB), in Berlin, Germany, alludes to historical building types, while partially wrapping all the buildings in a striped layer of veneer stone.

59 The "ruin elements" from James Stirling's 1984 Neue Staatsgalerie in Stuttgart, Germany.

of a larger image or pattern anticipates the tensions between material and image that we continue to grapple with today.

James Wines of SITE (Sculpture In The Environment) worked in a similar mode with his designs for Best stores. Wines' Peeling Building in Richmond, Virginia (1971), for example, presents the brick-clad suburban retail box type with a delaminating facade. 56 The facade of the brick veneer is laid with curled ends, as if it is peeling off from the volume of the building. The delaminating yellow brick of the front facade exposes a white facade behind it, reinforcing the illusion that the Best store is a multi-layered volume in a state of unraveling. The left-hand edge of the facade curls to form a partial cylinder. The upper-right corner peels diagonally, hanging precariously over the sidewalk. In order to build the peeling bricks, the masons thickened the outer layer and added steel reinforcement between the brick courses. 57

SITE went on to design dozens of stores for Best Products Company, each with a playful tectonic twist. The Best store in Houston, Texas, had an unfinished parapet and a pile of falling bricks accumulating over the entrance doors. The Best store in Sacramento, California, had an open "notch" corner that would slide out to reveal the entrance. The tilting facade of the Best store in Towson, Maryland, consists of the front facade inclined and bearing on a single point, with three notched openings floating overhead. Each store design used standard building materials while finding innovative ways to turn their thinness and everyday materiality into a signature for the store. Best Products Company went bankrupt in 1996, and most of SITE's designs for Best stores have since been demolished or otherwise altered beyond recognition. The demolition and alteration of SITE's projects for Best stores exposes the acute temporality of commercial architecture.

The articulation of thin veneers and ironic facades became part of postmodernism's tectonic language in the 1970s and 80s. James Stirling and Michael Wilford designed the Social Science Center (WZB) in Berlin to house a research institute dedicated to the study of the social sciences. The building is an addition to an existing monumental building, the 1894 Imperial Insurance Office by August Busse that fronts the Landwehr Canal. The planimetric design for the WZB is a collage of recognizable plan forms, such as a hexagonal campanile, a linear stoa, a basilica church plan, and a semi-circular theater.

Despite their diverse and idiosyncratic plan geometries, the facades of the new building

60 A material record of a moment in time, concrete is a medium ideally suited to recording the traces of its own making. The bush hammered concrete of I.M.Pei's Everson Museum of Art (1987) in Syracuse, New York.

61 The process of bush hammering concrete involves labor intensive work of jack hammering concrete to expose the aggregate. The Everson Museum of Art employed a special technique of impact hammering concrete in diagonal patterns.

are highly consistent, with horizontal banding alternating between pink and blue stucco. Punched windows are articulated with projecting red stone hoods. The facades meet the ground with a stone-clad base of alternating bands of red and yellow sandstone. The sandstone cladding extends from the ground up to the first floor in most areas. In other areas, it extends further before stopping abruptly in a staggered pattern. The stone cladding is deliberately incomplete, appearing as a kind of instant ruin. 58

Like many buildings in Berlin, the facade of the historic Busse building was heavily damaged during the war. The heavily rusticated stone is pockmarked with gouges and bullet holes. Michael Wilford describes the decision to retain the original building with its "disfigurations," as preserving the traces of history. The new buildings employ archetypal floorplates, as a kind of "instant archeology" that was popular during the postmodern period. The incomplete stone cladding is also a reference to the ruined state of the building, as well as a commentary on the materials and means of construction. Wilford recalls the tight budget and the use of stucco common in German buildings, but he also embraces the stucco as an opportunity to bring color and character to the new building. Wrapping the diverse collection of forms in a consistent skin, Wilford refers to the facade as "rather like wallpaper" that has been "pasted" onto the building mass to unify the buildings and give them a singular identity.[26]

Stirling and Wilford's articulation of the facade materials reveals the thinness of the sandstone cladding. The open joints and exposed stone support clips reinforce the veneer reading, as do the jagged edges of the "ruin" motif.[27] Wilford also calls attention to the window surrounds as a sun shading device as well as a means for making the facade appear thicker. The design of the WZB seems to celebrate the thinness of contemporary rainscreen masonry construction, while also giving the occupants a sense of solidity and security that the deep window surrounds convey.

The three examples of postmodern material tectonics represented by Venturi, Scott Brown & Associates' ISI building, Wines' Best stores, and Stirling and Wilford's WZB building reveal an attitude about contemporary construction, utilizing veneer materials with a matter-of-fact attitude. Making manifest the material realities of contemporary construction and celebrating the superficial, the veneer, and the super-graphic

26 An interview with Michael Wilford is on the WZB website. https://wzb.eu/en/news/anything-but-grey-how-we-designed-the-wzb

27 The Social Science Center in Berlin shares details and design approaches with other projects in the office at the time. The Performing Arts Center at Cornell University has a similar collection of archetypal forms as its conceptual framework. The Neue Staatsgalerie in Stuttgart shares the same open joint stone detailing as well as a moment where the stone panels appear to be falling off the building, leaving a curious and irregular opening in the facade.

62 Louis Kahn, Yale Center for British Art, New Haven, CT, 1966.
63 Jiakun Architects, West Village, Chengdu, China, 2015.
64 Lina Bo Bardi, SESC Pompeia, São Paolo, Brazil, 1986.
65 Machado Silvetti, PAAM, Provincetown, MA, 2005.

66 Marcel Breuer, Becton Engineering Center, New Haven, CT, 1970.
67 Marcel Breuer, Atlanta Central Library, Atlanta, GA, 1980.
68 Paul Rudolph, Boston Government Service Center, Boston, MA, 1971.
69 Herzog & de Meuron, 216 Chrystie Street, New York, NY, 2017.

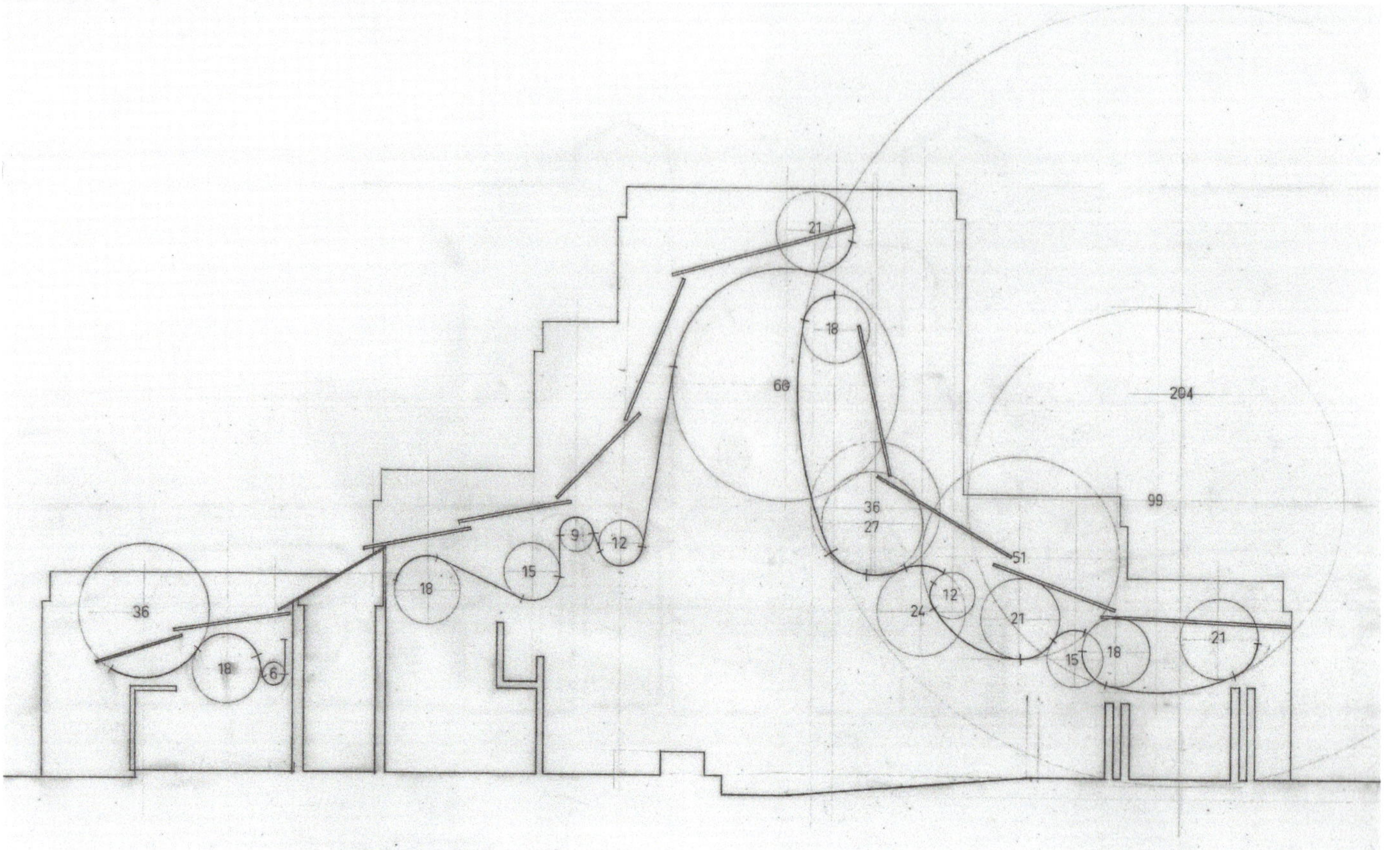

70 Jørn Utzon conceived of the Bagsvaerd Church (1976) ceiling as a billowing cloud and translated that concept into a concrete profile of convex arc segments.

qualities of contemporary culture, postmodern architecture challenges certain claims made about "authenticity" and working "in the nature of 60 materials," made by generations of modernist architects who sought expression in the new qualities of an evolving material culture.

BÉTON BRUT

The material system most commonly associated with notions of authenticity is site-cast concrete. Concrete has been admired since its earliest applications as a material that bears the traces of its making. Formwork leaves an imprint on the surface of concrete construction, giving rise to experimental patterns and textures in board-formed concrete as well as post-produced concrete finishes like bush-hammered concrete. 61 The term *béton brut*, which gave rise to the label Brutalism, refers to the rough-finished concrete that highlights the process of forming and casting. Le Corbusier referred to béton brut as a mode of building that was known for its unapologetic and unadorned qualities. His projects for the monastery at La Tourette (1960) and the Carpenter Center for Visual Arts at Harvard (1963) celebrate the rough and unpainted finishes of board-formed concrete. The craft and labor of forming concrete are preserved, frozen like a three-dimensional portrait — a material record of a moment in time.

Concrete is more a process than it is a material. Construction photos of Jørn Utzon's Bagsvaerd Church in Denmark, completed in 1976, show the processes of assembling the curved wooden forms that make up the building's unique roof profile as well as the placing of steel rebar mesh that follows the curvature of the section. 71 Photos document the process of applying the concrete mix as a shotcrete system with a pneumatic hose. Once the concrete is cured, the wood formwork is removed, and the billowing concrete "cloud" forms are revealed. Gaps between the curves are glazed to create a clerestory window at the top of the section, bringing indirect light into the church. 70 72 The process of concrete construction consists of building multiple versions of the building. First, the wood building is made as formwork, then a steel rebar mesh version of the building is overlaid, and finally the liquid concrete building is poured into the first two. In the process, the steel form is subsumed by the concrete form and the wood structure is removed and discarded.

The process of making three buildings that become one is part of what makes concrete construction a unique process and a site of

7.1 The construction of Bagsvaerd Church involved the forming of the sine wave geometry in wood, laying rebar, and covering the forms with a spray-on concrete.
7.2 The completed interior of Bagsvaerd Church with the indirect light entering the sanctuary between the layers of billowing concrete.

7.3 Preparing fabric formwork for the concrete panels of the Hanil Visitors Center (2009), in Chungbuk, South Korea.

experimentation, with the mix, its aggregate, as well as the patterns and textures of the forms. The labor involved with the formwork is also a reason why site-cast concrete construction has become prohibitively expensive in many labor markets. The production of fair-faced concrete is a science and an artform that combines pigments and admixtures, formwork, and form liners, as well as techniques for vibrating concrete to remove any bubbles trapped in the mix. The imperfections and imprints of concrete construction reveal how the molasses-like viscosity hardens into permanent stone-like solidity and are part of the procedural beauty of *béton brut*.

Spanish architect Miguel Fisac developed innovative concrete forming techniques with both precast and site-cast concrete construction. Fisac's Mutualidad de Papel, Prensa y Artes Gráficas (MUPAG) Rehabilitation Center in Madrid (1973) showcases his experiments with flexible formwork to produce a pillowed facade panel. 7.5 He writes: "After a decade making exposed concrete, I realized that something was not right, because the concrete took on the texture of the planks, as if it were wood; so I decided to give it an expression of its own, because if it is a material you pour on site when it is still soft, it should have a final appearance resembling that fluidity."[28]

Fisac's MUPAG facade panels capture the liquid nature of concrete pressed against the plastic membrane stretched over a frame and subdivided by steel wires. Architect Andrew Kudless notes how Fisac's bulging material experiments have been interpreted as "grotesque" aberrations, as well as "monstrous and imperfect."

"Like Gaudi and Otto, Fisac was not interested in purely rational structuralism, but in the ability of emergent material forces and new construction techniques to literally inform form. These forms, grotesque to some, offer a new approach to the aesthetics of architectural form. Impure, imperfect, and complex, the undulating facades of Fisac's pre-cast facades point to a certain resonance between form, growth, and behavior that is beyond the domain of the designer."[29]

Expanding on Fisac's fabric formwork studies, Canadian architect Mark West, has developed a body of work with his material research group, Centre for Architecture Structures and Technology (CAST) at the University of Manitoba, that finds formal invention in the hydraulic behavior of liquid concrete. His experiments with fabric-formed beams and branching columns evoke the work of Antoni Gaudí and suggest a biological inspiration for the forms. Like organic forms, no two beams are the same.

28 Miguel Fisac quoted in *Arquitectura Viva*, https://arquitecturaviva com/works/centro-de-rehabilitacion-mupag-madrid-

29 Andrew Kudless, "Bodies in Formation, The Material Evolution of Flexible Formworks," in ACADIA Proceedings, 2011, p. 102.

The Hanil Visitors Center (2009), designed by BCHO Architects with Mark West of CAST, experiments with fabric-formed concrete to produce the concrete rainscreen panels that give the center its distinct appearance.

75 Miguel Fisac pioneered experiments with fabric-formed concrete in the 1970s, like the facade of the MUPAG Rehabilitation Center, in Madrid, Spain.

West collaborated with BCHO Architects on the Hanil Cement Visitors Center and Guesthouse in South Korea. 71 The visitor's center for a cement factory was designed to showcase various techniques of using concrete, including fabric-formed concrete and recycled concrete in gabion walls. The fabric-formed vertical profiles are 23-foot (7m) long, site-constructed tilt up piers. The formwork consists of extruded tubular profiles draped with an impermeable membrane. 73 Plywood end dams cap the mold on both ends. The weight of the liquid concrete in the fabric formwork causes the fabric to ripple slightly at the ends of the piers. These subtle wrinkles and "imperfections" in the fabric are transferred into the concrete piers and serve as an indexical trace of the material process of casting the piers. "The vocabulary is given by the materials," says West. "Casting is a collaboration with materials."[30] Referring to the irregular parts of the concrete piers that West describes as "unpredictable," he argues for being deliberate about what you can control and what you do not need to control. "That's where the magic happens."[31]

Studio Gang's SOS Children's Villages Lavezzorio Community Center (2008) project creates a distinctive identity using the color difference between subsequent concrete pours. 76 The community-based project provides social services and neighborhood amenities in the Auburn-Gresham neighborhood of Chicago's South Side. The exterior wall was conceptualized as a "strata wall," made up of the different concrete pours. The color difference between concrete pours is highlighted in the stratified facade articulation, with alternating bands of light and dark gray concrete pours. 77 These "lift lines" or "cold joints," are traces of the material process of concrete construction, and become the signature image of the community center.

Concrete as a process lends itself to experimentation and technique, as architects and builders employ the formwork, pigments, and finishes to develop unique concrete finishes. Louis Kahn, 62 I.M. Pei, 60 and Paul Rudolph 68 each developed a signature concrete style that was inseparable from the craftsmanship of construction. Lina Bo Bardi employed a loose-fitting form on the SESC Pompeia water tower to produce a unique overflowing texture at the joints between forms. 64 Chinese architect Liu Jiakun experimented with woven bamboo form liners to imprint the walls of the West Village with a hand-crafted texture. 63 Steven Holl's crumpled paper concrete at the Kennedy

[30] Conversation with the author, August 8, 2024.

[31] Conversation with the author, August 8, 2024.

7|6 The "strata wall" of the Lavezzorio Community Center (2008) was conceived by Jeanne Gang as a kind of tectonic mural, made up of different concrete mixes, symbolizing a diverse community coming together to create a neighborhood institution.

7|7 Concept models of the layered mix of concrete pours for the Lavezzorio Community Center (2008).

Center REACH in Washington, DC, takes the appearance of an entirely new material.

INDETERMINATE MATERIALITY

The Snøhetta designed 2016 addition to the San Francisco Museum of Modern Art (SFMOMA) rises like a cliff behind the original Mario Botta designed structure. The addition is largely opaque, with buff-colored facades, and creased with horizontal undulations that recall the strata of a geological formation. 7|8 The composite museum, made up of the two buildings from different eras, speaks to the period in which each structure was built. Botta's museum, with its wedding cake massing and lipstick oculus was a self-conscious postmodernist icon. Snøhetta's addition, built 21 years later, reflects a completely different attitude about the city, about architecture, and material culture. Where Botta's building recalls the architecture of the historic city and the craft of the masonry, Snøhetta's monolithic addition evokes a towering landform beyond the scale of architecture or the city. Founding partner Craig Dykers describes the bowed and rippling facade as inspired by the natural beauty of the Bay Area: "We are creating a series of terraces that act like the San Francisco landscape."32

32 Craig Dykers quoted in *Architectural Record*, https://www.architecturalrecord.com/articles/2380-sn%C3%B8hetta-releases-details-of-sfmoma-expansion

During the design process, the facade evolved from a stone-like texture in the early renderings, to a precast concrete, to a glass fiber reinforced concrete (GFRC), and finally to a fiberglass reinforced polymer panel (FRP) on a unitized curtain wall. The lightweight fiberglass panels use shipbuilding materials and digital fabrication techniques to arrive at a cladding system that is lighter, cheaper, and faster to install than the earlier precast approach. The fact that the design concept began as one material and then evolved into another through an R&D process makes the project an instructive case study in contemporary design and construction.

The renderings produced during the 2011 competition emphasized the formal qualities of the rippling white volume but were ambiguous about its materiality. There are no joints shown in the renderings and a close-up of the Howard Street entrance shows a cement-like texture map. Once the project was developed through schematic design and design development, the architects drew and priced several alternate facade systems. The amount of irregular geometry on the facades suggested that the cladding material would require many unique panels or would need to undergo

7 8 The 2016 addition to the SFMOMA was designed by Snøhetta as evocative of a geological formation of rippling stone-like material. Over the course of the project design phases the facade evolved into a fiberglass panel system.

79 The north-east facade of the SFMOMA looks like an abstracted geological formation punctuated by discrete windows, terraces and a connecting bridge.

80 The development of the facade involved mockups and testing to demonstrate the viability of the fiberglass panel mounted to an aluminum curtain wall.
81 The custom fiberglass panels piggybacked on an existing unitized curtain wall system to achieve cost, material and schedule savings.

an optimization process. The process of design development including material research and full-scale prototyping is the standard method for managing the risks of uncertainty involved in an unprecedented facade design.

In the case of the SFMOMA, the use of FRP was serendipitous. Composite fiberglass fabricator William Kreysler was visiting Snøhetta's office to discuss the interior ceiling soffit. Kreysler recalls inquiring about how the facade was intended to be built and heard that the architects were unsure.[33] Kreysler, who has a background in fabricating fiberglass boat hulls, offered to explore the use of fiberglass as an exterior cladding material for the museum. FRP had not been used as a building material for a structure that large because it had not been approved by the National Fire Prevention Association's (NFPA) fire testing process, required for buildings over four stories. With the support of the owner, and through a design assist process, Kreysler successfully passed the NFPA 285 process, enabling them to use their proprietary Fireshield 285 system on the SFMOMA project.

The FRP system involves the production of large-format fiberglass panels over unique molds. The molds are custom-shaped EPS foam blocks that are CNC milled using tool paths derived from the architect's 3D model. Because each unit is unique, the 710 panels required 710 unique molds. The layup of the fiberglass involved applying layers of coatings and fiberglass sheets to the inside "tub" of the mold, to achieve a 3/16 inch (4.8mm) thickness. The face layer includes the buff-colored sand grains that are eventually sand blasted to achieve the stone-like look. While the outer face of the panel is contoured to produce the undulating effect, the interior face of the panel is planar and attaches to the rectangular frame of the curtain wall unit.

The curtain wall frame system, produced by Enclose, allowed the facade to be assembled quickly and precisely. 81 The unitized aluminum frame system is engineered to snap together with gasketed stack joints and has integrated pressure-equalized cavities and weeps.[34] The facade anchoring system was developed to account for slab construction and erection tolerances. By piggy-backing the FRP panels onto an existing building system, the design team isolated and minimized the risky elements of the exterior face and utilized a standard building envelope system as the primary building enclosure. 82

The finished museum building matches the initial renderings, albeit with a higher density of

33 The process of how Kreysler and Snohetta developed a custom FRP facade together is described in an article in *The Architects' Newspaper*, "Kreysler & Associates Erects an FRP First in San Francisco," https://www.archpaper.com/2015/08/kreysler-associates-erects-an-frp-first-in-san-francisco/

34 The horizontal stack joints in a unitized curtain wall are wide enough to account for slab deflections, where the vertical joints are narrower. The SFMOMA facade initially had a horizontal orientation to reinforce the horizontal stratifications. This was changed to vertical after it was noted that the vertical joints would be easier to hide than the wider horizontals.

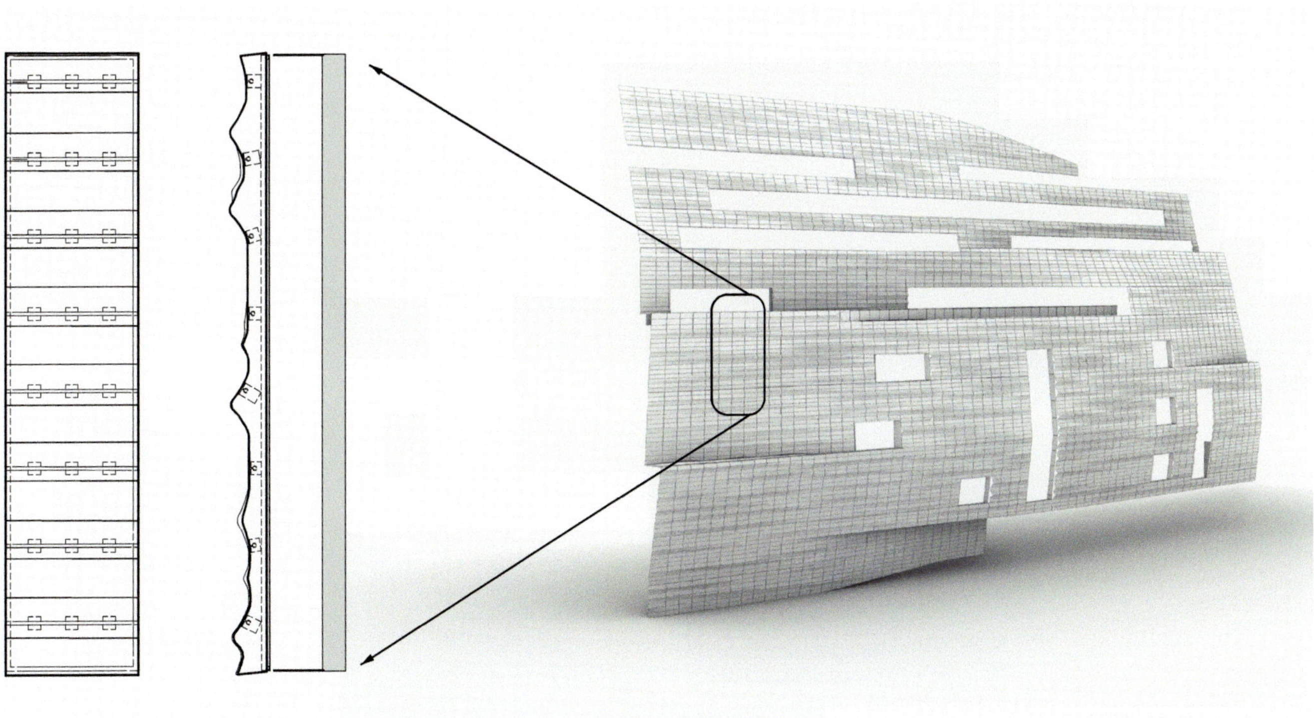

8.2 The entire facade of the Snøhetta-designed wing consists of unitized curtain wall panels that subdivide the structure into five-foot-four-inch (1.7m) modules. The R&D process that resulted in the fiberglass reinforced panel (FRP) facade is representative of the state of contemporary material systems and processes today.

vertical joints. Frontal views of the building's north-east elevation show the building's undulating horizontals that cast deep shadows across the facade. The horizontal ribbon windows at the setbacks slip into the solid-looking surface. The vertical joints are the telltale detail. They slice through the building's monolithic form with the regularity of an egg slicer and align perfectly with the curtain wall joints in the glazed areas. The five-foot-four-inch (1.7m) module, typical to curtain wall, runs through the building facades and soffits, hinting at the fact that the building is no monolith, but rather a carefully detailed panel system, designed to produce a monolithic effect. 35

35 For more on monolithic architecture, see Chapter [illegible], "Monoliths."

The material research and development process yielded a new, code-compliant cladding system and limited the risk to the owner and the builder. The integration of Kreysler's lightweight FRP panel with the unitized curtain wall system allowed the building envelope to be installed quickly in one pass, without a secondary structural layer, and the lightness of the system produced savings in both the steel tonnage and foundation costs. The architects were able to achieve their design intent through a process where the building envelope evolved from a preliminary material concept to an entirely different material system.

EXPANDED

The expanded process of design, from concept to construction, considers materials and their potentials, as well as the processes that they undergo, from extraction, cutting, fabrication and installation, as well as the labor involved, the skills required, and the code constraints met. Increasingly, digital tools are used in fabrication and surveying to achieve unique and unprecedented results. It is important to remember that materials have no predetermined application. Materials have qualities and they have potentials. They are subject to processes and transformations, and they are sites for invention and innovation. Conceptualizing a building as a material process, consisting of forms, materials, and labor, is to imagine architecture as a series of material operations and time-based procedures that result in a building as the carefully considered material outcome. That is tectonic imagination.

1 The Hemeroscopium House (2008), designed by Ensamble Studio, is an explicit assemblage of discrete parts that work together to form a balanced whole.

Chapter 4 Part to Whole

The wood siding of the Newton House consists of vertical planks of red mahogany [2]. At a three-quarters of an inch (75mm), the joints between planks are unusually wide. Unlike conventional siding, the planks are not interlocked or lapped. They are open, exposing a layer of horizontal, black-stained wood boards beneath. The Newton House employs a rainscreen system, creating a ventilated cavity behind the boards and allowing the material to weather and dry out. The eye-catching detail of layered boards is accentuated by the tapering of the ends of the boards, attenuating their proportions, accentuating the width of the joints between them, and calling attention to each individual board. Arranged together, they define the plane of the building's south elevation. Yet, the gaps between the boards and their tapered ends make them discreet. They hold the surface that constitutes the facade — the architectural whole — yet retain the autonomy of the individual board — the architectural part.

Designed by Nader Tehrani of NADAAA, the Newton House is a modest addition to a historic home that does not derive its form from the original but speaks to new forms of craft and modes of construction. The articulation of the wood siding exemplifies an approach to aggregative architecture. For Tehrani, aggregation is integral to the very act of architecture, or "the painstaking assemblage of elements that attempts to form something larger than the sum of its parts."[1] In his essay on the topic, Tehrani describes two approaches: the figurative and the configurative. The figurative consists of assemblages of parts that are legible primarily as a unifying figure — examples include building typologies such as the basilica or the rotunda. On the other hand, the configurative, "involves the systematic deployment of parts, privileging the unit of construction, whether it be the brick, the beam, or the module of inhabitation. Aggregation does not determine a priori a legible final figure, yet it precisely anticipates form by way of the materials, methods, and rules of assembly."[2] Tehrani offers Moshe Safdie's Habitat 67 building as an example of the configurative. Safdie's Habitat consists of stacked prefabricated housing units, which reflect the constitutive parts from which it is assembled. Another way to read the tension between the figurative and the configurative is that the configurative prioritizes the process and components of construction, whereas the figurative prioritizes the perception of the building as a complete whole. The tapered boards of the Newton House exemplify a configurative approach where the boards deliberately call attention to the constitutive

1 Nader Tehrani and Justin Fowler, "Aggregation," in *Material Design, Informing Architecture by Materiality*, ed. Thomas Schröpfer (Basel: Birkhäuser, 2011), p. 48.

2 Nader Tehrani and Justin Fowler, "Aggregation," p. 50.

2 The tapered wood boards of the Newton House (2010) were designed by Nader Tehrani to remain discrete, defining the wall, the architectural whole, while retaining their autonomy, the architectural part.

elements and part-to-whole relationships that are endemic to architecture.

Architecture is inherently heterogeneous. It is made up of thousands of discrete parts that must be assembled, formed, joined, bolted, adhered, welded, friction-fit, nailed, and fixed together to form what is often ultimately recognized as an architectural whole. It is the thoughtful, intentional, and legible organization of these parts, and a way of thinking about architecture as assembled, which translates buildings into discursive artifacts comprised of the specific materials and their qualities. This approach to architecture retains the logic of construction and assembly as an expressive driver for design and anticipates the legibility of the completed work of architecture. The question of how architecture is made, and how all the materials are together articulated to take the form of spaces and buildings falls within the realm of "tectonics."

TECTONIC CULTURES

The celebration of architectural components and their proper expression found their most vocal advocate in Kenneth Frampton, whose 1995 book, *Studies in Tectonic Culture: The Poetics of Construction in Nineteenth and Twentieth Century Architecture*,[3] reads architectural modernism through the lens of the constructed detail and its discursive capacities. Frampton tracks the concept of space in architecture discourses to explore how the processes of assembly and construction are made legible. From his introduction: "Without wishing to deny the volumetric character of architectural form, this study seeks to mediate and enrich the priority given to space by reconsidering the constructional and structural modes by which, of necessity, it has to be achieved. Needless to say, I'm not alluding to the mere revelation of constructional technique but rather to its expressive potential."[4] Frampton proposes a theoretical and philosophical framework within which to read the embodiment of these principles over 150 years. His writings provide a powerful lens through which to read the architecture of the recent past and offer a framework for the assessment of architecture and construction today. In the years since Frampton's landmark book, the interest in tectonics has ebbed and flowed as new generations of architects and critics have taken it up in light of new tools and techniques of design and construction. Contemporary architecture finds new relevance for questions of tectonics and techniques that are both challenged and re-affirmed by contemporary

3 Kenneth Frampton, *Studies in Tectonic Culture: The Poetics of Construction in Nineteenth and Twentieth Century Architecture*, (Cambridge, MA: MIT Press, 1995).

4 Kenneth Frampton, Studies in Tectonic Culture, p. 2.

5 The break is less between modernism and postmodernism, but rather the pre-digital, the digital and the post-digital. The shift has been described as the "digital turn" by architecture critic and historian Mario Carpo. *The Second Digital Turn, Design Beyond Intelligence*, (Cambridge, MA: MIT Press, 2017). Art critic Dave Hickey characterized architecture as experiencing a fundamental shift with the introduction of what he calls "custom steel."

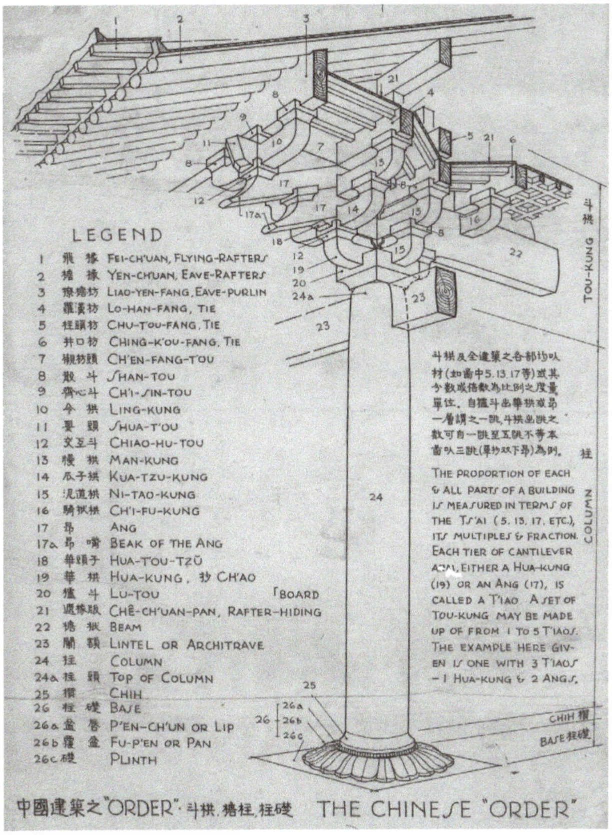

3 The Chinese Order, as documented by Sicheng Liang and Huiyin Lin in 1946, codified the proper usage and articulation of building components to achieve a sense of order and tectonic correctness.

work. The following two chapters argue for a re-examination of part-to-whole discussions in architecture, as well as its opposite: the monolith.

The most obvious difference between Frampton's *Studies in Tectonic Culture* text, and the present moment is the massive impact of digital technologies that have changed every aspect of architecture, from its conception, documentation, fabrication, and construction.⑤ If Frampton argued that the translation of architectural ideas from concept to construction could be evaluated by how those ideas were materialized, the specifics of assembly and their legibility remain critically important. Yet a discussion of the craft of architecture, what Frampton might refer to as "baukunst," or the art of building, has been significantly impacted by the continuously evolving means of design and production that incorporate digital tools, altering notions of craft and construction, and reframing a tectonic understanding of architecture.

As we continue to embrace a more complete understanding of the sheer materiality of architecture and the embodied energy of what we make, designers and builders are reassessing the impacts of their practices. An awareness of the life cycle of materials is what Kiel Moe calls "the details of the details."⑥ Shifts in understanding both how architecture is first conceived and later perceived warrant further analysis. For the purposes of this chapter, we will look at projects that advance the understanding of architecture as the intentional assembly of parts, which often results in varying interpretations of its composition as a broad spectrum of differences that synthesize to become whole.

Contemporary practices like Antón García-Abril and Débora Mesa's Ensamble Studio embrace the architectural ready-made. For example the industrially-sized beams and trusses they used for Hemeroscopium House are a built commentary on mass production. For other projects, like Sou Fujimoto's 2013 Serpentine Pavilion, assembled from thousands of distinct white extruded members, the articulation of parts produces specific perceptual effects, namely of a blurred boundary of the pavilion.

Unique shaped steel members, conceived on the computer and fabricated with CAD-CAM processes, and erected with digital surveying equipment, he argues, inaugurated a paradigm shift in architecture that is distinct from the previous means of construction. Comment made during a lecture at Harvard GSD, September 12, 2019.

⑥ Kiel Moe argues for the "radical act of literal description," to better describe architecture as a material and energetic artifact. Kiel Moe, *Unless: The Seagram Building Construction Ecology*, (Barcelona, New York, NY: Actar Publishers, 2020), p. 42.

BUILDING SYSTEMS

The discussion of part-to-whole relationships is not specific to modern or contemporary architecture. Historic treaties such as Vitrivius' *De Architectura* or *Ten Books on Architecture* (30 to 15 BC), or the Chinese *Yingzao Fashi*, or *Treatise on Architecture Methods or State Building Standards* (1103), serve

4 Kengo Kuma's Yusuhara Bridge Museum (2011) draws on building traditions to create a contemporary expression of the beam and bracket system.

5 The proliferation of the wooden joint leads to the engulfing of the Prostho Museum (2010) in a "cloud" of carpentry.

as templates for the proper usage and articulation of materials and other building components. In the case of the Chinese *dougong* bracket system for timber column/beam interfaces, for example, the manual codified the articulation of a hierarchy of parts and their expression.[7] 3

The intricate assemblies of individual brackets joined to form the elaborate structures of column capitals, beams, and eaves in Chinese Song Dynasty structures produce the architectural expression appropriate to its function. Hundreds of individually crafted components fit together to produce a cohesive whole, complete with a hierarchy of stacked brackets that form a canopy, like an abstracted architectural tree. The articulated beam ends and the stacking and interlocking logic of the parts were derived from carpentry joints developed over time through traditional carpentry craft. Looking at the underside of a Song Dynasty roof structure, all elements work together to make connections, transfer forces, and articulate the significance of the structure. Each component part contributes to the overall structure without being subsumed by the whole.

Contemporary examples of timber construction that employ carpentry details reminiscent of traditional carpentry joining techniques are illustrated by two projects designed by Kengo Kuma. The Yusuhara Wooden Bridge Museum (2011) employs a single column to support a tiered structure evocative of the bracketing system of traditional Japanese construction. 4 The contemporary bracket structure is made of glulam beams, and they are assembled with modern techniques, but the overall effect is of a meticulously crafted cantilevered structure transitioning from a single post to an occupiable building.

Kuma's Prostho Museum (2010) turns a Japanese wooden joint system, called a *chidori*, into a volumetric lattice that engulfs the museum. 6 The cubic lattice measures 80 inches (200cm) and is joined without adhesives or mechanical fasteners. The glass enclosure is embedded within and obscured by the lattice, resulting in an ambiguous sense of the limits of the building. 5 Both the Bridge Museum and the Prostho Museum depart from traditional detailing in significant ways, forming a hybrid construction that integrates steel and wood. Kuma describes his interest in "systems in which the possibility of an endless combination of small particles produces total freedom in the overall design — a similar concept to organic systems, in which small cells combine to form the whole."[8]

[7] Francis D.K. Ching, Mark Jarzombek, Vikramaditya Prakash, *A Global History of Architecture*, (Hoboken, NJ: Wiley, 2011), p. 401.

[8] Kengo Kuma, *Kengo Kuma Complete Works*, (London: Thames & Hudson), p. 185.

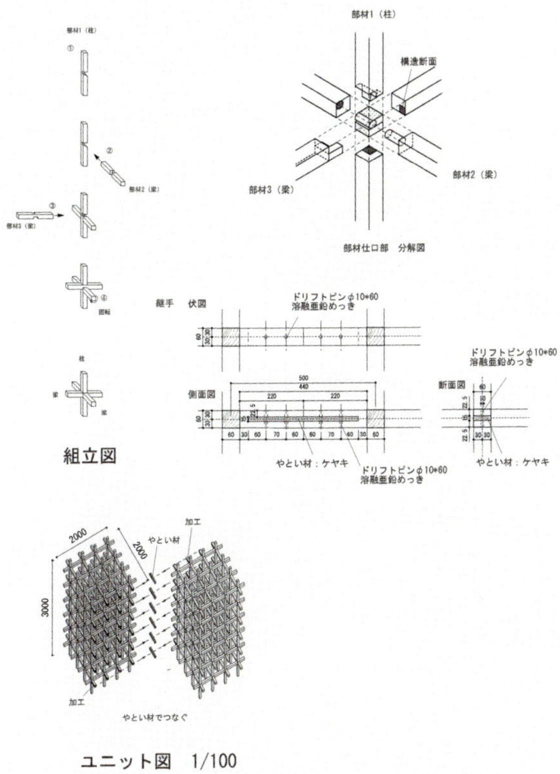

6 The Japanese wooden joint system, called a *chidori*, forms the basic tectonic unit that is multiplied in the Prostho Museum (2010).

7 The 2013 Serpentine Gallery, by Sou Fujimoto, created a surreal white lattice of powder coated steel members that combined to produce tectonic effects.

CLOUDS AND CATHEDRALS

Fujimoto's 2013 Serpentine Pavilion, a temporary structure constructed in London's Kensington Gardens, could be understood as an extension of the beam and bracket system of traditional Japanese architecture. 7 Fujimoto describes the structure as a "cloud" that melts into its context, which shifts the emphasis from Kuma's continuity with traditional construction methods to an assembly designed to produce perceptual effects. Fujimoto's use of painted ¾" x ¾" (2 x 2cm) steel members (27,000 of them) welded together departs from the materially specific traditions of the beam system and aims for an expressive quality. The three-dimensional lattice of 16" x 16" (40 x 40cm) or 31" x 31" (80 x 80cm) cubes remains open at its edges by employing half modules, with members hanging or projecting. The edges appear incomplete, and anticipatory, as if waiting to be extended further. The cantilevered elements expose the "beam ends" in ways analogous to the Kuma lattice. Fujimoto's Cartesian cloud provides a means of reinterpreting traditional dougong assemblies as cloud-like. Rather than read the lattice as an abstracted version of a tree, we might re-read them as non-hierarchical tectonic clouds where the boundaries between inside and outside are blurred, and the structure produces ephemeral effects.

The cloud is an unusual referent for architecture. It is everything that architecture is not. Perceptually and materially, a cloud is fleeting, ephemeral, blurry, and atmospheric. Fujimoto's tectonic cloud appears as a contemporary meditation on the elusive figure of the antithesis of architecture, an anti-architecture. Pavilion architecture and temporary structures serve as important prototypes for testing architectural ideas at scale. Similarly, Thomas Heatherwick's Seed Cathedral, for the 2010 Shanghai Expo, was a paradoxical structure, ephemeral and translucent, consisting of thousands of clear acrylic "quills" arrayed in a radial pattern. Seed Cathedral was a truly ambiguous building, where the density of quill parts made it unclear where the building began and ended, and the transparency of the quills amplified the reading of the building as an atmosphere.

Decades prior, Thorncrown Chapel (1980), designed by Fay Jones, epitomizes an architecture of blurred boundaries and expressed parts. The chapel, located in Eureka Springs, Arkansas, is surrounded by dense woodlands. 11 Jones set out to create a non-denominational chapel in harmony with its natural surroundings. The chapel consists of thousands of light-frame members arranged to

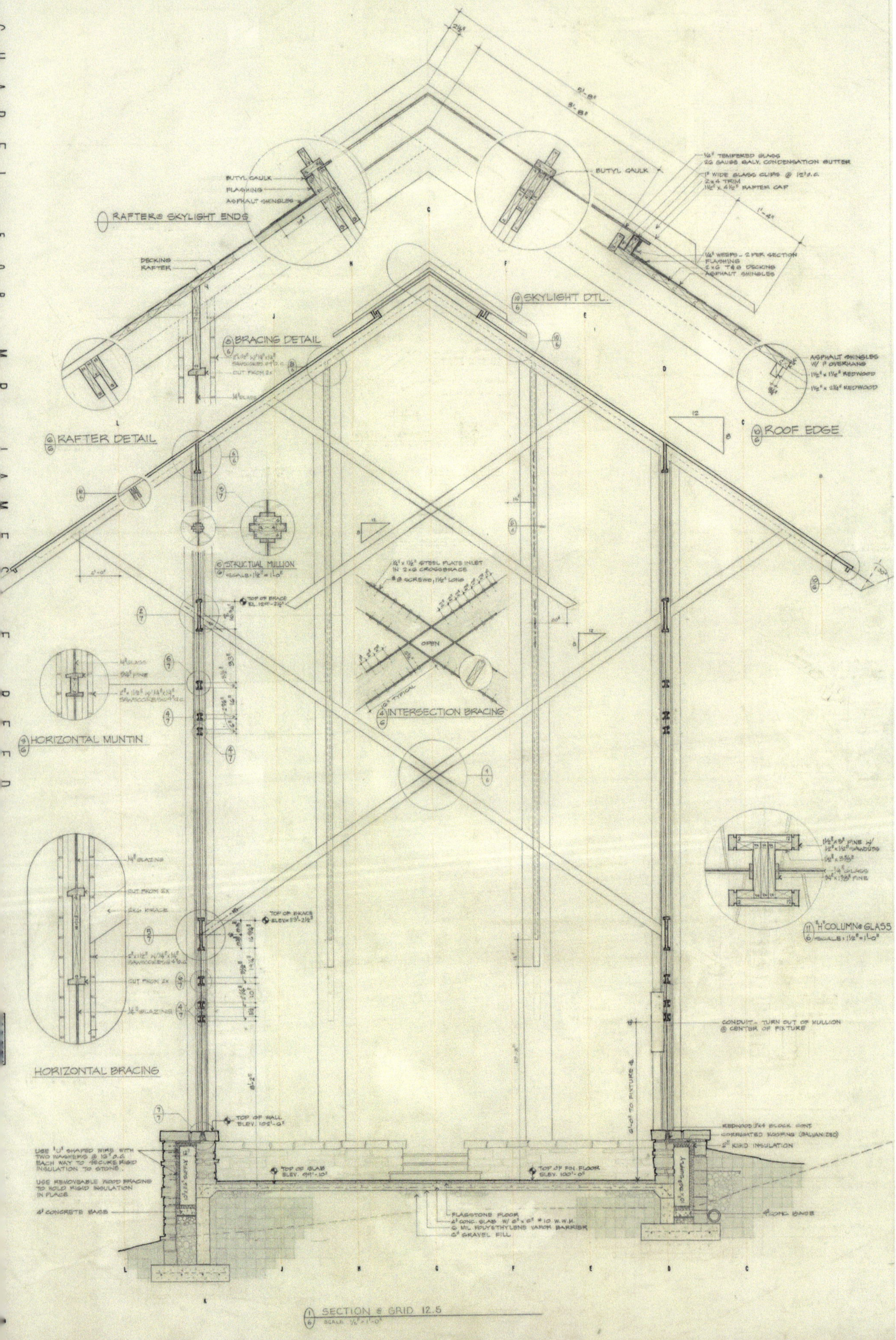

8 Fay Jones' intricate Thorncrown Chapel (1980), in Eureka Springs, Arkansas, uses the constraint of the site to limit the member sizes, which in turn produces the striking effect of blurring boundaries.

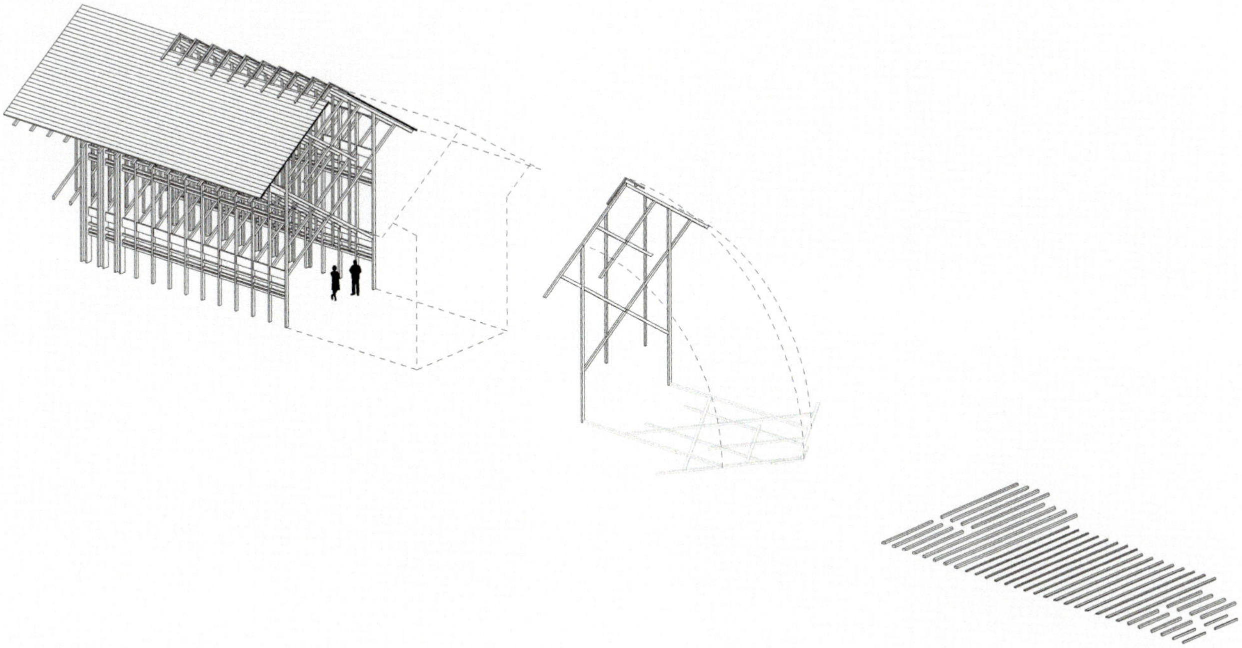

9 The assembly sequence of the Thorncrown Chapel includes bringing the members to site, assembling frames on the ground and standing them up to form the nave of the chapel.

provide a vertical, light-filled space. Jones insisted that no structural member be "larger than what two men could carry through the woods."[9] The Southern Pine "two-by" (2x4s, 2x6s, and 2x12s) framing members were assembled to allow the individual elements to cross and extend beyond the joints. The center of the cross bracing is articulated as a "hollow" diamond-shaped joint made of painted steel that allows the joists to occupy the same plane, as opposed to giving one member priority over the other. 8 The central nave is defined by a series of densely layered and repeated frames. The abundance of relatively slender members arranged in sequence produces the effect of what was referred to as "a great lattice storm" of diagonals.[10] The facade of the chapel is produced through layering, with several members suspended in front of the building enclosure, emphasizing the idea of the ambiguous building boundary. Cantilevered elements with tapered-end profiles accentuate a dense thicket of rafters and columns. 9 Approaching Thorncrown Chapel through the woods, the architecture appears to dissolve into its context, and standing inside the chapel produces the uncanny experience of not being indoors at all. Jones manages to marshal the materials of architecture in service of specific effects: dissolving architectural boundaries, mimicking the grain and density of tree trunks, and producing an architecture of branches and thickets that seems at home in the Ozark landscape. Jones' prolific use of "two-by" lumber allows it to blend into the wooded landscape of the site, while producing what could be described as a contemporary "carpenter's gothic," an articulated assemblage of wooden members that dissolve the mass of the structure into an immersive environment of tectonic effects.[11]

[9] *Fay Jones*, by Robert Adams Ivy, Jr, (New York, NY: McGraw Hill, 2001), p.35.

[10] *Fay Jones*, by Robert Adams Ivy, Jr, p.35.

[11] Carpenter's Gothic, sometimes referred to as "rural gothic" is a North American architectural style used to describe gothic revival architecture. I am grateful to Andrew Holder, who uses the term to describe a contemporary trend of articulated part-to-whole projects.

CARTESIAN COMPOSITIONS

In the early part of the 20th century, the Dutch De Stijl movement prioritized abstraction and cartesian composition as a means for challenging traditional forms of representational painting, sculpture, and architecture. Axonometric drawings by Theo van Doesburg depict orthogonal elements like blocks and planes, composed in space in a seemingly weightless world. The articulated edges and ends of elements ensure that each is understood as distinct and discrete. Planes, lines, and volumes are suspended in van Doesburg's axonometric compositions and rendered in primary colors and shades of gray. 13 The artists and architects of De Stijl sought an abstract language that could be

10 Thorncrown Chapel under construction reveals its powerful repetition of members to produce a thicket of roof trusses, analogous to the tree canopies beyond.

11 Despite its glass enclosure, the completed Thorncrown Chapel maintains the sense that its boundaries are blurred.

universally adopted, and the highly distilled palette of colors and materials were meant as a modern alternative to the conventional brick and timber construction systems of the time.

Gerrit Rietveld's iconic Red and Blue Chair (1917) comprises standard square-section linear elements for the frame, rectangular section elements for the armrests, and planar elements for the seat and backrest. 14 All the components are extended past their intersections such that each member is legible as a pure, orthogonal line or plane. The three-dimensional cartesian grid suggests the possibility of its extension, like a vector, expanding infinitely in all directions, implying that the chair is a discrete instance within a universal grid. The oblique planes of the seat and backrest are achieved through the misalignment of orthogonal frame elements, allowing the diagonal planar elements to read as exceptional geometrical instances and a tacit acknowledgment of the ergonomics of the human body.

The same logic of x-y-z axes and orthogonal frameworks underlies the geometry of the Schröder House (1924), also designed by Rietveld. 12 The house translates the abstract compositions of van Doesburg, and Rietveld's own furniture pieces, to the scale of architecture and the city. Situated at the end of a block of brick row houses, the house contrasts with its context through its abstract composition of white, gray, and primary-colored planes. Although the house is constructed of brick walls and timber elements, the finishes of plaster and paint mask the actual construction systems, giving the appearance of abstract planes instead. Like the axonometric paintings of van Doesburg, the Schröder House articulates its component parts, allowing the edges and ends of its elements to slip past one another, seemingly effortlessly. The walls of the house do not turn a corner. Instead, they are articulated as deliberately misaligned planes. Rietveld chose to express the thickness of materials and articulate the distinct profile of each element. The yellow I-beam post supporting the balconies extends beyond the roof line. The post is connected to a black beam that supports the plane of the roof, which slips past the yellow post, while the roof plane seems to slide over the beam. Van Doesburg's writing describes the new architecture as, "anti-cubic, that is to say, it does not try to free the different functional space cells in one closed cube. Rather, it throws the functional space cells, centrifugally from the core of the cube."[12] Later, he describes the architecture as "floating" and working "against the gravitational forces of nature."[13]

12 Theo van Doesburg, quoted in Kenneth Frampton, *Modern Architecture, A Critical History*, (London: Thames & Hudson, 1990), p. 145.

13 Theo van Doesburg, p. 145.

12 Intent to realize the abstract compositions of De Stijl, Gerrit Rietveld built the Rietveld Schröder House in Utrecht in 1924.

13 Theo van Doesburg's abstract axonometric compositions exemplify the gravity-free space of universal geometrical space.

14 Gerrit Rietveld's Red and Blue Chair (1917) articulates De Stijl's cartesian composition at the scale of furniture.

PARTS IN EQUILIBRIUM

In a video clip produced by Ensamble Studio, a close-up shot of a cardboard model is assembled piece by piece. A foot-long cardboard I-beam is gently set down on the edge of a cardboard plane. The beam is held gingerly between two index fingers. Next, the video cuts to a timelapse of a construction site where a crane lifts a real concrete I-beam into position. The precast beam is massive, approximately 8 feet (2.5m) tall and 65 feet (20m) long. The video cuts back to the cardboard model, and a U-shaped channel profile is placed at the end of the concrete I-beam. The channel profile is supported by two columns. The supports are roughly at quarter points, allowing the channel to cantilever significantly. Cutting back to the timelapse, a real concrete beam is lifted into position, mimicking the model footage. Cut again. In the model sequence, a similar U-shaped profile, facing upward, is gingerly balanced on the three points: the previously installed channel, a V-shaped frame, and a short wall at the end of the channel. The upright U profile cantilevers one-third of its length. Back to the construction timelapse, the concrete U-shape is hoisted into position, seemingly effortlessly. 16

The jump cuts, between the choreography of the scale model and the actual construction site continues until two Vierendeel steel trusses are placed to complete the helicoidal composition. 17 The final piece to be installed is a cubic counter-weight, which appears in the scale model to be the size of a sugar cube. In the construction footage, the counterweight is an 8-foot-tall (2.5m) and 8-foot-wide (2.5m) granite block. The camera pulls back, and we see the architect, Antón García-Abril, carefully placing the counterweight cube on the model. Satisfied that it is stable, he stands up, the video fades to white, and the credits roll. The video of the construction sequence takes less than four minutes. García-Abril notes that he spent three years on the design of the house, one year to engineer the structure, and seven days to erect it.

Ensamble Studio's Hemeroscopium House employs many of the techniques of part-to-whole articulation, with each of the seven primary structural elements clearly articulated as sequentially stacked in counterbalance. 1 The massive granite block, weighing twenty tons, was added to the composition to serve as a counterweight at its highest point, a didactic reminder of the balancing act of the house and its ready-made components. The enclosure of the house seems almost secondary to the heroic

15 The cantilevered ends of Ensamble Studio's Hemeroscopium House (2008) recall the abstract part-to-whole expression of Theo van Doesburg's axonometric compositions.
16 A video clip produced by Ensamble Studio draws the comparison of the assembly of a physical model and the construction of the Hemeroscopium House.

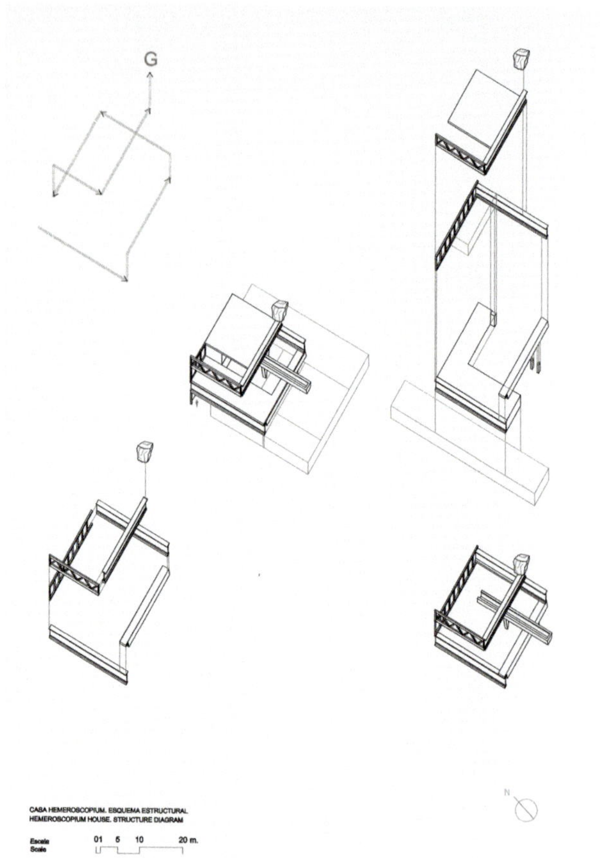

17 A series of axonometric drawings illustrate the careful shoring and balancing that the Hemeroscopium achieves with its cantilevered sections and granite block counterweight.

structural work performed by the full-story beams and trusses. Large glass infill panels span the floor-high beams to create floor-to-ceiling windows. Even the floor slabs, constructed of precast concrete planks, are understood as secondary or infill elements, occupying the gaps between the primary structural members. 18 19

The detailing of the Hemeroscopium House is consistent with the hierarchy of elements. The primary members call attention to their connections. Just as the hands in the video show the careful placement of elements, the dimensions of the beams and trusses are sized to fit together perfectly. The top and bottom chords of the Vierendeel truss extend over and under the concrete I-beam at the second-floor level. The granite block is fastened to the concrete beam with four threaded rods that bolt through the concrete beams and are secured with a steel plate to the underside of the concrete flange.

Naming the Hemeroscopium House after a mythical Greek place of the setting sun, García-Abril describes the project as a space of heightened perceptions. The cantilevered U-shaped section contains a pool. Its end is dammed with a glass plane, allowing the concrete U-profile to remain legible and giving the swimmer the surreal experience of floating above the horizon, overlooking a canopy of trees and the Madrid cityscape beyond.

14 Ensamble Studio, *2G*, Issue #82, July 2021, p. 6.

A rigorous syntax of detailing ensures that the distinct I-beam profiles are expressed on their cantilevered ends. The vocabulary of industrially produced components is an explicitly arranged composition of parts. The intent is to showcase the discrete structural elements and demonstrate how they achieve equilibrium.

Unlike the abstract, immaterial components of De Stijl compositions, Ensamble employs recognizably industrial components. While the floating planes of van Doesburg and Rietveld seem to defy gravity in the weightless space of axonometric representation, the materially evident and counterbalanced beams of Ensamble call attention to the physics of their mass and weight. In his introduction to Ensamble's recent book, Philip Ursprung notes that in an age where lightness and fineness have been celebrated, the work of Antón García-Abril and Débora Mesa shifts toward the heavy, the grounded, the geological, and the earthbound.[14]

Ensamble Studio conceptualizes each project as a confluence of materials, systems, processes, and protocols. The carefully edited video of the model and construction process shows how elements arrive on site, and are sequenced in their assembly, conceptualizing construction as a choreography of

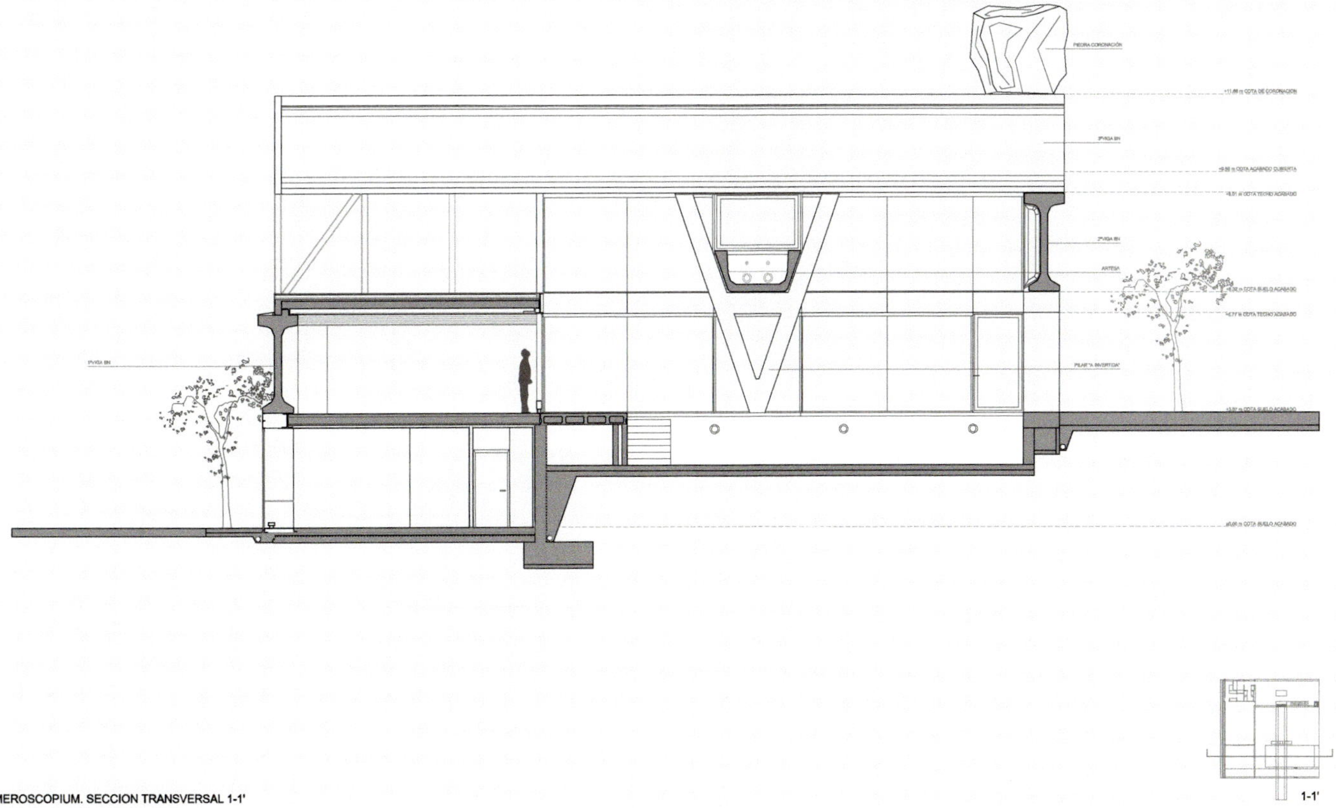

CASA HEMEROSCOPIUM. SECCION TRANSVERSAL 1-1'
HEMEROSCOPIUM HOUSE. CROSS SECTION

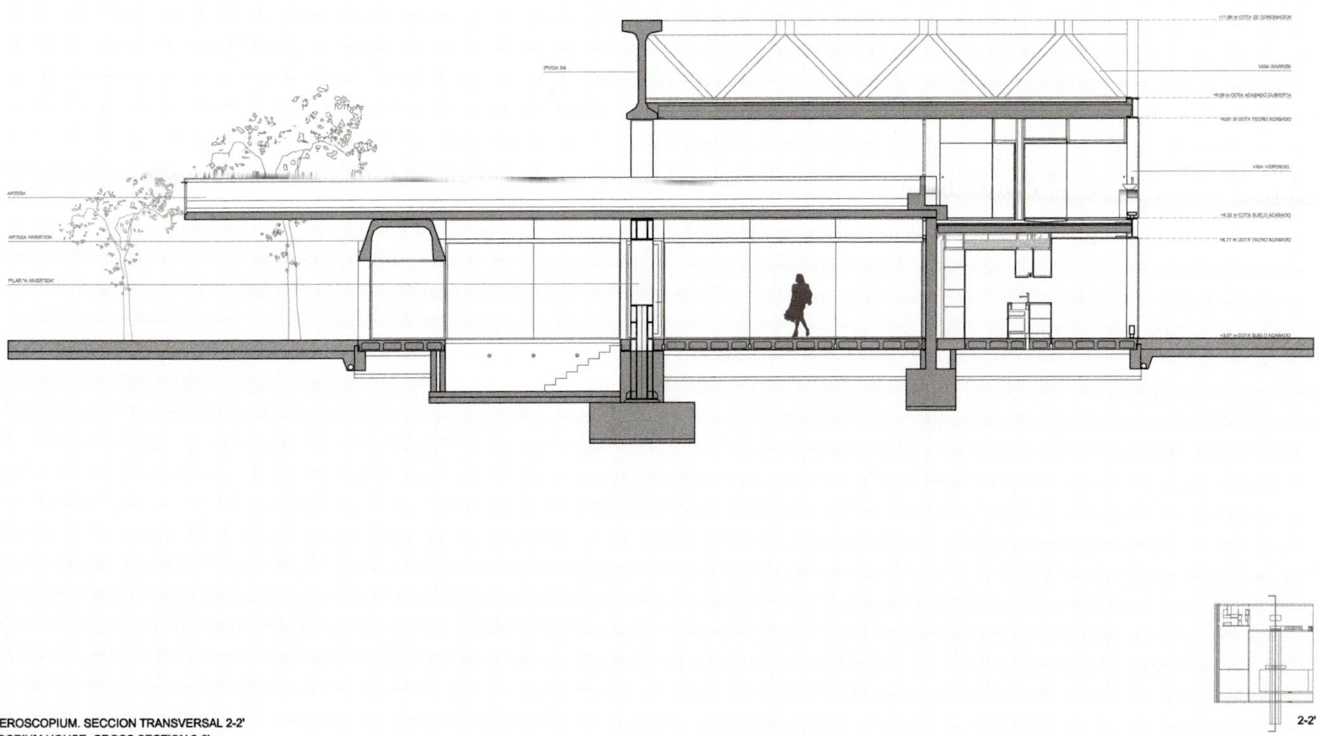

CASA HEMEROSCOPIUM. SECCION TRANSVERSAL 2-2'
HEMEROSCOPIUM HOUSE. CROSS SECTION 2-2'

Escala
Scale
0 1 2,5 5 10 m.

18 Transverse section through the elevated pool with counterweight block beyond.
19 Longitudinal section through the elevated pool showing the cantilever and back span.

20 The Centre Pompidou (1977), designed by Richard Rogers and Renzo Piano, exhibits its component parts through the articulation of building systems.

21 The New York Times Headquarters (2007), by Renzo Piano, uses its part-to-whole strategy to present the tall building as the opposite of monolithic – as light, airy and transparent.

cranes and carriages, rigging, and scaffolding. It also shows how architects are designing through the procedures and sequences of construction. Ensamble's part-to-whole approach articulates the parts to reveal the sequence of assembly.

MACHINE PARTS

The concept of an industrial "kit of parts" is shared with the historical movement known as High Tech architecture. The most visible examples of High Tech are the Centre Pompidou in Paris by Richard Rogers and Renzo Piano (1977) 20 and the Hongkong and Shanghai Bank Headquarters by Norman Foster (1985). In each case, the architectural vocabulary is one of repeatable and expressive machine-like components. Both projects are influenced by the visionary architecture of the Archigram Group, which popularized the imagery of industrial componentry and futuristic assemblages. Conceptual projects from the 1950s and 60s, like Plug-In City and Walking City, imagined a hyper-modern future of robotic structures capable of rapid assembly and disassembly.

Centre Pompidou's systems, including structural, plumbing, mechanical, and conveyance are all color-coded, and expressed on the building's exterior. Critics at the time described the articulated ductwork and structure as a new kind of ornamentation that employed the functions of the building as the new cornice lines and entablatures. Showcasing systems on the exterior is highly didactic: all parts are shown, and nothing is "hidden." This exposure also makes them susceptible to the elements, and the long-term maintenance has proven labor-intensive and expensive.

The use of building components that evoke machine parts corresponds to a moment in the 20th century, when architects strove for a technological future that felt slightly out of reach. Its imagery is aspirational technology — more an image than an instrument. Looking back at the density of components and the evident display of engineering, High Tech architecture displays an impatience for progress — it was yesterday's vision of tomorrow.

The New York Times Headquarters (2007) on 8th Avenue in New York is a more contemporary version of the High Tech architecture of the 1970s and 80s. 21 Piano's design for the tower continues a tradition of articulated parts and discrete elements. The tower is made up of four primary planes, one for each of the primary building faces. The extension of the planes, in plan and section, works to deemphasize the building mass, and is

2.2 The installation of a panel of the terracotta screen at the New York Times Headquarters (2007).

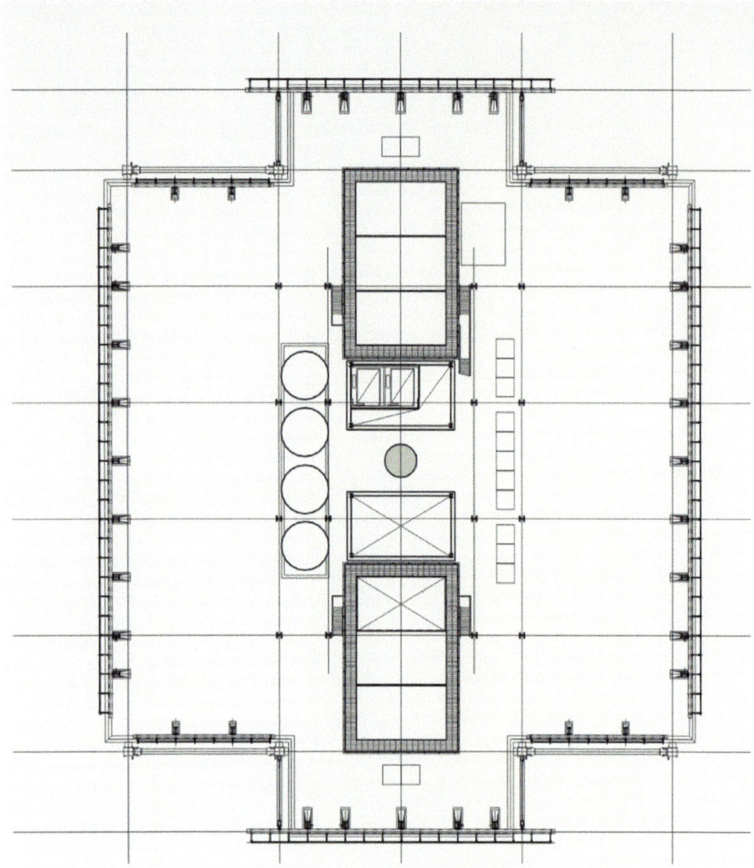

2.3 The cruciform plan of the New York Times Headquarters (2007) and the overhanging facade elements means the tower reads more like four planes than a singular mass.

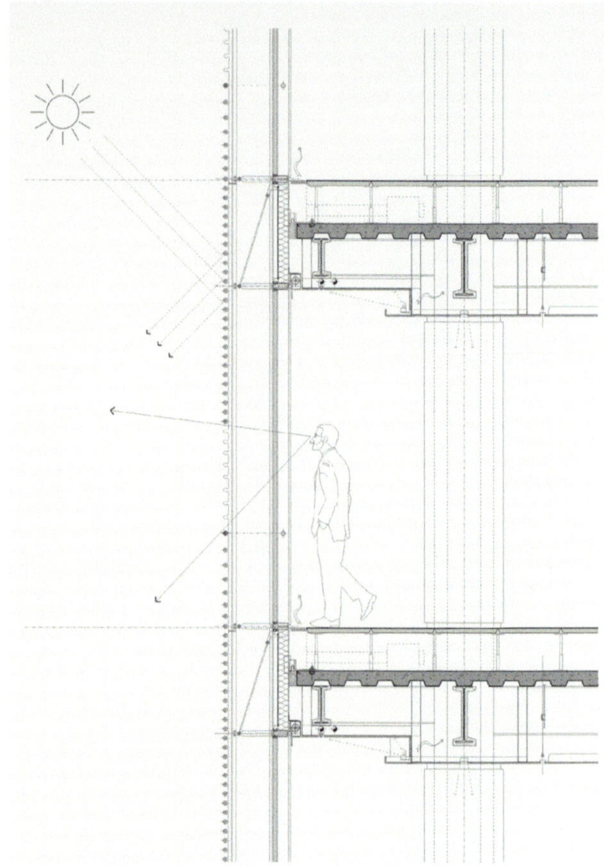

2.4 The section through the curtain wall of the New York Times Headquarters (2007) shows the terracotta screen masking the glass curtain wall.

reinforced by recessing the corners, resulting in a cruciform plan. 2.3 Each articulated plane is composed of a lattice of horizontal members that form a screen-like veil suspended in front of the building mass. The screen is supported by a series of outriggers that project beyond the building envelope, and the gap in between is used as a maintenance zone. 2.2

Piano's use of the lightweight screen blurs the building's edge and creates a thickened zone around the tower. It also acts as a device that mitigates solar heat gain by blocking the sun's rays before they reach the glass curtain wall of the building envelope. 2.4 The light-colored lattice is made of terracotta extrusions with a circular cross-section. The horizontal "rungs" of the lattice vary in density to allow for unobstructed areas in the vision zone, as well as areas of higher density in the spandrel zone. For Piano, the planar reading of the screen was essential to the idea that the headquarters for the newspaper conveys permeability and transparency. The horizontal ribbon window openings do not reach the edges, ensuring the continuous reading of the facade as a plane. Seen from a distance, the tower appears to be made up of four translucent screens that appear delicate and diaphanous. The articulated parts of the tower resist the reading of the tower as a solid mass and offer instead individual members, lightweight layers, and complex visual effects.

CONCRETE COMPONENTS

Boston City Hall (1968) is loved by architects but hated by almost everyone else. The term Brutalism, derived from the French *béton brut*, has been applied to concrete buildings of a certain era, circa the 1960s, of which Boston City Hall is a prime example. 2.5 Won through a design competition by Gerhard Kallmann and Michael McKinnell, the architects aspired to a monumental and enduring architecture that they felt was appropriate for the new seat of municipal government. Kallmann and Mckinnell's concrete architecture was intended as an "architecture of substance," and an explicit rejection of what they saw as the flimsy "candy wrapper" buildings, as they called the curtain wall architecture of late modernism. The new city hall building was constructed of a combination of site-cast concrete piers supporting tiers of cantilevered precast concrete facades suspended over the brick-paved plaza. While the building looks forbidding and fortress-like, there is also something lattice-like about it. The repetitive windows are framed by

25 Boston City Hall (1968), designed by Gerhard Kallmann and Michael McKinnell, combines repetitive precast concrete elements and site-cast concrete elements to form a structure that appears monumental yet still made up of legible parts.

26 Studio Gang's 11 Hoyt Tower (2021), in Brooklyn, New York, is made of repetitive precast concrete "donuts" that form an undulating field of rippling three-story strips.

projecting window surrounds that look like beam ends.[16] In fact, the cantilevered members are reminiscent of Kuma's stacked wood lattice from the Yusuhara Wooden Bridge Museum.

Studio Gang's 11 Hoyt Tower (2021) in Brooklyn employs a precast concrete panel system for its facade that conveys variation and repetition. The undulating wave pattern of concave "bay windows" animates the surface of what would typically be a highly repetitive building type: multifamily residential. 26 A review of the elevations of the tower reveals that the facade consists of approximately 60 percent of typical planar panels that establish a "field." The atypical projection panels comprise a small percentage of the facade, but their distribution brings a middle-scale movement to its composition.

A typical precast facade unit is often referred to as a "donut" because it is a rectangular concrete panel with a window-sized hole in it. 29 Precast "donuts" typically span from column to column, so their weight is carried directly by the columns without gravity loading the slabs. At 11 Hoyt, the system of arraying repetitive precast "donut" panels with intermittent "bay window" units allows it to be expressive and economical. 27 The joint patterns between panels reveal the modularity of the units, but the primary reading is of a field of windows punctuated by three-story tiers of undulating projecting bays. The careful calibration of parts creates an elegant effect of an articulated pattern of repetitive panels, alternating between continuity and modularity, typical and atypical, background and figure.

Diller Scofidio + Renfro's Broad Museum (2015) in Los Angeles oscillates between repetitive modularity and a veil-like continuity of the surface. The concept for the museum relies on the contrast between the collection storage volume called "the vault," and the thick exterior wall draped over it, which the architects called "the veil." The public galleries consist of a full floor of column-free space between the inner volume and the outer skin. The desire for diffuse natural light in the gallery space gives the veil a geometrical constraint, demanding that it act as a multi-directional sun screening system that permits only indirect light to enter the galleries. The translation from the conceptual premise of the project to the construction system involved the use of sophisticated daylighting tools to study the qualities and angles of light on the roof plane as well as the four elevations of the project.

Equipped with the specific parameters of the project, the architects opted for a repetitive glass

16 Classical Greek architecture is said to have been a material translation from timber construction to stone, with the triglyph of the entablature representing the beam ends of the timber joists.

2|7 The wave-shaped bay windows of 11 Hoyt Tower form a continuous surface, but the joint lines between panels reveal its modular construction.

2|8 Diller Scofidio + Renfro's sponge-like design for The Broad Museum (2015) is made of repetitive glass fiber reinforced concrete (GFRC) modules, or parts, that are legible by their joint lines, yet the local inflection around the boardroom window suggests that the parts are working, and deforming, together like a whole.

29 One of the hundreds of modular prefabricated bay window parts of 11 Hoyt Tower (2021).

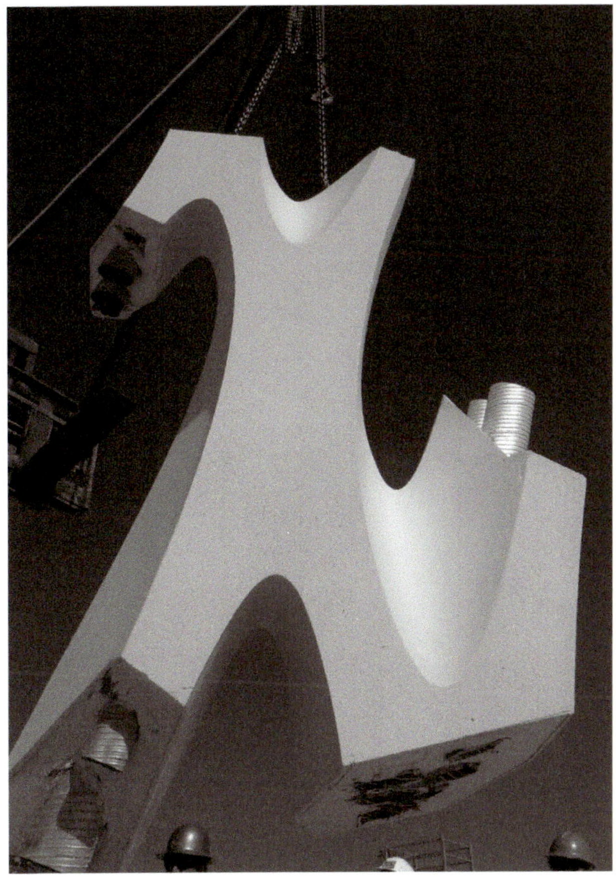

30 The individual component parts fit together to form The Broad's unique lattice-like facade—the architectural whole.

fiber reinforced concrete (GFRC) diagrid module that was attached to a steel-frame exoskeleton. 2,500 individual panels were formed and installed on the frame. 30 The glass facade of the building sits behind the veil and consists of large floor-to-ceiling glass lites. The roof is punctuated by north-facing daylight monitors. From the interior, the GFRC diagrid appears to continuously wrap the ceiling and walls, creating a single expression for both systems.

Exceptions to the system, like the irregular diagonal geometries of the sloped soffits and the figural concave depression, emphasize the overall form of the building and nominate The Broad to the category of monolith. 28 Approaching the museum along South Grand Avenue, the building appears like a giant sponge with the lower corners sliced off and a conspicuous depression in its primary facade. 31 Closer, the regularity of its modules becomes more apparent, and the cellularity and repetition of its parts become its primary reading. Appearing at times as a singular object subdivided and at others as an assembled piece-by-piece aggregation, The Broad seems to enjoy the double reading as both a part-to-whole strategy of aggregation of discrete units, while also appearing, at times, to be a monolith.

PREFABRICATION AND PROCUREMENT

Questions of modularity, panelization, and construction sequence are never far from tectonic thinking. Conceiving of a design from the point of view of its eventual execution is a common feature in part-to-whole design approaches. It is impossible to conceive of the Hemeroscopium House without considering the elements of which it is composed, and the sequence of installation, balancing, and equilibrium that it achieves. Increasingly, questions of procurement and provenance surround design decisions as architects consider the embodied energy of materials employed in construction.

Considerations of module sizes, lifting capacity of construction equipment, and transportation constraints become drivers for design when decisions are made that test the limits of any of these variables. This tendency is reinforced by several factors. Firstly, the cost of on-site construction and the cost premium for union labor have created a demand for more and more off-site prefabrication. Quality control for components assembled in the shop environment is also a driver for prefabricated units whenever possible, as shop assembly is always more precise than field work. Prioritizing prefabrication results in maximizing module sizes,

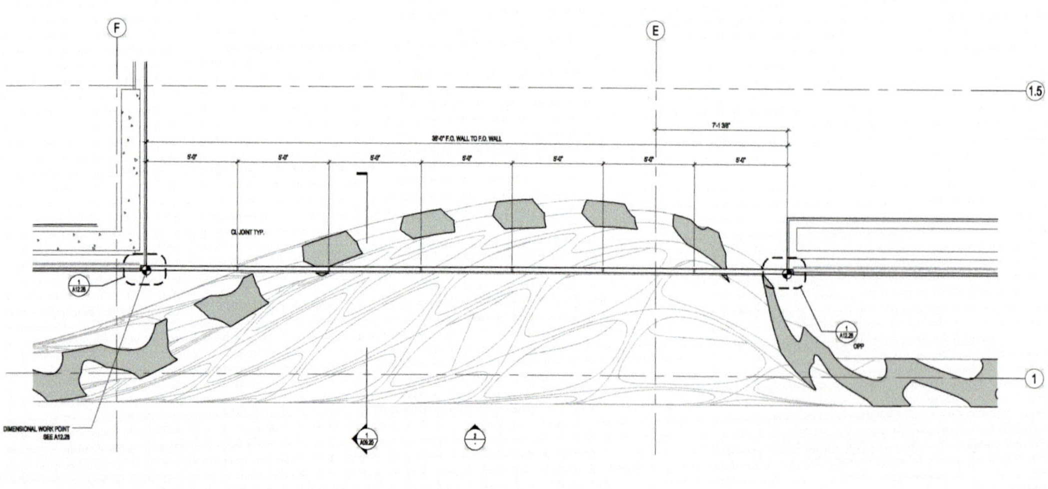

3 1 The typical and atypical components of The Broad's (2015) GFRC facade.

3 2 290 Mulberry House (2013) in New York, was designed as a modular brick-faced precast concrete facade.

3 3 An L-shaped precast concrete module awaiting installation at 290 Mulberry House (2013).

which tests the limits of equipment capacity and transportation constraints. Optimizing a module for efficient erection sequencing and crane time can limit the maximum weight of a prefabricated module. And transportation considerations can result in module sizes governed by the bed size of a truck or Department of Transportation (DOT) clearances under bridges and through tunnels. These constraints, applied to modular construction, often circumscribe the largest prefabricated panel sizes and are often legible in the finished building.

MODULAR THINKING

290 Mulberry House (2013), by SHoP Architects, exemplifies modular thinking applied to a brick facade design. 3 2 The nine-story apartment building takes its inspiration from the brick material palette of the surrounding neighborhood on Houston Street. The architects proposed a contemporary brick facade with an undulating pattern. The variability of the brickwork design and its modular repetition made it a strong candidate for a prefabricated precast panel. This allowed for a highly regular pattern of face bricks cast into the concrete panel that was shop fabricated and field installed. The architects developed a unique process whereby special bricks with puzzle-piece profiles were first set into a custom liner mold. Rebar cages were then installed over the brick and finally a concrete mix was poured over the mold to create the composite panel. The L-shaped panels were designed around the parameters of the floor-to-floor heights and window sizes, as well as transportation and equipment limitations. 3 3 The resulting facade is a striking texture of brick walls that seem to fold in and out of plane, catching the light to dramatic effect. The architects refer to this as a "more and less"[16] process of optimization, acknowledging constraints of site and labor, leveraging digital fabrication tools and methods to produce unique designs, and finding complex yet cost-effective solutions conceived with the construction process in mind.

Thinking through material systems, modularity, and construction sequences was also the driver for 212 Stuart Street (2022) by Höweler + Yoon. 3 4 The idea of a multi-story pier as a typical cladding unit was chosen to achieve the effect of a solid building with a vertical fluted grain and a depth of material articulation. The precast concrete panels allowed for a range of expressive profiles while their vertical pattern created a variable coursing of the facade, changing the perception of the scale of the buildings. The height of the piers created a coursing

16 SHoP Architects, *ShoP Architects Out of Practice* (New York, NY: Monacelli Press, 2012), p. 154.

34 212 Stuart Street (2022), designed by Höweler + Yoon, reads like a "coursed block" of multi-floor packages that increase in height as they rise.

35 The multi-story "fluted" precast piers of 212 Stuart Street (2022) are two or three stories tall, maximizing their dimensions within the parameters of fabrication and transportation.

pattern that increased with the height of the tower, beginning as two stories, then three stories, four stories, up to six. 36

The two- and three-story panels project forward while the infill window wall system recedes, giving the impression that the glazing is a shadow gap between the vertical piers. The sizes of the precast panels were calibrated to the load capacity of the construction crane, as well as the dimensions of a flatbed truck. The project was conceived of with its construction sequencing in mind, resulting in a building where assembly is key to its conception, and its parts are legible in its completed state. 35

The outside corners of the tower express the fluted piers as 12 inch (30cm) thick elements that project past the corner and allow the glass window wall to fill the gap. The corner is consistently made up of a planar precast element adjacent to a glass infill element, so the corners are always open, with a biased orientation. Successive courses of piers alternate the corner bias, so the view of the corner looking upward reveals a stack of alternating and overhanging solid panels.

Designs involving prefabrication inevitably incorporate sequencing and part-to-whole strategies, which are visible in completed projects to varying degrees. Haus Gables is a single-family house in Atlanta, Georgia, conceived as a mass timber, cross-laminated (CLT) structure. 37 Designed by Jennifer Bonner of MALL, the house is a play on the roof form semiotics of suburban houses that the name Haus Gables refers to. It is also a design research project for the architect, who is also the client. Interested in understanding the material system's properties, Bonner designed the house around a CLT blank module of 50 foot x 9 foot (15.25m x 2.75m) as well as the parameters of shipping container dimensions and truck bed sizes.[17] The 87 panels for the house were procured in Austria by KLH and shipped to the US. The installation team of four builders erected the entire house in 14 days.

The large CLT blanks can be both structural members and interior finish surfaces, eliminating the need for the multi-trade laminate construction we associate with typical "two by" stick construction. Bonner likens Haus Gables to a doll house and refers to her architectural drawings as assembly instructions, akin to IKEA furniture. 38 Bonner's use of CLT blanks that are both structure and finish imagines a design process where construction and assembly methods are the drivers for the conceptualization of the house from the outset — construction method preceding conceptual design.

17 "We believe the CLT blank can and must be conceptualized beyond its sustainability," write Jennifer Bonner and Hanif Kara addressing the potential of CLT and thinking through modular construction components as parameters for designers working with mass timber products, *Blank, Speculations on CLT* (New York, NY: Applied Research and Design Publishing, 2021), p. 27.

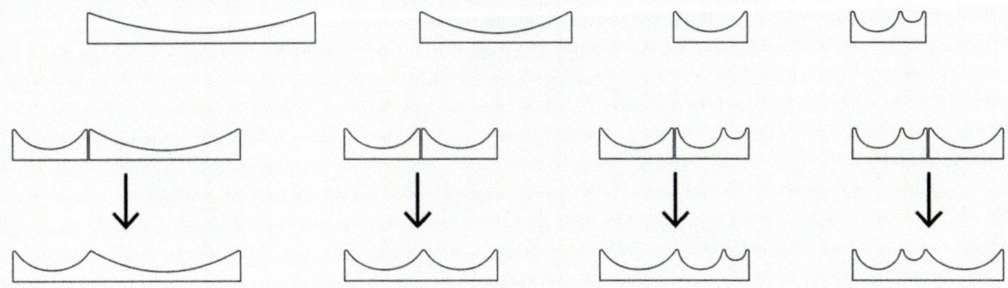

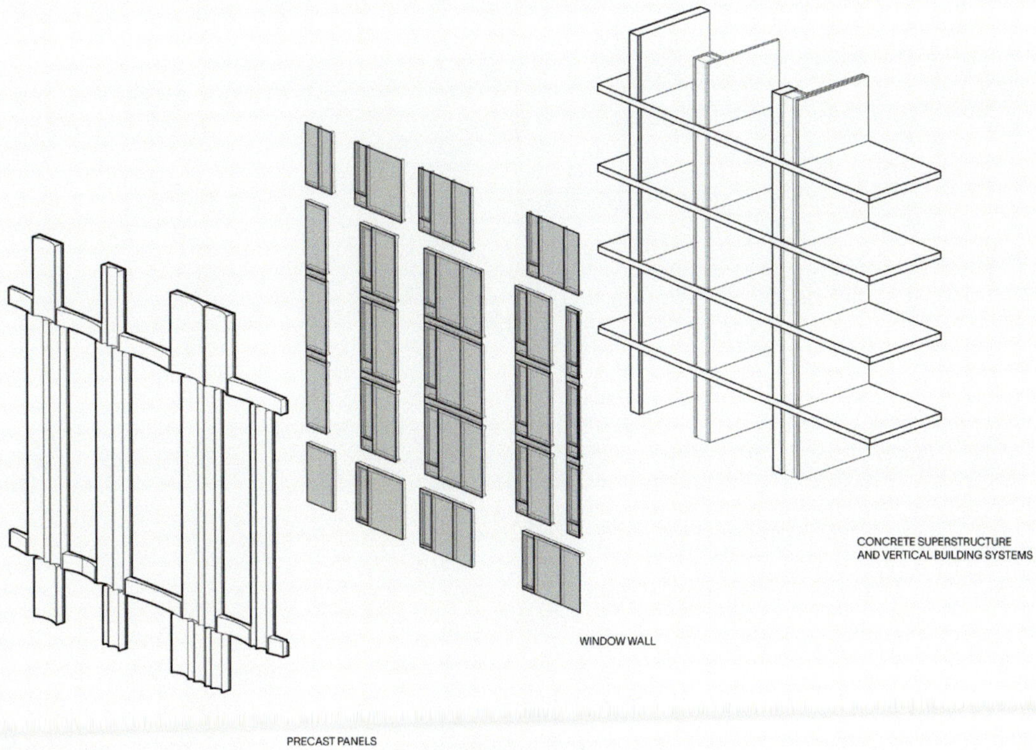

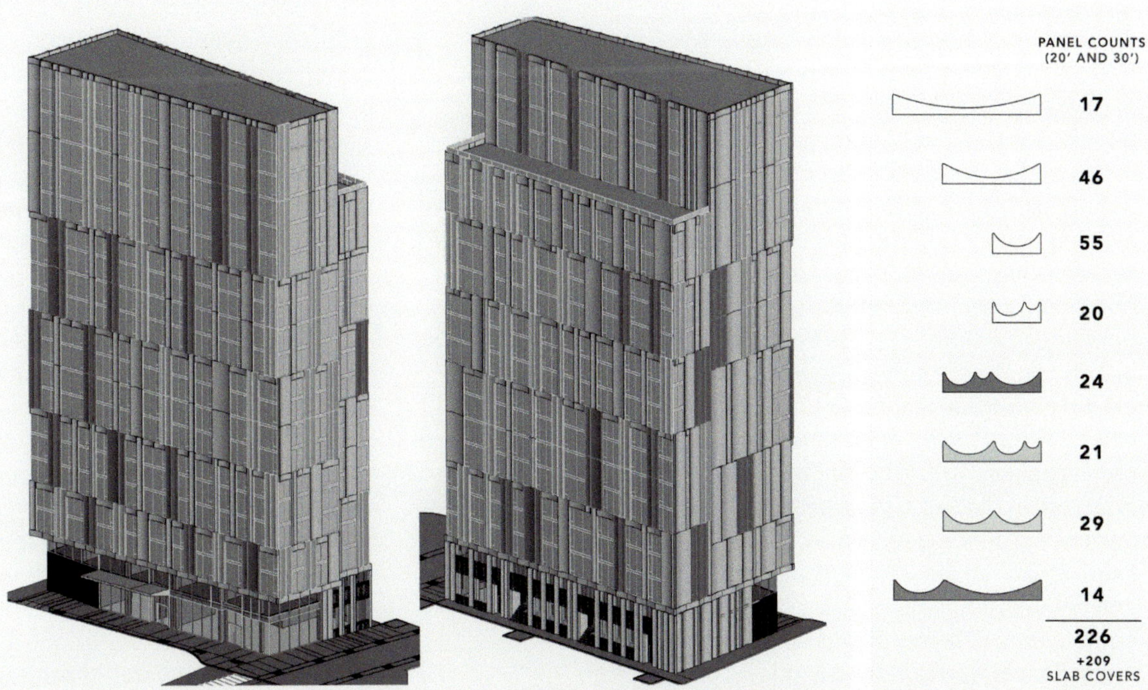

3.6 The design of the 212 Stuart Street (2022) facade was developed around parameters for efficient fabrication, transportation and installation, with precast modules optimized for the hoisting capacity of the on-site tower crane.

37 Jennifer Bonner's Haus Gables (2018) was conceived of as a commentary on the vernacular roof profile typologies of American single-family houses. It was also designed to be assembled from cross laminated timber (CLT) panels that make up its walls, floors and roof.

ASSEMBLED EFFECTS

Jean Nouvel's Torre Glòries (formerly Torre Agbar) in Barcelona is consistent with this expressive approach and represents both a part-to-whole and a monolithic approach. The Barcelona skyscraper combines an iconic, somewhat phallic, building massing with a densely layered facade of operable glass louvers and multicolored metal panels. 39 The combination of components in the facade design achieves the goal of solar control by reducing the incident radiation, but it is also intended to produce a unique, shimmering polychromatic effect. The structure of the tower is a reinforced concrete shell with a pattern of irregular punched openings. 40 The openings are arrayed in a checkerboard pattern, evoking random pixels, glazed with 4,500 individual windows. The concrete walls are clad on the exterior in corrugated metal panels painted in bright shades of red and blue and arranged in the same pixel pattern. The outermost layer of the building envelope is a second facade of fixed horizontal glass louvers in a range of patterns and finishes that diffuse and reflect light. 41 The combination of layers of glass in frames, painted metal panels, and punched windows produces an extraordinary, mirage-like effect of a shimmering surface of ambiguous extents and material makeup. Nouvel describes the tower's facade design as evocative of the effects of water and light. "This tower is a fluid mass that burst through the ground like a geyser under permanent, calculated pressure."[18] The pixelation of the pattern is a surprisingly effective technique to produce a blurring effect. Typically, a punched window facade would create a normative and recognizable pattern on a facade. Nouvel's punched windows at Torre Glòries are camouflaged by their interspersed distribution and the fact they share the same dimensions with the colored metal panel pattern. The overlay of glass louvers in a gridded frame adds another layer of pixelation, making the Torre Glòries impossible to scale.

The part-to-whole strategies of both the New York Times building and the Torre Glòries employ legible component parts to achieve optical and perceptual effects. For Piano, the delicate scale and texture of the terracotta rods produce a diaphanous screen over the shaft of the tower that recasts a massive building typology as something rich and layered. For Nouvel, the combination of colors and patterns blurs the extent of the normative figures of windows and slabs to produce a far more ambiguous effect, a multi-story polychromatic reading. Pixelation is itself a part-to-whole effect, rendering a smooth

18 Torre Glòries (formerly Torre Agbar) houses the headquarters of Barcelona's water utility, hence the relevance of the water geyser metaphor. Quoted in an Instagram post @ateliersjeannouvel on April 3, 2022. Also see Scott Murray, *Contemporary Curtain Wall Architecture* (New York, NY: Princeton Architectural Press, 2009), pp. 140–147.

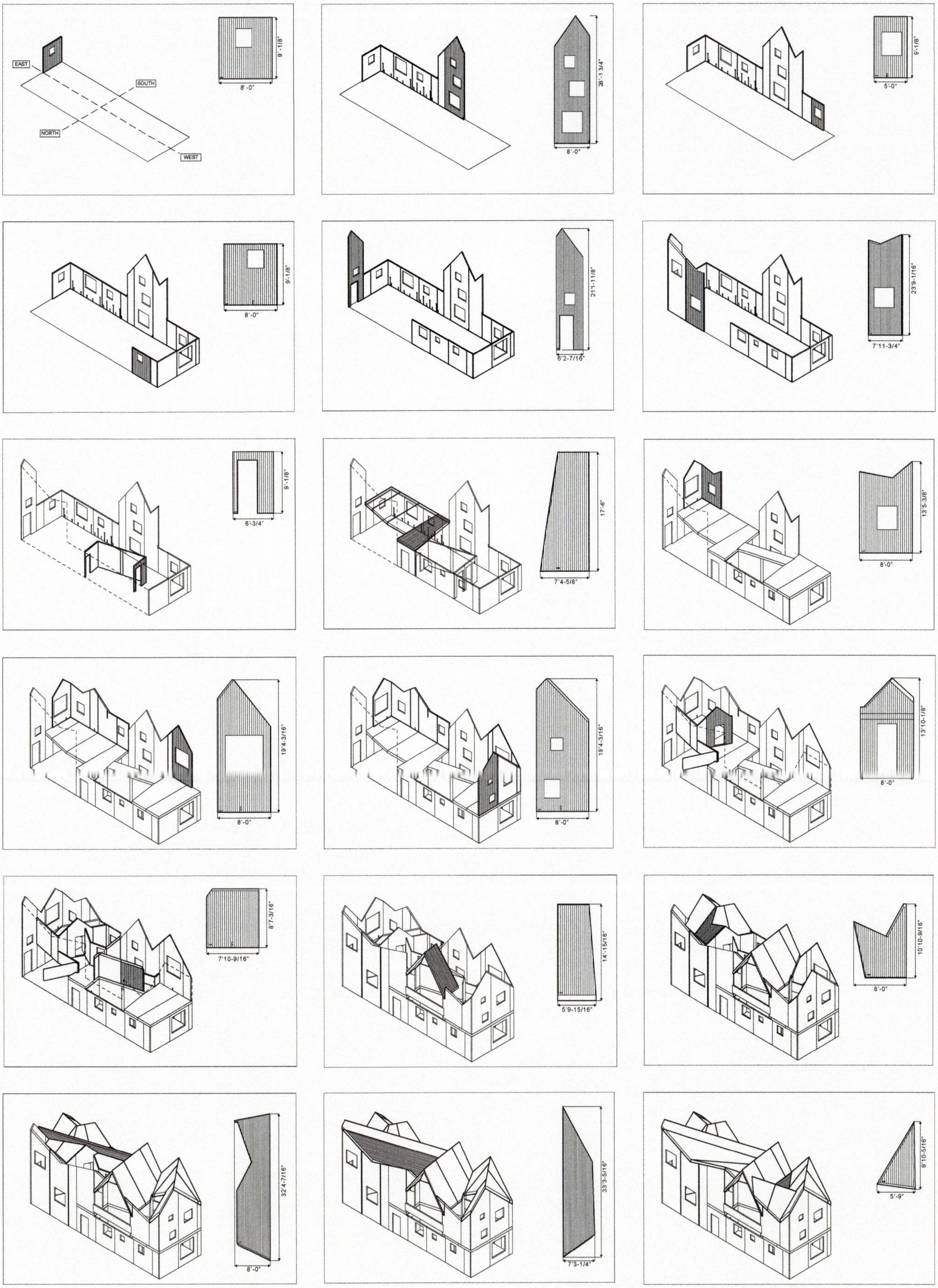

38 The Haus Gables CLT part drawings show each panel individually and how it fits into the house as a whole. Bonner has likened this assembly sequence drawing to the instructions that come with IKEA furniture and the do-it-yourself culture of "some assembly required."

3|9 Torre Glòries (formerly Torre Agbar, 2004), in Barcelona, Spain, was designed by Jean Nouvel to create a pixelated effect of blurred edges and shimmering surfaces.

4|0 The combination of punched windows, translucent glass layers, colored metal panels and glass louvers produces a striking composite effect, wrapping the tower in a tectonic assembly whose architectural intent is an ambiguous image.

gradient as a series of bounded units within a range of color values. Nouvel's clever alignment of the window, painted panel, and glass louver patterns effectively masks the normative figure of the windows within a layered, part-to-whole technique in the service of architectural and material effects.

IMAGE

Part-to-whole relationships are not confined to a particular moment in time nor a particular scale or type of project. They reflect an attitude about design with materials, systems, as well as assembly and construction, such that concepts are significantly informed by the articulation of their constituent parts and their processes of construction. For Frampton, the way of building and the expressive potential of tectonics underscore how architecture can be a means for "being present," as experienced by an embodied subject, and resistant to certain tendencies that he sees as "gratuitously figurative" and image-based. Frampton writes: "Thus the tectonic stands in opposition to the current tendency to deprecate detailing in favor of the overall image."[19] The prominence and power of the image have not subsided since Frampton's writing and may be useful as a means to understand the part-to-whole approach to architecture.[20] Strategies for making architecture distinct from its image basis include the articulation of its parts and an emphasis on its material qualities and demonstrative detailing. Strategies like Nouvel's pixelation of parts seem to acknowledge the pervasiveness of the image and its contemporary currency. Utilizing the pixel rather than the part, Nouvel's approach embraces the culture of imaging, icon-making, and branding to produce a highly self-conscious architecture of components and camouflage that appears at times to articulate its components, and at others to produce a blurry monolithic whole.

19 Kenneth Frampton, *Studies in Tectonic Culture*, p. 26.

20 For more on the contemporary status of the image as a cultural artifact that occurs currency, see David Joselit, *After Art* (Princeton, NJ: Princeton University Press, Point: Essays on Architecture, 2012).

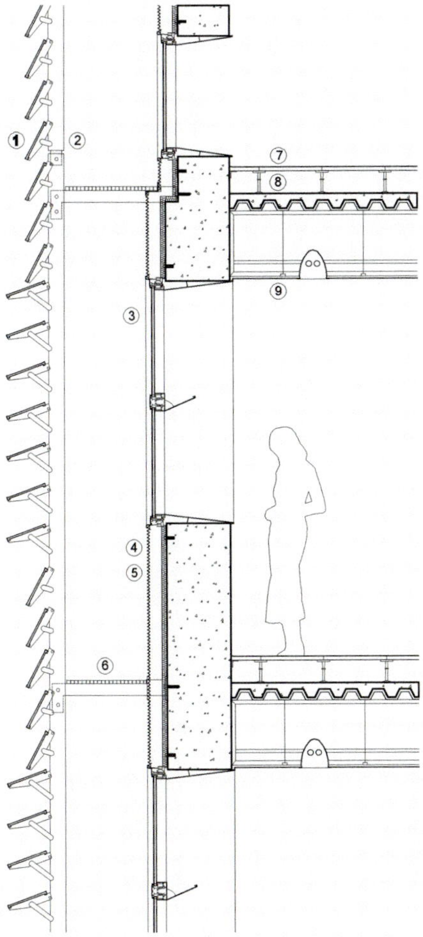

Section

1 Laminated glass louver
2 Anodized extruded aluminum rail
3 Aluminum window with low-e coated insulating glass
4 Painted corrugated aluminum sheet
5 Mineral wool insulation over reinforced concrete wall
6 Galvanized steel maintenance catwalk
7 Raised floor
8 Concrete slab on mtl. deck and steel beam below
9 Suspended ceiling

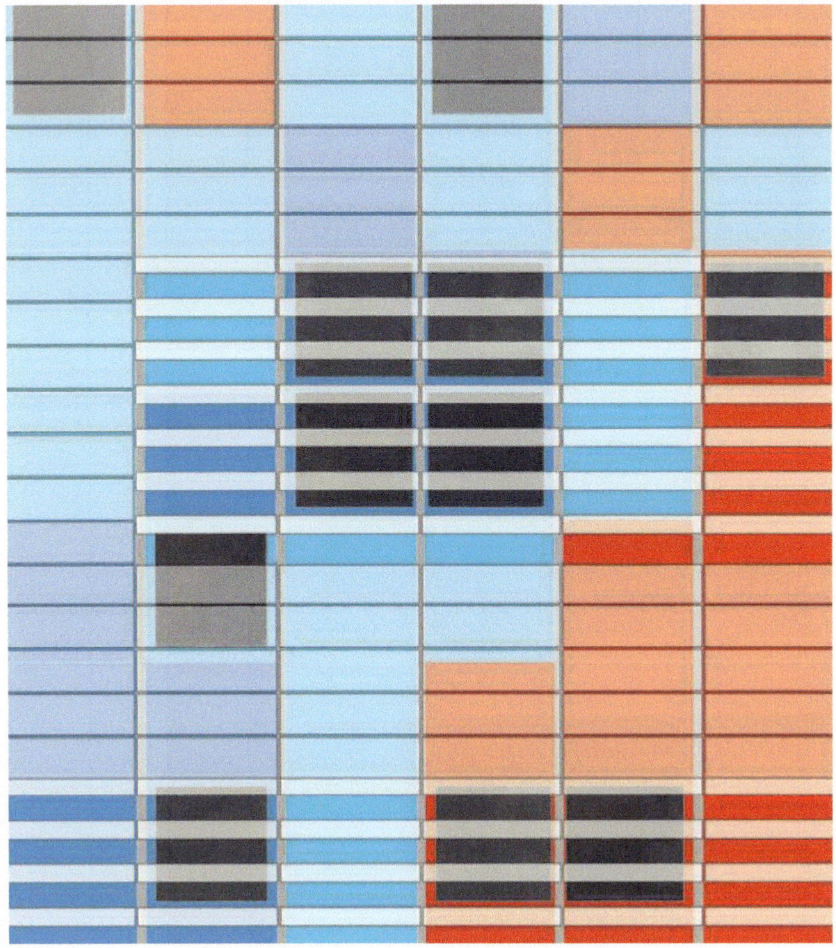

Partial Elevation

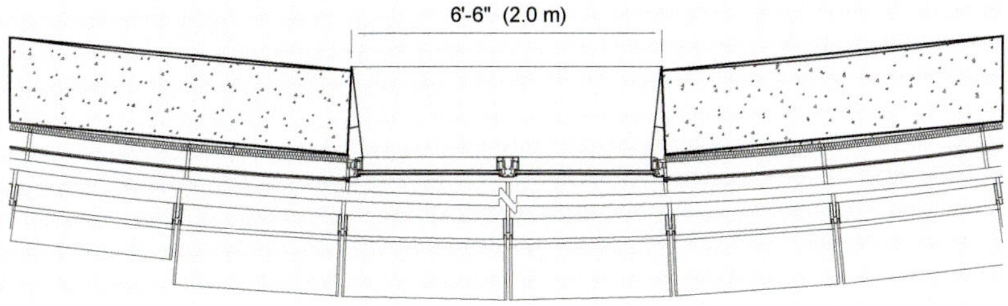

4.1 Wall type drawing of Torre Glòries (2004), produced by Scott Murray, documents the facade elevation, section and plan, and reveals the means to achieve the particular architectural effects.

1 The exterior facade of the Azuma House (1976), designed by Tadao Ando, appears mute and monolithic.

Chapter 5 Monoliths

Tadao Ando's Azuma House (1976) does not disclose much about its interior or inhabitants. The primary facade, facing a narrow street in Osaka's Sumiyoshi district, is completely blank, save for a single, centered, door-shaped opening. [1] Cast-in-place concrete is Ando's signature material and is often associated with monolithic architecture. Unlike other construction types that rely on multiple parts, studs, fasteners, laminates, veneers, etc., cast-in-place concrete produces the appearance of pure solidity. Once the formwork is removed, the house is a single solid block. Only the traces of its construction reveal the parts and components that contributed to its production.

Looking closely, a single horizontal joint line divides the facade into two similar rectangles. The regular pattern of form tie holes creates another layer of information, a point grid, roughly 2 feet x 2 feet (500mm x 500 mm), arrayed across the facade. Seen in a certain light, the facade is revealed to be far from flat, with subtle warps in plane, where the pressure of the concrete pushed against the plywood formwork. The jointing pattern of the plywood is also visible in the subtle relief of the concrete facade. The house is slightly wider than two sheets of plywood, so a sliver of plywood was added in the middle, forming a vertical band that looks like a keystone over the doorway. Upon closer inspection, the monolithic concrete house is less mute than it first appears. Careful observation reveals parts and processes that work together to achieve the monolithic effect of the facade. [2]

Monolithic architecture requires great effort to subsume a multitude of discrete components and processes into a singular expression. The counterintuitive nature of the monolith makes it a challenging ambition for contemporary architecture, where designers are taking a range of approaches — for example, the Perelman Performing Arts Center in Lower Manhattan and the MAAT Museum in Lisbon — and use very different formal and constructional logics to achieve a singular unifying effect.

In 1995, Rodolfo Machado and Rodolphe el-Khoury curated the *Monolithic Architecture* exhibition at the Heinz Architectural Center, presenting nine projects to illustrate what they called a "counter sensibility."(1) Machado and el-Khoury saw the expressive potential of the monolith as a way to counter the "conceptual habits and compositional clichés in the aesthetics" of the "avant-garde" practices of the moment. "Simultaneously solid and hollow, uniform and heterogenous, impregnable and permeable, foreign and familiar, the monolith that is not one stands

(1) Rodolfo Machado, Rodolphe el-Khoury, *Monolithic Architecture*, (Munich; New York, NY: Prestel, 1995). Remember, 1995 was only a few years after *Deconstructivist Architecture* (1988), and just after *Folding in Architecture* (1993), so to talk about monoliths, muteness, and abstraction, was counter to all the tendencies of the moment that were veering toward complexity, fragmentation and disjunction. Machado and el-Khoury are keenly aware of the "ism-ism" of the moment and the artificiality of new schools or movements. They write, "We present them [the projects in the exhibition] as a set of isolated, but parallel anomalies rather than as the converging symptoms of a new tendency," p. 14.

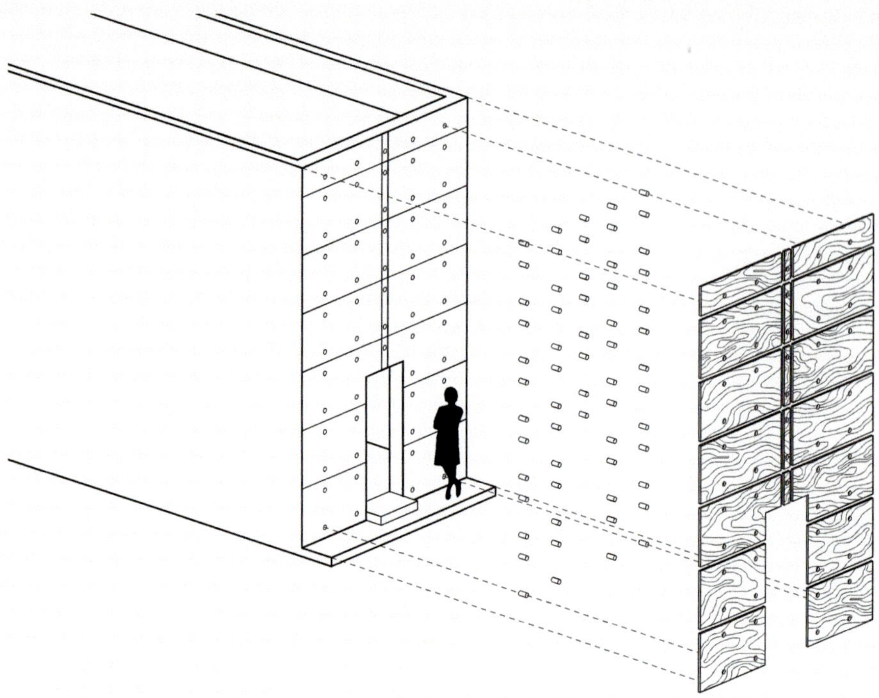

2 Isometric "process drawing" of the Azuma House reveals the large number of component parts, including plywood sheets and form ties, to achieve the monolithic appearance that is such a distinctive aspect of the design of the house.

out as a glitch, a disturbance within the dialectical logic and the corollary axiological dichotomies that continue to regulate the theoretical discourse in architecture and persistently sanction the emancipatory virtues of the formless as opposed to the totalizing and repressive authority of the 'formed,' the small to the big, the fragmented to the coherent, the transparent to the opaque, the irregular to the orthogonal."[2] The essays collected in the exhibition catalog and the projects they use to illustrate their argument remain relevant today, as architects apply new design tools and methods to address similar sets of concerns.

The *Monolithic Architecture* exhibition catalog included both complete and unbuilt projects. Herzog & de Meuron's Signal Box building (1994) in Basel had recently been completed, as had Simon Ungers' T-House (1992). The Yokohama Terminal building by Foreign Office Architects (FOA) was represented in the catalog with early renderings and model photos but the project itself was unbuilt at the time of the exhibition. These projects continue to be highly influential, partly because of the singularity of their intent and because they were proposed as "counter-trends" at a moment that tended toward an exuberant articulation of complexity.

2 Rodolfo Machado, Rodolphe el-Khoury, et al., eds., *Monolithic Architecture* (Munich; New York, NY: Prestel, 1995), p. 13.

CLAD BOX

The Signal Box 4 building typifies the monolithic approach through techniques of masking and smooth transformations within material systems. 3 The "clad box" strategy employed to produce the appearance of a singular volume was to wrap the entire volume in a single material with a single detail. In this case, the eight-inch (200mm) wide copper strip wraps the building as a rainscreen system. The building envelope is set back from the facade and includes opaque walls, windows, and doors. These normative elements are masked by the copper screen, hinted at by the subtle twisting of the copper louvers that rotate from vertical to diagonal, articulating porous zones behind the screen. 4 The openings are kept to the middle of the facade, and the corners are solid. A small-radius fillet allows the copper strips to bend around the corners and emphasizes the continuity of the ribbon-like bands. The six-story Signal Box building could not be mistaken for a solid, given the horizontal banding of its exterior screen, yet the design approach is emphatic in masking building parts to achieve the singular expression of the wrapped copper volume of ambiguous scale. At home among the sheds, poles, wires, and railroad infrastructure, the Signal

3 The Signal Box 4 building (1994), designed by Herzog & de Meuron, was included in the *Monolithic Architecture* exhibition (1995), and remains a highly relevant example of the "clad box" type.

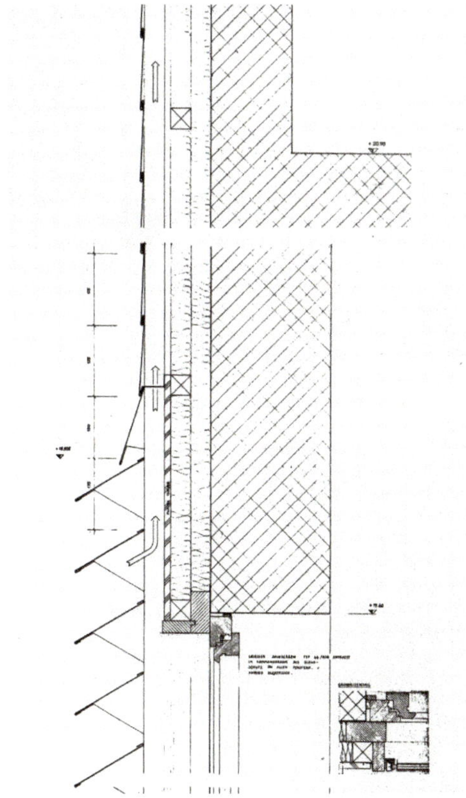

4 A detail wall section of the Signal Box 4 window head documents the lapped copper sheet rainscreen system and its transition from a typical "closed" configuration to an "open" configuration at the window.

Box, understated and mute, nearly disappears into its context. Following the completion of the Signal Box in 1994, the "clad box" approach has since been used to deemphasize parts and organize all aspects of a project around a singular form. Five contemporary examples serve as "clad box" case studies that achieve a monolithic expression.

The Suzhou Chapel by Lyndon Neri and Rossana Hu is perhaps most like the Signal Box in its approach. In this case, a perforated, creased metal panel screen forms a "veil" over a light-colored cubic volume with punched windows. The white cube of the chapel volume is distinctly articulated in contrast to a series of lower volumes clad in traditional Chinese gray brick. 5 The vertical pattern of creased metal panels is contrasted with the highly textured brick volume below. A recessed reveal separates the floating upper cube from the grounded landforms of the brick base. The monolithic expression of the cube is achieved by delaminating the rainscreen metal panel from the building envelope behind. The white cube encloses the vertical space of the chapel as well as the vertical circulation to the choir balcony above. The exterior screen masks programmatic and geometric variations within, as well as the building envelope components, the windows, frames, sills, and copings. In daylight, the randomly distributed square windows are hardly visible behind the screen. At night, the windows form a distinct lighting pattern behind the screen. The architects describe the perforated screen as "a metaphorical veil that alternatively hides and reveals."[3] Entering the chapel, the viewer finds an interior of warm wood finishes. The veil seems to mask the interior, achieving a counterintuitive theatrical disclosure upon entry.

A screening double facade is also utilized in Rahul Mehrotra's KMC Headquarters (2012) in Hyderabad. In this case, the screen comprises a vertical, vegetated living-facade that climbs an aluminum trellis system. 6 The cavity between the building envelope and the screen is wider than the Suzhou Chapel's, allowing a three-foot (1 m) maintenance cavity to access the green facade and the glazed building envelope. This facade strategy requires specific infrastructure to accommodate the hydroponic trays, integrated drip irrigation, and regular maintenance by the gardening staff. Mehrotra describes the screen as a "space of climate-mitigation and social connection."[4] The living screen masks the building functions within while producing a natural sun-shading system that humidifies the air through evaporative cooling. These passive strategies are well-suited for Hyderabad's hot and windy summer season. At the same time, the continuously

[3] Neri&Hu, Design and Research Office, *Thresholds: Space, Time and Practice* (London: Thames & Hudson, 2021), p. 231.

[4] Architecture is always simultaneously technical and social, material and political. Mehrotra's discourse around the possibility of a momentary visual connection across the planted vertical facade of the KMC Headquarters is exemplary of this condition. Rahul Mehrotra, *Working in Mumbai, RMA Architects* (Berlin: Architangle, 2020), p. 195.

5 The exterior view of the Suzhou Chapel (2016), designed by Neri&Hu, in Suzhou, China, wraps the structure in a perforated metal screen that veils and obscures the pattern of windows on the facade.

6 The KMC Corporate Office (2012), designed by RMA Architects, in Hyderabad, India, employs a vertical green facade strategy to envelop the building in a living screen.

evolving appearance of the facade is subject to the seasonal cycles of growth and flowering of the species of climbing plants. The green facade also gives the building a distinctive and unified appearance suitable for a corporate headquarters building, masking floor slabs and interior partitions. The monolithic expression of the KMC building appears like a cubic green volume emerging from the surrounding landscape while its passive skin qualifies as a new type: the performative monolith. 7

The Fundación Santa Fe de Bogotá (2016) by Colombian architect Giancarlo Mazzanti is a 12-story, brick-clad volume that appears to float over a low-rise plinth. 10 An addition to the 1983 building by George Nelson, the brick cube uses the material palette of the original in new ways. The cladding of the addition consists of a brick screen suspended on steel cables that allow the brick facade to appear solid, or checkered, in horizontal bands to screen the sunlight and views. 8 9 The building houses hospital functions with perimeter patient rooms as well as a rooftop healing garden. The horizontal ribbon windows correspond to the bed level of the patient rooms. A maintenance zone separates the building envelope from the brick rainscreen. A subtle, concave multi-story impression in the facade pushes the brick out of the plane as if pressed by a large sphere. The profile of the impression is outlined with a circular steel frame and braced by several tiers of horizontal members. The carved faces of the facade unify the interior programs and their cellular organization behind a single curving surface. Interstitial spaces between the orthogonal interiors and the concave exterior wall are occupied by atrium, garden, and circulation spaces. The building is raised on exposed concrete cross bracing, which creates the impression that the cube is hovering, while the diffuse pattern of bricks dissolves the solidity of the volume in surprising ways.

The final two projects in this series of "clad box" approaches to contemporary design are both theaters. The Dee and Charles Wyly Theater (2009) and the Perelman Performing Arts Center (2023), both by Joshua Ramus of REX, employ a refined monolithic facade enclosing highly-functional performance spaces. The Wyly Theater in Dallas, Texas, appears like a surreal shimmering cube, hovering over the ground in the low-density neighborhood of Dallas' Theater District. 11 Its vertical grain of bundled cylinders gives it an unusual look, like a silvery cube covered in icicles that are dripping and running down the facade. The lack of conventional windows contributes to its surreal effect, making it hard to scale. A pair of blurry-edged horizontal

7 The living screen of the KMC Corporate Office facade partially obscures the programs housed within, while offering a constantly changing organic presence to the exterior.

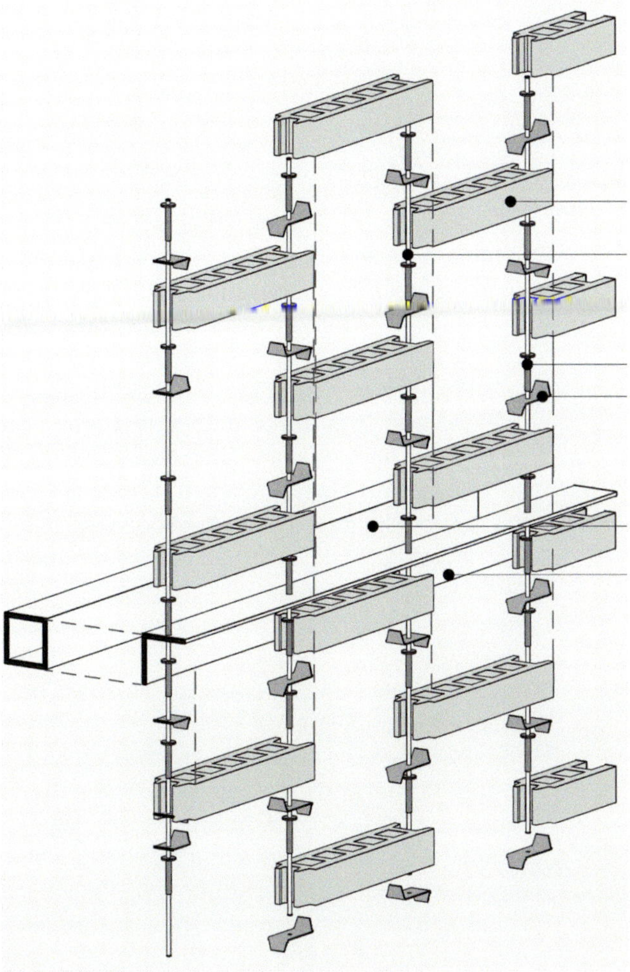

8 The brick screen facade of the Fundación Santa Fe de Bogotá is achieved by suspending the hollow brick units on stainless steel cables that form an exterior screen for the entire facade.

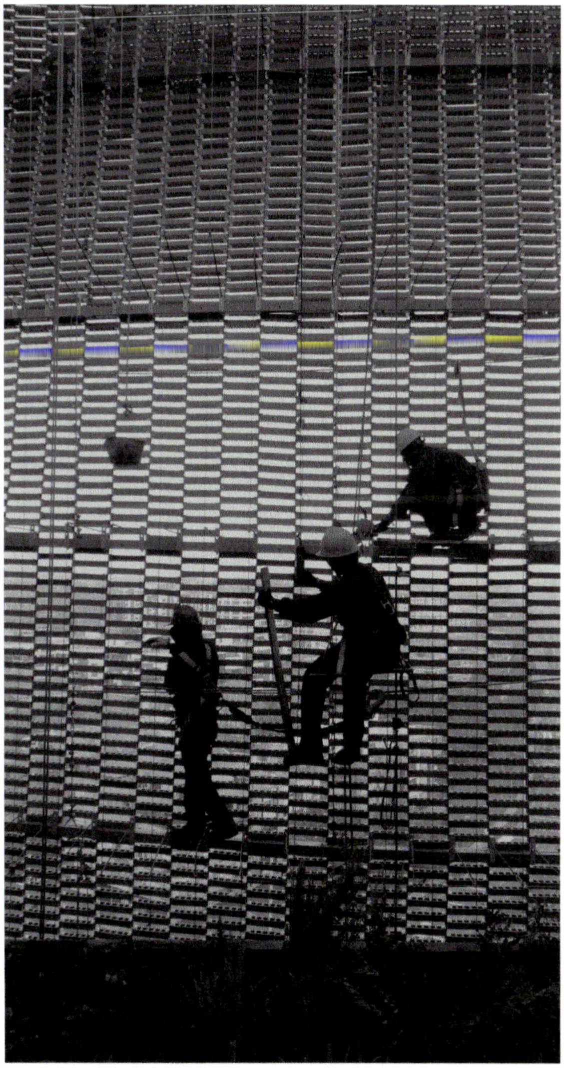

9 The construction of the brick screen facade of the Fundación Santa Fe de Bogotá, involved installing thousands of bricks in the suspended masonry screen.

10 The brick screen of the Fundación Santa Fe de Bogotá (2016), designed by Giancarlo Mazzanti, masks the various hospital programs within, while presenting a singular, though inflected, facade to the exterior.

11 The Dee and Charles Wyly Theatre (2009), designed by OMA/REX, presents a surreal silver bundled facade to the exterior, while hinting at performance and rehearsal programs through local modifications to the extruded rainscreen facade elements at the windows.

stripes across the top of its north facade hint at a space behind the evocative screen. The cladding is made of cylindrical extruded aluminum profiles, and the horizontal stripes are areas where the extrusions are compressed to allow for views out of the program spaces behind.

The Perelman Performing Arts Center in Lower Manhattan is similarly mysterious and mute, offering no windows in its blank stone facade. 12 The richly patterned alabaster stone cladding of the upper volume is contrasted with a black-granite-clad base that is undercut on the south face providing an entry into the underside of the hovering volume above. The building houses multiple flexible and reconfigurable theater spaces within that are subdivided into four quadrants with circulation spaces between them and around the perimeter. 13 The facade's alabaster stone has been book-matched to produce symmetrical patterns along its centers and at its corners. The thin layer of stone is sandwiched between glass layers to create a translucent stone panel that diffuses exterior daylight to the interior glows like a lantern at night. The facade learns from the Beinecke Library at Yale, where one-and-one-quarter-inch (32mm) thick marble slabs create a unique space of diffuse light. The meticulous patterning of the book-matched stone produces a Rorschach-like pattern that seems to tattoo the exterior skin of the building, binding the disparate interior spaces into a singular expression and providing a uniform and legible pattern on the perimeter.

A review of these five "clad box" projects demonstrates a range of motivations, techniques, and effects that employ monolithic strategies to different ends. Applying Machado and el-Khoury's criteria, these five buildings exhibit qualities that qualify them as composite, hollow, and epidermal monoliths. Although many appear solid, they are inhabitable interiors that employ the paradoxical qualities of "hollow solidity." The projects employ surface techniques in their building envelopes to mask heterogeneous interior spaces. Machado and el-Khoury speculated on the capacity of some monoliths to appear to float and the degrees to which they appear grounded. All five contemporary examples meet the ground in deliberate ways that disengage them from the ground plane and reinforce the effect of singular, volumetric, weightlessness. The Suzhou Chapel, with the reveal of detail and material contrast, perceptually levitates over its brick base. The Fundación Santa Fe rests on concrete cross bracing, while Perelman Performing Arts Center is grounded and destabilized by undercutting.

1|2 The absolute muteness of the Perelman Performing Arts Center (2023), designed by REX, in Lower Manhattan. Clad in a taut stone glazed curtainwall, the design removes all scaling elements like windows and doors and provides an intriguing and patterned urban object as its outward appearance.

1|3 The construction of the Perelman Performing Arts Center (2023) reveals the elaborate structure and interior organization required to achieve the designed discreteness of the facade strategy.

Contemporary monoliths tend to turn inward and away from their contexts. Chapel and theater programs lend themselves to the monolithic approach in that their programs are predisposed to an interior focus. Normative figures such as windows and doors challenge monoliths that often appear scaleless until a window or a door provides a scaling element. As with the nine projects included in the *Monolithic Architecture* exhibition, contemporary monoliths often rely on critical details and perceptual techniques of masking, unifying, rescaling, and screening to achieve their desired effect.

STRANGE OBJECTS

Scaling is a unique condition of monolithic architecture, as many of these examples mask the elements that would reveal the scale of a building, such as the spacing of floor slabs or the articulation of a normtive figure, such as a door or window. At a domestic scale, the monolithic approach has been applied to single-family houses and community buildings to make them appear familiar yet strange. Machado and el-Khoury included one residential project in the 1995 exhibition, the T-House, by Simon Ungers. 1|4

The T-House (1992) differentiates itself from its context with its abstract geometry and distinct cladding of corten self-oxidizing steel plates. The lower volume is embedded in the site like a landform building, while the upper volume cantilevers out dramatically. Windows and doors are rendered abstract, like a pattern of repeating rectangular openings that approximate a band within the upper volume. The T-House was prefabricated in large panels and brought to the site on large flatbed trucks. 1|5 The construction of the large steel-plate facades was not achievable through conventional residential construction techniques where attaching the cladding in situ would create significant challenges. As a prefabricated structure, the steel panels could be welded together in the shop and brought to the site in several large pieces. By welding the panels together, the architect eliminated the presence of fasteners, the expression of repetitive joint lines, and the articulation of the normative components of flashing, coping, vents, and sills.

The View House (2009) by Johnston Marklee sits on a flat open site in Rosario, Argentina. Its sculptural massing and concrete construction make it appear, at times, both stable and unstable. The curved carveouts on opposed faces make the house appear as a large, volumetric sculpture. 1|6

14 The T-House (1992), designed by Simon Ungers, in Wilton, New York, was included in the 1995 *Monolithic Architecture* exhibition. Its sculptural massing and steel plate cladding helped qualify it as a residential scaled monolith.

15 In order to achieve its monolithic qualities the T-House was conceived as a prefabricated steel plate assemblage of parts that were welded together on site. The large modular parts and the apparent lack of joints reinforce the sense of a singular monolithic block.

Large floor-to-ceiling windows occupy the broad faces of the irregular facades, while flush and inset punched windows mark the other faces. The form of the house is derived from extruding and carving curved profiles, producing the dynamic effect of a constantly changing object in the round. Diagrams show the house as a series of frames that look out at the surrounding landscape, cropping, and editing views through the shaping and framing of the windows. 17 Reto Geiser describes the house which, "through the careful manipulation of light, volume, and views, responds to its context through an interior promenade arranged along a continuous sequence of precisely placed apertures and culminating in a mirador, while nevertheless appearing as a solitary figure in the open field."[5] The house as an instrument — a multi-scope, instead of a periscope — is understood as inhabiting the center of a visual array. Its mute mono-material approach is consistent with the desire to deflect attention away from itself and into its surroundings.

The design of a building as a recognizable vernacular form, yet modified to become atypical, is the approach taken by Marlon Blackwell in his Thaden School Bike Barn. Set in an open site in Bentonville, Arkansas, the red-painted barn appears familiar yet strange. 21 The asymmetrical profile is a departure from the traditional gambrel roof form, as one side is extended to an asymmetrical peak and fitted with sloped skylights like an industrial shed. The western edge is manipulated to create an overhanging area fronting the athletic fields. The barn's interior reveals the modified wood truss system with a symmetrical interior geometry. 22 The cladding material is locally grown cypress boards spaced to allow for ventilation and light to filter through. Seen from a distance, the barn looks like a kind of geometrical zoomorphic figure.[6] The overhang resembles a tail, while the peak could be the head. Blackwell is careful to employ geometry and details to transform the common into the unfamiliar, suppressing the visibility of the windows and doors to allow the vernacular barn to masquerade as a fantastical figure. Commenting on working in the Ozarks for 30 years, Blackwell describes a way of working: "in the place, of the place, for the place, but also out of place." His work is inspired by its cultural and material context, both the ordinary and the extraordinary. Blackwell states, "Our task is the task of recreating strangeness."[7]

The Tournai School (2017) by Aires Mateus offers another example of the "strange objects" approach to monolithic architecture. The building

[5] Reto Geiser, "Specific Indeterminacy," in *House is a House is a House is a House, Architectures and Collaborations of Johnston Marklee* (Basel: Birkhäuser, 2016), p. 23.

[6] Andrew Holder's categories of contemporary architectural techniques include the category of the "creaturely," which would seem to apply to the Thaden Barn. K. Michael Hays, Andrew Holder, and Harvard University, eds., *Inscriptions: Architecture before Speech* (Cambridge, MA: Harvard University, Graduate School of Design, 2021).

[7] Marlon Blackwell, quoted in a video "Building Between" presented during a lecture at Harvard GSD, April 4, 2022.

1|6 Johnston Marklee's View House (2009), in Rosario, Argentina occupies a large open field, from which it seems to survey the surrounding landscape.

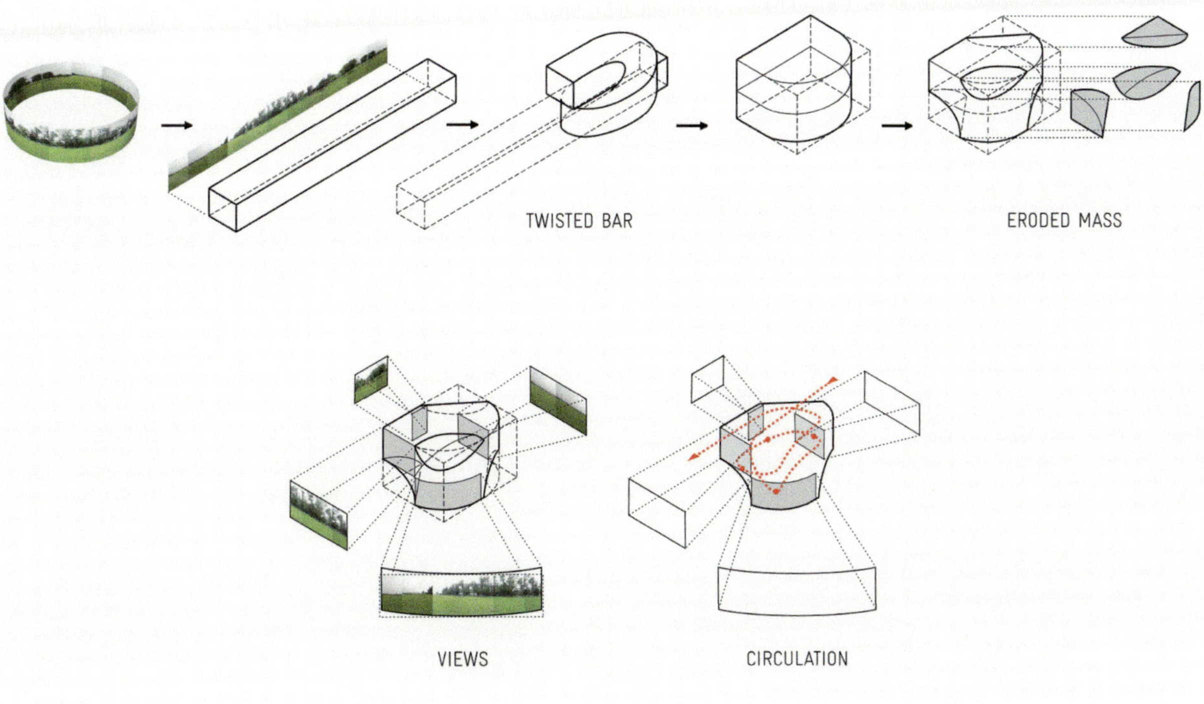

1|7 Concept diagrams of the View House illustrate how the house is conceived as a viewing device, surrounded by panoramic vistas and designed to selectively frame specific views.

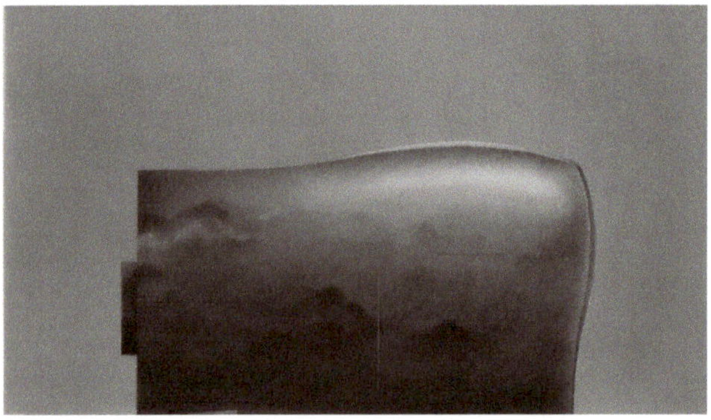

18 The elevation of the Tokyo Opera House (1987) competition entry by Jean Nouvel hints at its contents through the subtle bulging of its facade.
19 The section of the Tokyo Opera House reveals the two performance halls nested inside that seem to press on the exterior, creating the deformation seen from the outside.

20 The blank and sculpted facade of the Tournai School (2017), by Aires Mateus, reveals little of its use or contents. The "muteness" of the exterior qualifies it as a contemporary example of monolithic architecture.

presents a blank stucco facade wedged between two existing brick buildings. 20 The facade has no openings and is undercut by a house-shaped figure that carves the corner of the volume and presents a seemingly gravity-defying suspended wedge-shaped form. The school is designed to fill the space between the buildings and connect three different frontages. Its sides define two existing courtyards with L-shaped edges. While the facade on Rue du Glategnies appears object-like, most of the building acts more like infill fabric at the urban scale. The sculptural approach to massing and the lack of normative figures like windows and doors make the scale of the Tournai School hard to apprehend. Its use of the gable roof form as an opening makes the building appear at once familiar and strange. The use of stucco as a uniform exterior finish makes the Tournai School appear from the street like a solid block.

Both the Fundación Santa Fe de Bogotá and the Tournai School are sculptural forms that are impacted or carved with legible voids. The Santa Fe hospital has a spherical indentation on its primary face, while the Tournai School has an opening in the form of the negative figure of a gabled house. These legible subtractions or incisions suggest that the monoliths interact with similarly scaled figures implied only by the traces they have left. Another monolith in this category is the Auditorio Infanta Doña Elena in Águilas, Spain, which is shaped by two concave cuts in its largely opaque white massing. The auditorium was designed by Estudio Barozzi Veiga from Barcelona. Sited along the concave line of the bay, the auditorium fronts both the water and the city. The two faces addressing the urban context are planar and largely opaque. 23 The south and west faces are marked by large openings in their concave surfaces. A large, stepped plaza abuts the auditorium on the west.

The auditorium building contains two primary multipurpose halls, one with 1000 seats, and the other seating 300. 25 The lobby, circulation, and pre-function spaces benefit from the large figural openings that seem to punch into the concave surfaces. The balance of the building consists of largely opaque white stucco surfaces. The fact that the auditorium programs do not need natural light explains both the opacity of the building and, again, demonstrates the theater typology typical to the monolith.

The 1986 Tokyo Opera House competition design by Jean Nouvel also included in the *Monolithic Architecture* exhibition, employed a similar, if inverse, strategy. 18 Nouvel used the

2|1 The designed "strangeness" of the Thaden School Bike Barn (2021), by Marlon Blackwell Architects, appears both familiar and unfamiliar.

2|2 The interior of the Thaden School Bike Barn reveals the symmetrical wood trusses that are modified to change the profile of the structure on the exterior.

2|3 The Auditorio Infanta Doña Elena (2011), designed by Estudio Barozzi Veiga, in Águilas, Spain, presents a largely blank concave facade to the waterfront promenade along the bay.

theater hall program to produce a highly opaque sculptural form. The building's multiple concert halls and their associated programs were expressed on the exterior as subtle bulges in the facade, as if the exterior facade was an elastic membrane stretched over an irregular interior volume. The convex expression of the internal halls on the exterior form implies the volumetric arrangements within. In both the Auditorium in Águilas and the Fundación Santa Fe, the individual program elements are entirely subsumed within the exterior form, but the primary geometrical expression is a doubly curved concave indentation in the facades.

The essay included in the *Monolithic Architecture* catalog by Machado and el-Khoury begins with a quote and a diagram from Antoine de Saint-Exupéry's *The Little Prince*. The drawing of a snake that swallowed an elephant speaks to the idea that monoliths are surprising and ambiguous. They defy expectations and provide moments that play with perceptions of scale and stability, associations, and ambiguities. The Thaden School Bike Barn resembles *The Little Prince*'s drawing of the snake. It contains a recognizable and familiar barn within it, but the barn seems to be absorbed and digested by an architectural skin that renders it unfamiliar — or, to use Blackwell's terms, a "recreated strangeness."

MONOLITHIC TOWERS

The skyscraper emerged as a typology without precedent in the late 1800s. Enabled by new technologies of steel-frame construction and the elevator, early skyscraper designs wrapped their structural frames in traditional masonry or terracotta skins of eclectic historical allusions. For example, the Woolworth Building (1913), by Cass Gilbert, employs a vertically articulated terracotta cladding to give the building a familiar gothic appearance. Gilbert used a tripartite composition of base, middle, and top to further articulate the tower in hierarchical parts.

The Shanghai World Financial Center (SWFC) (2008) combines several attributes of monolithic architecture in a single supertall tower. 2|6 The 1607-foot (490m) tower combines hotel, office, and observation deck programs within its tapered shaft. While typical office floor plates are deep and broad to maximize leasable area, hotel floor plates tend to be shallow to take advantage of perimeter views. 2|7 The building's designer, Bill Pedersen of Kohn Pedersen Fox (KPF), chose to articulate the programmatic stacking within the exterior form of the building. By tapering the tower along its shaft, the lower zone of the tower contains the

24 The concave faces of the Auditorio Infanta Doña Elena, seem to gesture to a focal point outside of the object, and the composition of rectangular openings of varying sizes makes the scale of the building hard to determine.

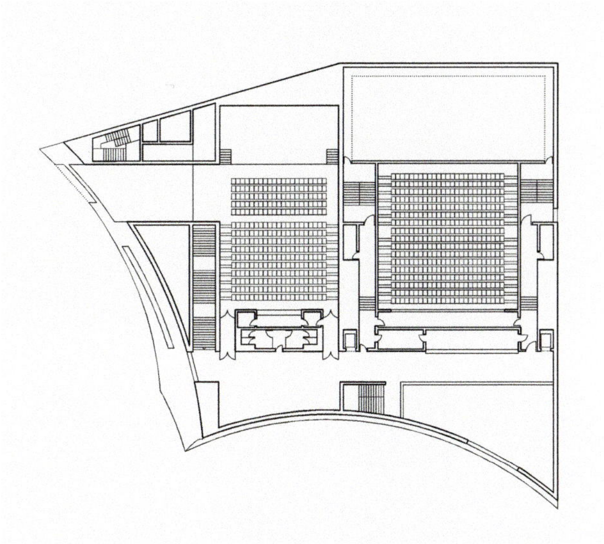

25 The plan of the Auditorio Infanta Doña Elena reveals the side-by-side arrangement of two auditoriums surrounded on two sides by prefunction spaces.

optimal floorplate for its office functions, while the upper floor plates contain optimal floor plates for hotel functions. The transitions between the two programs are blended by a cutting plane that follows a curve with an extremely large radius. 28 Thus, each floor plate of the SWFC is unique, diminishing from bottom to top along a slow curve. The service core at the center of the tower also decreases in size as the tower rises, as elevator banks drop off. The programmatic spaces are framed by these two profiles: the stepped interior profile of the service core and the curved exterior profile of the curtain wall.

The tower's exterior wall is clad in a highly reflective glass facade that obscures the activities within. To mask the scale of the building further, Pedersen made the spandrel module the same dimension as the vision module and articulated the horizontal mullions with an extruded horizontal cap. The result is a highly banded horizontal pattern of repetitive proportions that reaches the top of the tower, obscuring the difference between vision from spandrel glass and office uses from hotel uses. The tower achieves a singular expression of great simplicity even as it stretches skyward for more than 100 stories. The monolithic approach to the SWFC subsumes the individual parts (mullions, windows, structure) in the service of a larger sculptural whole while also utilizing geometrical carving to blend the programed areas, unifying the disparate programs (office, hotel, restaurant) and systems (service cores, elevator shafts, lobbies) into a singular expression.

8 Stan Allen and Princeton University, eds., *Landform Building: Architecture's New Terrain* (Zürich: Lars Müller, 2011), pp. 22-23.

INFINITE SURFACES

Beyond a certain scale, structures merge with infrastructure or appear more like landscapes than buildings. At the threshold between architecture and infrastructure, another type of monolith takes the "landform" as its model. Architect and critic Stan Allen has described the landform as a concern with "field-like organizations" that "distribute and channel flows of energy, information, and people on-site."[8] The "infinite surface" approach to monolithic architecture was anticipated in the 1960s in drawings by the Italian group, Superstudio, whose Continuous Monument showed a neutral white grid extending to the horizon, sparsely inhabited by small groups of people and fragments of landscapes. 29 The extended grid hinted at a scaleless architecture of infinite extendibility, forming a new alternative topography that could compete with the city and the territory.

26 **Shanghai World Financial Center (2008), designed by Kohn Pedersen Fox, creates a 100-story monolithic tower as a skyline icon for Shanghai.**

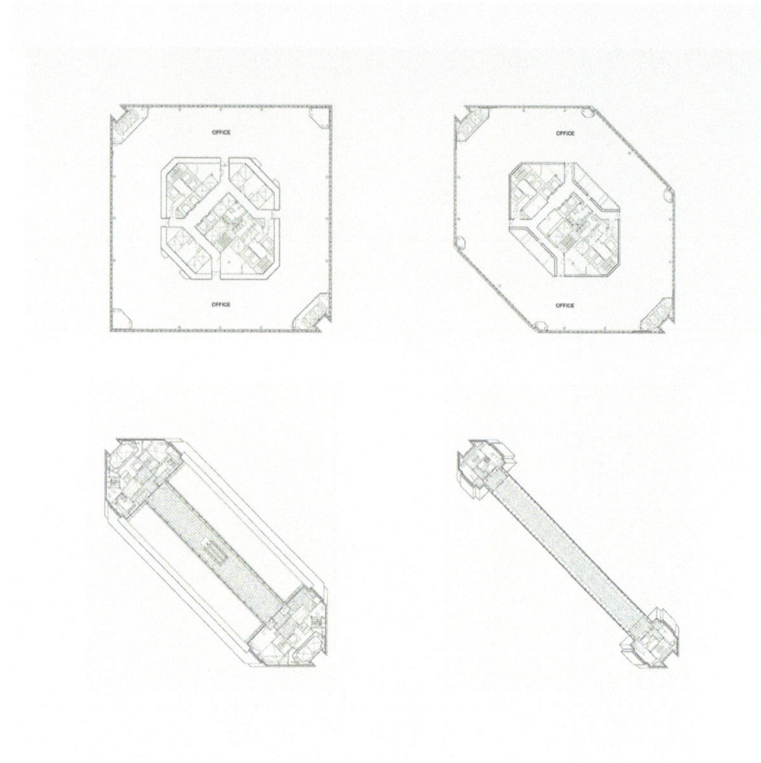

27 **The typical plans of the Shanghai World Financial Center (2008) document the typical office floor plate at the lower portion of the tower with a square proportion, as well as the narrower floor plate at the upper portion, where the elevators have dropped off and the tapered geometry of the exterior is approaching the proportion of a diagonal line.**

While Superstudio's radical proposals remained unrealized, they appear to have inspired another generation of landform projects. Foreign Office Architects' (FOA) proposal for the Yokohama Ferry Terminal won a 1995 competition and was completed in 2002. The project takes its cues from the singular topography of Superstudio's collages. The competition renderings included in the *Monolithic Architecture* exhibition depict an extended grid stretching nearly to the horizon. 30 In a departure from Superstudio, FOA's grid deforms, wrinkles, and folds to enclose spaces. And, unlike the gridded abstraction of Superstudio, FOA's design for the Yokohama Ferry Terminal used landscape as a performative and organizational surface, folding to control the flow of vehicles and pinching to produce enclosure. Allen notes that the emergence of the landform as a design driver coincided with "the newfound potential of the computer to model complex surfaces."[9] This technique of using a single surface to produce floor, wall, and ceiling was latent in the early competition renderings and eventually manifested as a detailing technique for the actual building.

Determined to stay true to the concept, even as the construction constraints necessitated the evolution of the design, FOA principals Farshid Moussavi and Alejandro Zaera-Polo insisted that the detailing consist of a single material per surface. On the upper deck, all surfaces are made of faceted wood boards to absorb the new ground's compound geometries. 31 On the lower levels, the ceilings are made of faceted steel plates folded to produce stiffness. The envelope consists of an undulating glass wall without mullions, relying instead on the folding of the glass lites to achieve the necessary stiffness. The choice of materials and their attitudes toward detailing allow the quarter-mile (430m) long structure to appear coherent as a singular surface described by the architects as "a field of movements with no structural orientation."[10]

While the Yokohama Ferry Terminal translates an undulating landscape into faceted surfaces, the MAAT Museum (2016) in Lisbon, by Amanda Levete Architects (AL_A), translates the single surface building into a smooth and swelling landform clad in ceramic tiles and stone pavers. 32 The museum takes the form of a swell in the landscape that crests toward the bank of the Tagus River to create a shallow, arch-shaped opening under its overhang. The rippling geometry of the museum was conceived of with 3D modeling software, and the construction of the complex geometry was achieved using repeating tiles with

[9] Stan Allen, *Landform Building*, p. 23.

[10] Foreign Office Architects, *Phylogenesis, FOA's Ark* (Barcelona: Actar, 2004), p. 228.

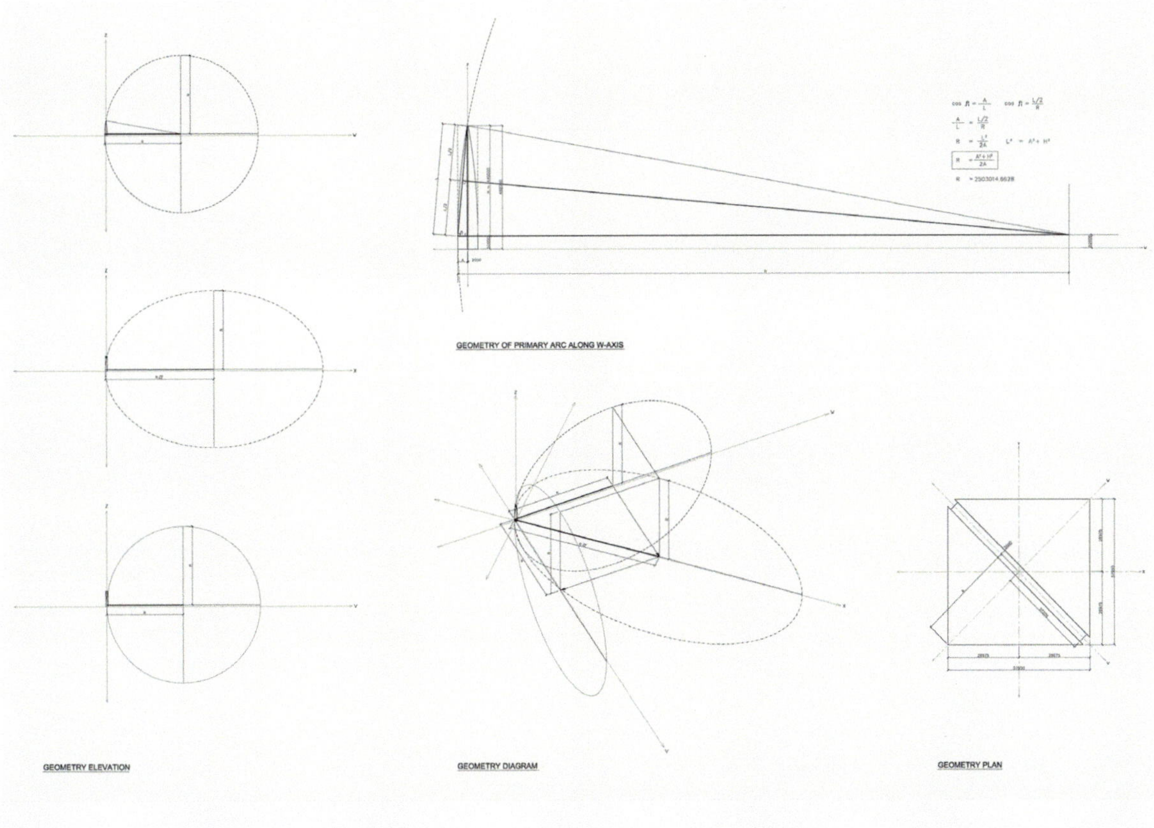

28 **The geometry of the Shanghai World Financial Center is defined by the arc of two very large circles that carve the tower to produce its distinctive sculptural form.**

variable joints. The architects intended for the building to reproduce the rippling lighting effects of the adjacent waterway. Levete evoked the tradition of Portuguese glazed tiles in her use of a contemporary glazed tile for the facades of the new museum. The trapezoidal tiles, which the architects call a "split hexagonal grid," approximately 12" x 30" x 1¼" (260 x 600 x 30mm) in dimension, are arranged in parallel strips that form a ruled surface while the glazed tiles have different degrees of inclination to create a subtle, faceted surface. The entire facade was clad with three types of glazed tiles, optimizing the variable geometry of the curvatures by varying the gaps between tiles. The open-joint rainscreen system allows the tiles to absorb slight changes in the geometry while keeping the building enclosure separate from the exterior cladding. The perceptual effects of the glazed tiles clad over the swooping curves of the landform structure give the building its distinct shimmering optical effect while contributing to its monolithic expression by blurring the edges of the building and its surrounding landscape.

The next generation of architectural monoliths will likely build on the conceptual strategies outlined by Machado and el-Khoury while leveraging new digital tools and construction techniques to produce architectural effects of solidity, singularity, muteness, scalelessness and "recreated strangeness."

11 Jeff Kipnis, "The Cunning of Cosmetics," in Herzog & de Meuron, 1993–1997, *El Croquis* 84, 1997, p. 26.

COSMETICS

Writing in 1997, architecture critic Jeff Kipnis proposed a theory he called the "cunning of cosmetics." He likened Herzog & de Meuron's Signal Box 4 to the siren songs in Homer's *Odyssey*. For Kipnis, the understated details and restrained use of a singular material are seductive, intriguing, and even perilous. Kipnis is quick to differentiate between ornament and cosmetics. "Ornaments attach discrete entities to the body like jewelry, reinforcing the structure and integrity of the body as such."[11] Cosmetics, on the other hand, are "indiscrete," blending with the complexions of the face, accentuating geometry and amplifying effects. For Kipnis, "cosmetics work as fields, as blush or shadow or highlight, as aura or air." Herzog & de Meuron's alluring architecture operates on surfaces with effects that transform and augment architecture alchemically. "HdM derives its critical edge from the assumption of architecture's basic adequacy and an ease with the controversial proposition that architecture has no other profound

29 The speculative drawings of Superstudio, such as A Journey From A to B (1969), imagined extensive environments between architecture, landscape, and infrastructure.

30 The conceptual drawings of the Yokohama Ferry Terminal (1995), designed by Foreign Office Architects (FOA) imagine a continuous and inflected urban surface of activities.

31 The completed Yokohama Ferry Terminal (2002) realizes an urban-scaled landform structure as an "infinite surface" type of architectural monolith.

3|2 The MAAT Museum (2016), designed by Amanda Levete, provides a contemporary version of the landform monolith, appearing to crest like a wave at the edge of the Tagus riverfront in Lisbon, Portugal.

project than to fabricate a new sensibility from its own palette."[12] For Kipnis, that means the new is not to be found at the margins, but rather in an "aggressive assertion of the center."[13] The palette at the center of Herzog & de Meuron's work is in the deliberate and determined arrangement of the components of architecture that, through details, distribution, and disposition, create surprising and seductive effects, causing the viewer to ask: What am I looking at? What scale is it? And, what else is not what it seems?

12 Jeff Kipnis, "The Cunning of Cosmetics," p. 26.

13 Jeff Kipnis, "The Cunning of Cosmetics," p. 26.

14 Jeff Kipnis, Instagram caption, @jeffkipnis, September 3, 2016.

Part-to-whole and monolithic strategies employ degrees of calibration and articulation. The wood components arranged in a grid, with overhanging edges and articulated ends, appear like a cartesian cloud in Kengo Kuma's Prostho Museum, while the planks of timber flooring that make up the Yokohama Ferry Terminal create an undulating surface that creases and folds to guide flows of people and vehicles to their destinations. Kuma's use of timber is discrete, while FOA's use of timber produces continuity. The cellular modules of The Broad Museum appear to repeat and aggregate until they form a uniform surface deformed to highlight the boardroom window. The careful calibration of modules, material finishes, and micro-faceting makes the MAAT Museum appear continuous with the surrounding landscape, glistening in the reflected light of the Tagus River. The articulation of building components into discrete parts or subsumed by a singular gesture is not predetermined or automatic. Achieving either outcome takes tremendous effort and is the result of a highly intentional design process. In each approach, the coordination of individual parts that form the larger whole requires careful attention to material and systems selection, module sizes and configurations, geometry at both local and global scales, and careful consideration of material tolerances, installation sequences, and access to expertise and skilled labor. Quoting Kipnis: "Architecture is not about how space is marshaled and assembled, but the implications thereof."[14]

1 The Zollverein School of Management and Design (2006), designed by SANAA, employs an array of square windows to perforate the concrete cube, introduce daylight, and disclose certain information about the building.

Chapter 6 From Window to Envelope

The geographer looks up from the map on the table and glances out the window. 2 A soft, natural light illuminates his face, right arm, and the partially unrolled maps on the table before him. A wardrobe sits in the corner of the room, immersed in shadows. The combination of highlights and shadows, known as *chiaroscuro*, is constant in the work of Dutch painter Johannes Vermeer, which nearly always includes a window as a source of dramatic horizontal light. The window in *The Geographer* is a divided-lite window, with smaller panels of translucent glass, held in place by lead strips.

Vermeer captures the solitary geographer mid-thought. The light floods the room, lighting the right side of his face, casting deep shadows on the rear wall, and creating shiny reflections of the geographer's divider, his thumbnail, and the surface of the terrestrial globe on the wardrobe. Light, rather than figure, is the true subject of Vermeer's paintings, and the window is the source of that dramatic light.

In the late 17th century, while Vermeer painted his solitary figures in domestic scenes, the world was experiencing a global trade revolution. His own port city of Delft was connected to the world through the trade networks of the Dutch East India Company. Dutch sailors traded silver, copper, silk, porcelain, and spices. Scientific instruments advanced cartography and global navigation, while microscopes and optics changed how the world was seen, studied, and represented.[1] Art historians have speculated that Vermeer may have used a camera obscura device to produce his flawless perspectival compositions, pointing to the low horizon line, the alignment of the window sill, and the apparent distortion of objects in the foreground that hint at mechanical representation. The camera, like the telescope, altered established perceptions of the world. So too, the windows and their elaborate frames.

1 Timothy Brook provides a beautiful account of the intimate interiors painted by Vermeer and the vast world beyond that is referenced in the details that are carefully depicted by Vermeer. Timothy Brook, *Vermeer's Hat: The Seventeenth Century and the Dawn of the Global World*, Paperback ed. (New York, NY; Berlin: Bloomsbury, 2009).

ARTIFACT

The window is an architectural element that plays many roles, introducing light into an interior space or framing views out. Windows can be opened for ventilation or closed to protect from the elements. They are often punctures — exceptions in the wall or solid plane in which they occur. Discrete or punched, windows are often built into load-bearing walls, where an arch or lintel transfers loads around its opening.

As a technical artifact, the window in Vermeer's *The Geographer* is a material construct, made up of wood members, glass lites, and lead cames. The

2 *The Geographer* (1669), by Johannes Vermeer, depicts an atmospheric interior scene with a solitary subject. The *chiaroscuro* effect of deep shadows and contrasting light is enabled by the window to the left. The window and its lighting effects are the other subject of the painting.

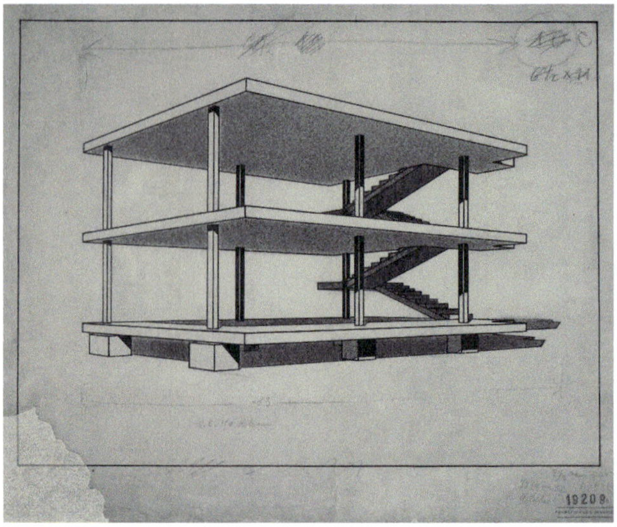

3 A model for mass produced housing, the Maison Dom-Ino (1914) by Charles-Edouard Jeanneret, also known as Le Corbusier.

elaborate detailing of the window in the painting is evidence of the labor and craftsmanship that produced it. A glassmaker was involved, heating silica to form the sheets of glass to be cut into rectangles to make up the divided lites. A metalsmith worked the lead cames around the glass to create the larger opening. A woodworker formed the fixed frame of the window, the casing, and the jambs, as well as the operable rails and stiles. All these materials and trades converge at a single window, allowing Vermeer's geographer to contemplate the expansive world outside.

The window is also a cultural artifact, tied to the time and place of its making. In the painting, the window is a source of light, providing a place to sit, read, or study. The window offers a view of social interactions happening on the street from the privacy of the home and is also reflective of larger cultural norms and values. Dutch windows in cities like Delft and Amsterdam have also been associated with the Calvinist tradition of putting a virtuous life on display, where everyday life indoors was displayed through large windows without curtains or blinds.

From the divided-lite windows of Vermeer's Delft to the triple-glazed, thermally broken windows of contemporary energy-efficient facades, the window has been the subject of significant evolution and transformation. Early glass fabrication limited the size, flatness, and visual clarity of windows. The 20th century saw the radical transformation of the glass window from a discrete punched opening in a load-bearing wall to increasingly larger windows that eventually became walls, spanning from floor to ceiling as part of a building envelope independent from the structural frame. The separation of structure from the enclosure, and the shift from its placement in a load-bearing exterior wall to the lightweight building envelope enclosing the entire structure, is central to narratives of modern architecture. By the second half of the 20th century, the curtain wall, a non-load-bearing glass envelope outboard of the building slab, had become a standard system for commercial architecture, and the evolution from window to envelope was seemingly complete.

ELEMENTS

In the introduction to his 2018 book, *Elements of Architecture*, Rem Koolhaas argues that despite the transformation of contemporary architecture beyond recognition, the basic building blocks are common to all architecture. Windows, facades, balconies, stairs, and corridors make up the architectural elements

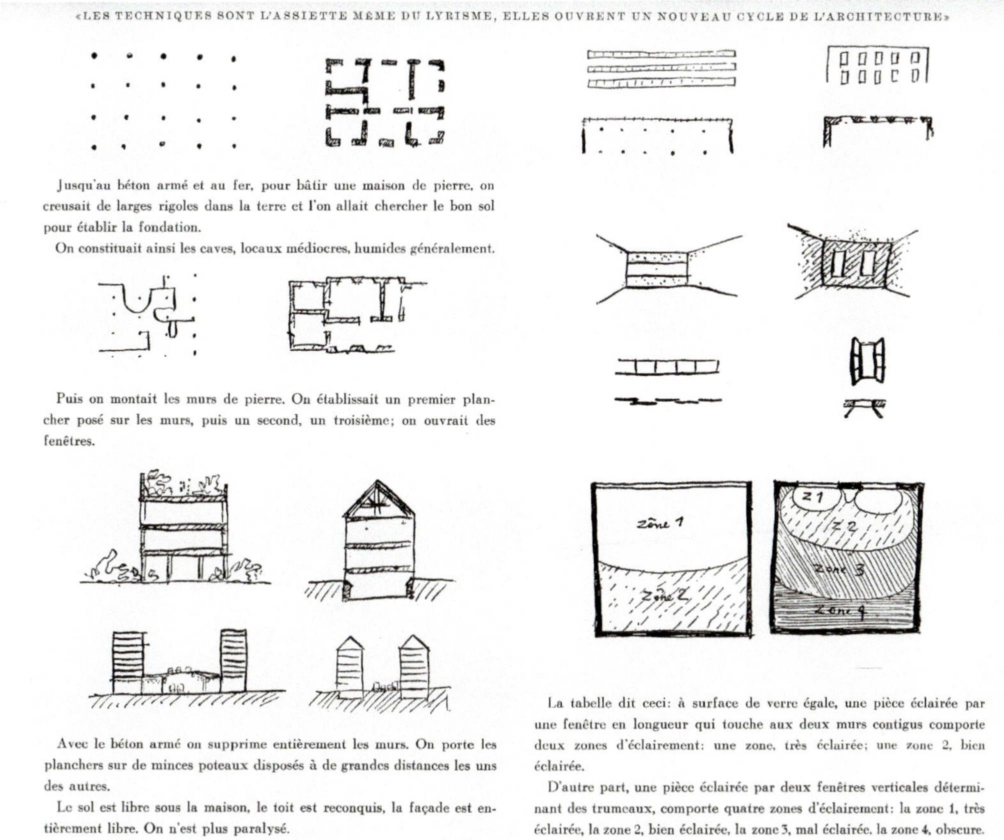

«LES TECHNIQUES SONT L'ASSIETTE MÊME DU LYRISME, ELLES OUVRENT UN NOUVEAU CYCLE DE L'ARCHITECTURE»

Jusqu'au béton armé et au fer, pour bâtir une maison de pierre, on creusait de larges rigoles dans la terre et l'on allait chercher le bon sol pour établir la fondation.

On constituait ainsi les caves, locaux médiocres, humides généralement.

Puis on montait les murs de pierre. On établissait un premier plancher posé sur les murs, puis un second, un troisième; on ouvrait des fenêtres.

Avec le béton armé on supprime entièrement les murs. On porte les planchers sur de minces poteaux disposés à de grandes distances les uns des autres.

Le sol est libre sous la maison, le toit est reconquis, la façade est entièrement libre. On n'est plus paralysé.

La tabelle dit ceci: à surface de verre égale, une pièce éclairée par une fenêtre en longueur qui touche aux deux murs contigus comporte deux zones d'éclairement: une zone, très éclairée; une zone 2, bien éclairée.

D'autre part, une pièce éclairée par deux fenêtres verticales déterminant des trumeaux, comporte quatre zones d'éclairement: la zone 1, très éclairée, la zone 2, bien éclairée, la zone 3, mal éclairée, la zone 4, obscure.

4 Le Corbusier's "Five Points of a New Architecture," lays out a polemical agenda for reinventing architecture around the possibilities provided by a new material system, namely steel reinforced concrete.

he sets out to research. He writes: "A surprising percentage of everything architects assemble in a so-called building is defined by a limited number of ancient categories: wall, floor, ceiling — to name the elements with which three-dimensionality is constructed. *Elements of Architecture* looks at the fundamentals of our buildings, used by any architect, anywhere, anytime."[2] By looking at each part, Koolhaas hopes to understand the constituent parts common to all buildings and perhaps shed some light on the architectural whole.

FREE FACADE

In 1914, at the age of 27, Swiss architect Charles-Édouard Jeanneret developed the design for a concrete frame house known as Maison Dom-Ino. 3 He was convinced that the design could address the demand for high-density housing and filed a patent application for his design. Maison Dom-Ino consists of six columns, three slabs, and a stair. The columns are arranged on a two-square grid, forming an A-A-B rhythm along its long axis, and a C-A-C rhythm on its short axis. Significantly, no enclosure is shown in the drawing. The Dom-Ino frame was explicitly independent of the enclosure, suggesting that the envelope was secondary and freed of the constraints of the structure. Jeanneret later adopted the pseudonym Le Corbusier, and architecture would never be the same again.

Le Corbusier describes the design of Maison Dom-Ino as an *ossature*, a skeletal framework, deliberately open and incomplete. The independent structural frame allowed the interior layout of the floor plans, and the facade to be "free," to be planned in a variety of ways. Le Corbusier imagined the Maison Dom-Ino as a generic unit that could be aggregated and varied within the parameters of the gridded system.

Le Corbusier writes, "The format of the 'dom-ino' frame, the very particular position of the posts, allow countless combinations of interior layouts and all imaginable light intakes on the facade. We had conceived the idea of a company, sister of the first, which would sell all the elements of the equipment of the house, that is to say, everything that can be manufactured in the factory in large series, according to standard measurements and meet the multiple needs of rational equipment: windows, doors, standard lockers serving as cupboards or furniture and forming part of the partitions."[3]

The Maison Dom-Ino was conceived of as a mass-produced structural framework to be infilled with a system of component parts that could be

[2] Rem Koolhaas et al., eds., *Elements of Architecture* (Köln: Taschen, 2018), p. XLV.

[3] Le Corbusier and Pierre Jeanneret, *Oeuvre Complète. 1: Oeuvre Complète 1910–1929* / W. Boesiger / O. Stonorov (ed.), ed. Willy Boesiger and Oscar Stonorov, 16. ed. (Basel; Boston, MA; Berlin: Birkhäuser, 2006), p. 23.

5 The Villa Savoye (1931) is regarded as the translation of Le Corbusier's "Five Points" in a single exemplary building. The taut white surfaces of the exterior wall with the horizontal ribbon window that cuts the facade in half, articulates the new agenda for architecture.

prefabricated off-site and assembled in a variety of combinations. Le Corbusier thought through the part-to-whole logics and means of production. His listing of "domestic equipment," the windows, doors, and cupboards, resonates decades later with the obsessive cataloging of fundamental components of Koolhaas' "elements."

The Maison Dom-Ino is the second project featured in his first *Oeuvre Complete 1910–1929* volume. The project is illustrated in two sections: a structural plan (showing the cellular makeup of the slab), a diagrammatic foundation plan, and a two-point perspective. The drawings show slabs composed of cellular pockets that optimize the use of concrete while providing a flat slab above and a smooth soffit below. Le Corbusier illustrates the combinatory logics of the Dom-Ino frame as well as possible interior layouts — labeled as *variante* — and facade designs. The plan variations do not take full advantage of the freedom offered by the column grid, as most of the columns are absorbed in partition walls. Similarly, facades represented in the perspective drawings show a combination of punched and screen-like openings. The radical potential of the Maison Dom-Ino would have to wait to be revealed in Le Corbusier's subsequent projects.

4 Le Corbusier and Pierre Jeanneret, *Oeuvre Complète*, p. 129.

The Maison Cook (1926) and the Villa Savoye (1931) would eventually allow Le Corbusier to disengage the house from the ground plane through the independent structural frame. Throughout the 1920s, he studied window size and proportion with large, glazed openings corresponding to double-height space in the Colonie De Weissenhof (1927) and the horizontally proportioned ribbon windows of Petit Villa at Lac Leman (1924). Le Corbusier's design for a painting studio for his friend Amédée Ozenfant (1924) also allowed him to experiment with large, steel-frame windows on two faces and clerestory windows to produce the effect of a cube of light.

Le Corbusier's early house projects allowed him to test components and configurations, leading to his "Five Points of a New Architecture." 4 The *fenêtre en longueur*, or ribbon window, was one of the key elements of a new architecture. Together with the *pilotis,* structural frame, the roof garden, the free plan, and the free facade, the ribbon window was a crucial ingredient in radically transforming architecture for a new epoch.

Le Corbusier published his "Five Points" as a series of diagrammatic hand-drawn sketches.[4] To describe the use of the structural frame, he contrasts a field of columns with the footprint of load-bearing walls with openings. To emphasize the freedom of

6 The Villa Savoye under construction reveals a concrete frame with masonry infill for the walls. The running bond pattern of brick on the filleted walls of the roof garden is the opposite of the clean, white, "machine" aesthetic that Le Corbusier aspired to.

partition walls, he shows a series of thin partitions snaking between the grid of columns. In section, he illustrates a house floating above a gardened landscape, contrasting with a traditional house section that includes a basement. He writes: "The ground is free under the house, the roof is reclaimed, the facade is entirely free. We are no longer paralyzed."[5]

The diagrams on the right concern the proportion of windows. 4 At the top, he contrasts three strips of ribbon windows with a facade of punched windows. Partial plans show the column grid that allows the windows to become a lightweight screen. Interior perspective vignettes contrast a single horizontal ribbon window spanning wall to wall with a dark, stippled room with two punched windows. The bottom-right diagram shows two square rooms in plan with light intensity zones hatched and labeled.

For the punched windows, Le Corbusier shows two pools of light adjacent to the windows and four different zones of light becoming progressively darker toward the interior. The ribbon window, in contrast, shows the same room, brightly lit for the top half of the plan, and the bottom half is labeled "Zone 2." Le Corbusier's intent is clear. The ribbon window provides superior daylight penetration, providing illumination and eliminating dark and dank corners associated with the interiors of the traditional house.[6]

Taken together, Le Corbusier's "Five Points" chart an interrelated set of configurations for a new architecture, exploiting the material properties of steel-reinforced concrete to find its own expressive potential — one that aligns with the aspirations of an emerging conception of lifestyle based on access to sun and the landscape beyond. In Le Corbusier's words, "Progress brings liberation."[7] Windows would become, for Le Corbusier, the site of intense polemics about daylight, hygiene, and structural integrity, and modernity would be manifested in the composition of the ribbon window and the free facade.

5 Le Corbusier and Pierre Jeanneret, *Oeuvre Complète*, p. 129.

6 Le Corbusier's ribbon window sketch anticipates daylight penetration in an age before digital simulations, so it is worth noting that he overestimates the daylighting performance of the ribbon window. Running a daylighting simulation today, would reveal that tall vertical windows would likely produce more effective daylight penetration, especially in buildings with high ceilings. For further discussion on this topic, see chapter 10.

7 Le Corbusier and Pierre Jeanneret, *Oeuvre Complète*, p. 128.

EVOLUTION

To trace the evolution of the window as a basic element of architecture, we must also trace the evolution of the wall in which it sits and the structural frame that displaced the wall as the primary structural support. The projects and polemics of Le Corbusier track the changing attitudes about windows, window walls, and, eventually, the curtain wall. The trajectory of modern architecture, from the opening in a masonry wall to the lightweight

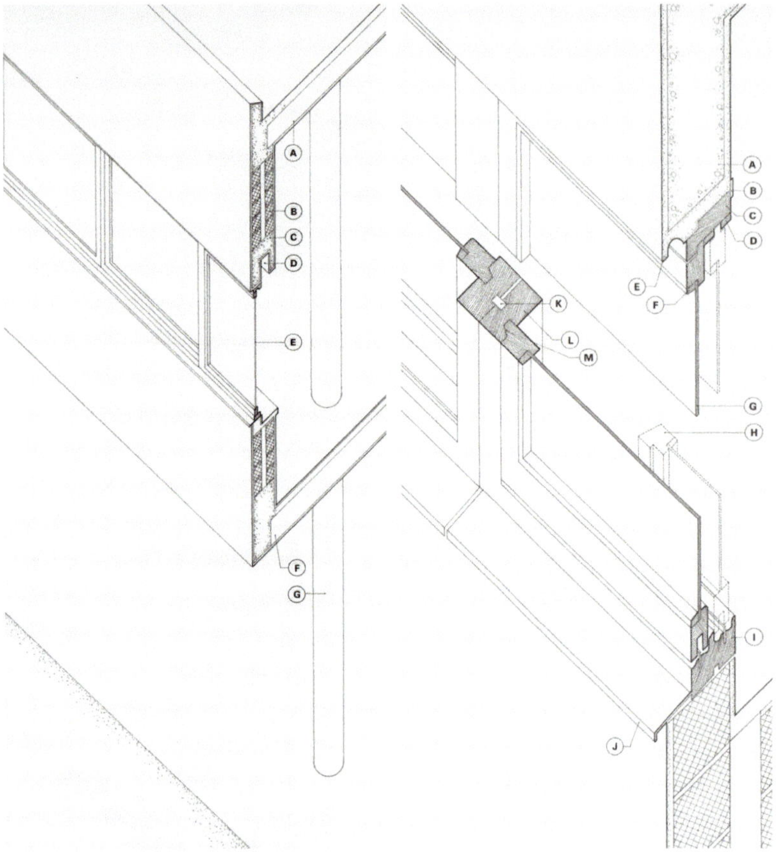

7 **Axonometric drawings of the exterior wall construction of the Villa Savoye (1931) produced by Edward R. Ford and published in *The Details of Modern Architecture.***

expressive membranes of contemporary curtain wall envelopes, passes right through Le Corbusier. A careful study of at the windows of his Villa Savoye reveal a pivot point between the window and the window wall. 5

The Villa Savoye in Poissy, a suburb west of Paris, was commissioned in 1928 by the wealthy French insurer Pierre Savoye and his wife, Eugénie. Conceived of as a country retreat, the house was set within a pasture surrounded by a grove of trees along the south bank of the River Seine. Le Corbusier understood the house in the tradition of the Renaissance villa, a second home in the country to complement an urban home in the city center.

Le Corbusier designed Villa Savoye from the point of view of arrival by car. Approaching along Rue de Villiers, a gatehouse gives way to a curving driveway through a bosque of trees. Rounding a bend, the house appears as a freestanding structure in the open pasture. The driveway proceeds off-center, aligned with the right-hand bay of pilotis, and passes under the house. The second-floor volume has an attenuated horizontal proportion, cut by a ribbon window extending across the facade's full width — a horizontal incision in the taut face of white stucco. The only articulation of the window is a small projecting sill running acros the length of the bottom of the ribbon window. 7

The ribbon window cuts horizontally across all four elevations, with minor variations due to the cantilevered ends and small breaks to accept the interior partitions. From the outside, the effect is a pristine white box suspended above the landscape with horizontal windows that split the facade into two equal bands. The window is a structural discontinuity in the facade, giving the house an abstract immaterial lightness that seems to defy gravity.

WHITE SURFACES

Construction photos of the Villa Savoye reveal a different story. The taut, white facade is revealed to be a layer of stucco and white paint, a thin veneer over a block wall. 6 The block coursing makes up the fascia under the ribbon window, and a concrete lintel carries the bricks above. Even the billowing curved walls on the roof garden level are constructed of block in a running bond pattern. Photographs of buildings under construction always show them in their most naked, exposed, and vulnerable state. Mark Wigley has noted the thinness of modern architecture's signature white surface in his book *White Walls, Designer Dresses: The Fashioning of Modern*

8 **Historic photo of Villa Stein at Garches with the car as a prop and a protagonist.**

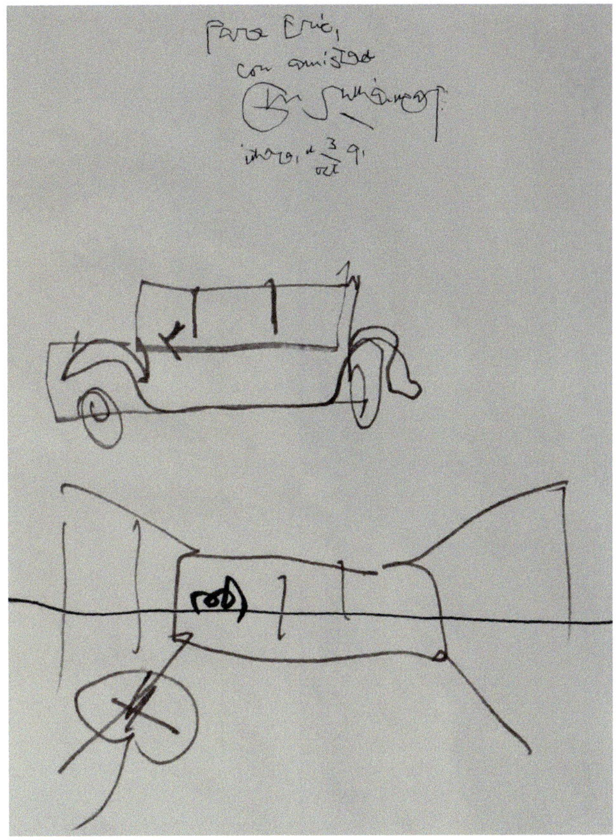

9 **Sketches produced by Jullian de la Fuente (1991), who worked closely with Le Corbusier, represent the ribbon windows of Villa Savoye as the same proportion as the car windshield.**

Architecture. On the before and after images of Villa Savoye, published in *L'Architecture Vivante*, Wigley notes "the extent to which the machine age finish of white-painted stucco is but a 'look' that veils the basically hand-crafted structure beneath."[8]

The proportions of the ribbon windows were unprecedented, but the details of the wooden frames appear rather conventional. 7 At the head condition, a drip edge profile articulates the header under the lintel, and at the sill, a bent metal sill projects beyond the plane of the facade to manage water and limit staining. Le Corbusier's choice to use a horizontal sliding configuration resulted in their occupying two different planes, as one window must slide past the other. This condition and the dimension of the window frames in the operable panel make it difficult for the window to recede. While the white stucco face unifies the building as a taut box, the components of the ribbon window work against the reading of the aperture as a singular negative cut.

Edward R. Ford's *The Details of Modern Architecture* carefully documents the construction of Villa Savoye to better understand Le Corbusier's detailing in the context of his contemporaries. Ford acknowledges the radical ambitions of Le Corbusier's approach of incorporating industrially produced and standardized parts yet notes his misconceptions about the building industry. For example, he claims that Le Corbusier failed to recognize the difference between the functional requirements of commercial and residential buildings, that he overemphasized the importance of standardized sizes in building components, and generally showed a lack of consideration for the impact of surrounding environments on buildings and their materials.[9]

Ford finds some of Le Corbusier's details counterintuitive and curious. Walter Gropius and Jan Duiker, two of Le Corbusier's contemporaries, used industrial window detailing borrowed from factories and applied these to schools and other institutional buildings. Le Corbusier insisted on wood-frame sliding windows for Villa Savoye. Whereas Duiker's Open Air School in Amsterdam turns the corner and creates a prismatic glass volume, Le Corbusier's ribbon window remains bounded in the picture plane of the facade. "The recessed notch at the head, the projecting sill, and the two planes of the sliding window make it possible to see the wall as a series of parallel planes...The wall thus becomes an analog or subsystem of the building itself."[10]

Returning to the arrival sequence (or the departure sequence), the first and last glimpses of the Villa Savoye were seen through the windshield

8 Mark Wigley, *White Walls, Designer Dresses: The Fashioning of Modern Architecture* (Cambridge, MA: MIT Press, 1995), p. 116.

9 Edward R. Ford, *The Details of Modern Architecture. Vol. 1*, 1st paperback ed., vol. 1 (Cambridge, MA: MIT Press, 2003), p. 247.

10 Edward R. Ford, *The Details of Modern Architecture*, p. 243.

10 Conceptual drawing of the Slow House (designed 1989) by Diller + Scofidio. The house was conceived as an instrument for seeing, just as Le Corbusier described the house as a "machine for living." Documenting the house, with its curvilinear geometry, the architects developed an instrument for drawing sections along a curved plan. Note the model car on axis with the front door, and the car windshield aligned to the front elevation of the house.

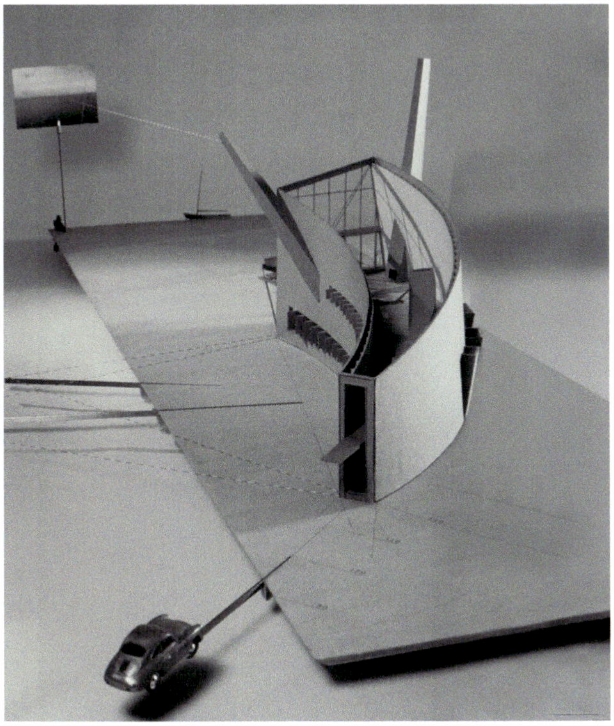

11 Conceptual model of the Slow House depicting the "boundary conditions" of the house as the arrival to the house by car, and the exiting of the house, through the panoramic view, represented by the postcard image of a seascape.

12 One of a series of sections of the Slow House depicting the activities occurring within.

of the car, or even perhaps as a reflection in the car's rear-view mirror. Le Corbusier made numerous comparisons between cars and buildings,[11] and he saw the automobile industry, with its assembly line efficiencies, as a model for modern architecture. He also used cars as props, which were prominently featured when he photographed his buildings.[12] [8] The extreme horizontal proportion of the ribbon windows at Villa Savoye is said to frame the horizon line of the landscape beyond. They also mimic the proportion of the automobile windshield and the rear-view mirror. Jullian de la Fuente, who worked in Le Corbusier's studio, has described the ribbon window of the Villa Savoye as having the proportions of a windshield. [13] [9] For Le Corbusier, the window was a protagonist in his polemics of machines and modernity.

[11] Le Corbusier, *Towards a New Architecture* (New York, NY: Dover Publications, 1986).

[12] The photographs of the Villa Stein at Garches show Le Corbusier's own Voisin parked in the driveway. There are no people in the photographs, but the car seems to be moving around the house like a mechanical visitor.

[13] Jullian de la Fuente, sketched the Villa Savoye as a car windshield in a lecture sketch in 1991.

WINDSHIELD TO PANORAMA

Speculating on the window's relationship to a view, Diller + Scofidio's unbuilt Slow House from 1989 turns the form of the house into a viewing device. [10] Designed for a waterfront site on Long Island, New York, the architects conceptualized the house as a sequence that begins with the horizontally proportioned car windshield, transitions to the vertical proportions of an oversized door, and then sweeps along a curve to terminate in a double height panoramic window facing the ocean view. The conceptual model of the house includes a miniature car on-axis with the front door, and a postcard-sized image of the view clipped to the base beyond the panoramic window. [11] The picture window at the terminus of the sweep was interrupted by the presence of a television suspended in front of the window. The TV was intended to play a video feed of the same view, but at a different time. The architects argued that the panoramic view, a primary driver of oceanfront real estate, should not be passively consumed by the occupants. The suspended display in the frame of the view acts as a minor irritant — a reminder that the view is not unmediated.

The Slow House proposed a transitional space between four significant optical frames: the windshield, the door, the panoramic window, and the television screen that partially obstructs it. Between these culturally loaded apertures, the architects arrayed domestic spaces, including bedrooms, kitchens, and bathrooms. [12] The corridor on the inside sweep of the decelerating curve also becomes a kind of vertical horizon that unfolds as the viewer approaches the panoramic window. The delay

1 3 The irregular arrangement of windows in the Zollverein School of Management and Design (2006) masks the scale of the building. The small-scale windows relate to the adjacent residential neighborhood, while the building massing relates to the industrial architecture of the coal facility beyond.

1 4 Abstract physical model of the Zollverein School of Management and Design (2006).

proposed by the Slow House is a deferral of the valuable view — a postponement of the pleasure of seeing the sea.

The project description in the exhibition catalog, *Scanning: The Aberrant Architectures of Diller + Scofidio*, describes the Slow House's relationship to the panorama: "This vista, a much-desired and commodified feature of domestic architecture, is held tantalizingly from direct view until one reaches the end of the structure's 100-foot curving walls: the Slow House, ultimately, is a passage to the view."[14]

The Slow House was never completed, but the house has been exhaustively represented in models and drawings. 1 0 The panoramic window at the end of the sweeping arcs is drawn as a glass face — an architectural lens mediating the interior domestic space and the expansive waterfront beyond. As the wider aperture capping the sweeping gesture of the house, the waterside facade was imagined as a window expanded to fill the opening. The dimensions of the opening necessitated a subdivision of glass modules. The height of the opening required a horizontal mid-span brace, depicted in the model with two tension rods to limit its depth. Views and their social and political dimensions have been a persistent interest that Diller + Scofidio have pursued across their body of work, including the ICA Museum in Boston, and the Juilliard School of Music at Lincoln Center.[15]

The window is not simply an aperture to bring in light. It is also a means by which we are reminded of our surroundings — an instrument through which to view the world. Vermeer's camera obscura would have been aimed at a domestic scene to project light from the window into the room and onto a surface that becomes the picture plane. Every window is an artifact, both technical and cultural, capable of establishing norms and conveying values.

NORMATIVE FIGURES

Though the early 20th century produced progressively larger windows to the point of introducing the all-glass facade as complete enclosure system, the window today remains a site of design innovation and invention. It is a "normative figure," one related to the scale of a body. Its placement can reveal the arrangement or organization of spaces behind the wall. The window is an indicator of scale, use, tectonics, and cultural attitudes.

The Zollverein School of Management and Design (2006), by Kazuyo Sejima and Ryue Nishizawa

14 Aaron Betsky, ed., *Scanning: The Aberrant Architectures of Diller + Scofidio* (New York, NY: Whitney Museum of American Art; New York, NY: Distributed by Harry N. Abrams, Inc, 2003), p. 103.

15 See also, Guido Incerti, Daria Ricchi, and Deane Simpson, *Diller + Scofidio (+ Renfro), the Ciliary Function: Works and Projects, 1979–2007* (Milano, Italy: New York, NY: Skira; Distributed in North America by Rizzoli International Publications, 2007).

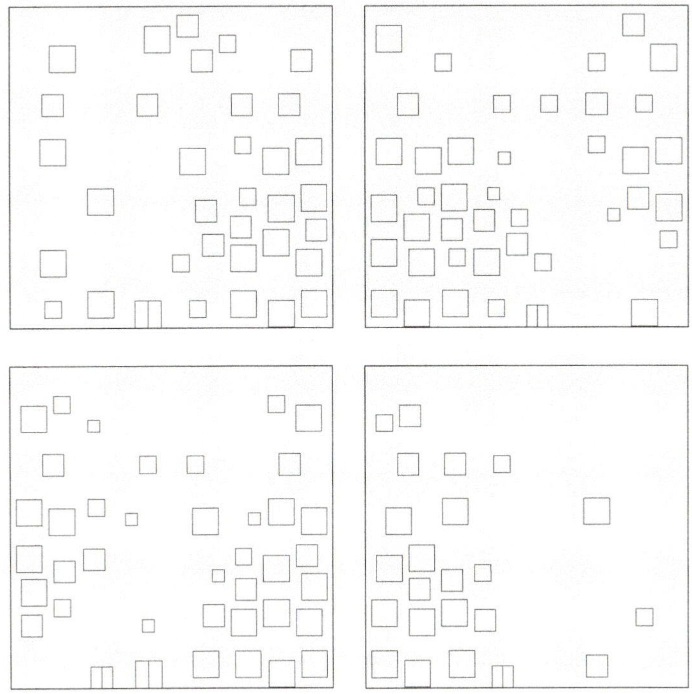

15 **Abstract elevations of the Zollverein School of Management and Design (2006).**

of SANAA, stands at the edge of an industrial complex in Essen, Germany, with a decommissioned coal factory on one side and a three-story residential neighborhood on the other. 1 The abstract design of the facade, with its irregular pattern of punched windows, is a nuanced response to its local context. 13

The school contains 54,000 square feet (5,000sm) of program, including seminar rooms, offices, a library, and a small auditorium, all contained within a compact 115-foot (35m) concrete cube. The architects chose to pack the building's program into a cubic envelope and wrap it with a singular material, a plain concrete shell. Given the diversity of the program and the varied demands for daylight the architects' facade strategy masks the programmatic differences in a composition of square windows of varying sizes.

SANAA's elevation drawings of the project are deliberately abstract. Four squares floating on the page. 15 Within each square is a scattering of smaller squares of different sizes, most of which float in the field of the larger square. The squares are casually misaligned, shifting up and down, left and right, like a school of square fish in an abstract aquarium. Some of the squares seem to be concentrated in the corners. Large areas without squares contribute to the ambiguity of scale. It is hard to discern where the interior floor slabs might be. Occasionally, a square touches the bottom edge. That is the building entrance.

Approaching the Zollverein School from the factory side, the concrete cube appears to be free-standing, surrounded by a small grassy field with wildflowers. 1 Square windows appear like die cuts in the abstract facade of the building. The glass surface is slightly set back from the face, and the relative lighting condition on the interior renders the glass a dark shade of gray. Seen from the street, the window in the upper left-hand corner appears backlit by the sky, the result of an open-air roof terrace, challenging the apparent solidity of the cube.

The abstract building fenestration recalls the sharp cut of the ribbon windows in the Villa Savoye. Even the open-air terrace seen through the windows is like the Villa Savoye. Similarly, the buoyant windows of the Zollverein School do not reach the edges, keeping the corners intact. While the pristine surface of Le Corbusier's ribbon window is interrupted by the projecting sill, this is not the case with SANAA's windows. In fact, there is also not even a visible window frame in the Zollverein School. The abstraction of the facade contributes to the sense of irreality, as if looking at a cardboard model. What we see is

16 A large-scale picture window in the Zollverein School, as seen from the interior. Note the aluminum frame all around the window in this view, and its total absence from the exterior views.

17 The abstraction of the design of the windows in the Zollverein School, and the elimination of the window frame and its associated trim, drip edge, and metal flashing is achieved by one critical detail: the area drain in the ledge.

a glass plane set back from the face, revealing the apparent thickness of the concrete wall, which appears paper thin — or rather like the thickness of a piece of cardboard from the scale model. 14

Upon closer inspection, the windows do have aluminum frames, but the concrete wall is pocketed to allow the frame to sit within a recess, concealing it from the outside. 16 The daylight opening — the dimension of the visible opening inside the frame, is aligned with the concrete opening on the facade, thus eliminating the frame-within-a-frame effect, typical in a punched window, and the same effect that frustrated Le Corbusier's ribbon window at Savoye.

Returning to the questions of frame and sill, flashing and drip edges, the purpose of the aluminum sill from Villa Savoye, the bent metal trim that projects beyond the picture plane of the facade, was a practical detail to keep the water from staining the facade. 7 How did SANAA avoid the projecting sill on the Zollverein School to achieve such abstraction? The window's eight-inch (20cm) setback creates a horizontal shelf for water to accumulate. The architects treated each sill like a small roof, complete with its own area drain. 17 The concrete ledge pitches to the drain, and each ledge is individually plumbed with a one-and-a-half-inch (40mm) pipe through the wall. 18

06 The integration of plumbing leaders into the cast-in-place concrete adds a challenge to the concrete scope of the project. The Zollverein School was a pioneering project in employing geothermal heating and cooling through thermally active surfaces, which involved running flexible one inch (25mm) plastic piping through the concrete walls and floors, so the addition of more piping may not have been as exotic as it might at first appear.

The integration of the ledge drain into the concrete formwork may appear extravagant.06 Yet, in the pursuit of a taut and abstract building facade, it is applied across the entire building. It is the detail that gives the building its unique and surreal effect. It is as if, walking across the grassy field, we have suddenly come upon a huge cardboard model of a building — like one of many models produced as part of SANAA's signature design process. 14

What does it mean to make a built work of architecture resemble a model? This is not the same as Le Corbusier imagining prefabricated, standardized architecture analogous to the automotive industry. It is also distinct from Peter Eisenman's "de-material" embrace of "cardboard architecture." The SANAA project of abstraction and the manipulation of scale is informed by a deep understanding of both construction systems and human perception. It is vested with the disciplinary knowledge of Villa Savoye as well as the material expertise and craftsmanship of concrete formwork. Sejima and Nishizawa's design for the 10-story institutional building, in dialog with a varied neighborhood of residential and industrial structures, produces a unique structure that appears uncannily scaleless and is executed through exacting attention to the elements of architecture — specifically the window and the wall. The tectonic imagination of

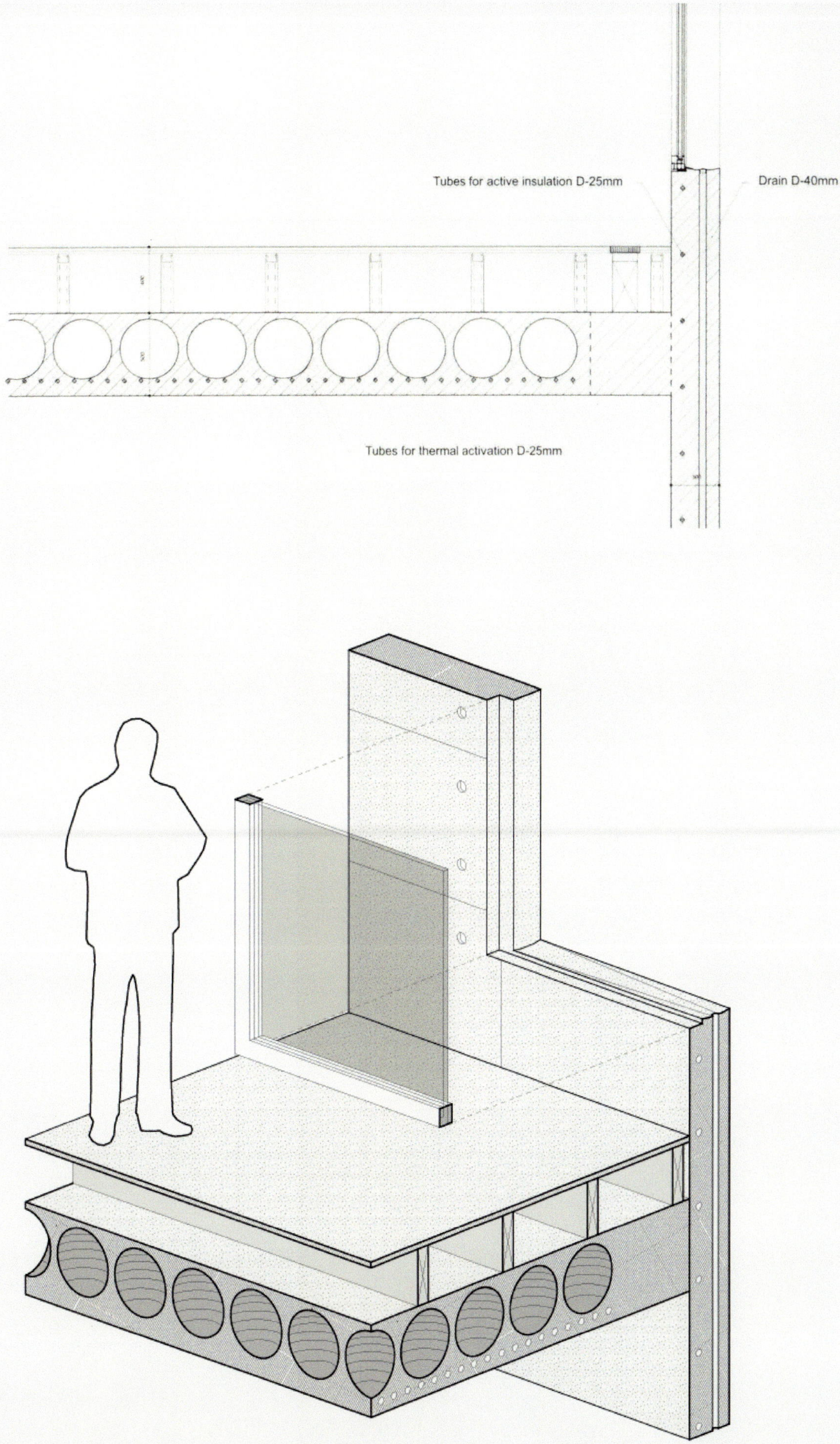

1.8 The wall section and isometric drawing of the exterior wall shows the pocket in the concrete wall to conceal the dimension of the window mullion. It also shows the area drain in the sill to manage the rainwater and avoid the sill flashing.

9.1 Simmons Hall (2004), designed by Steven Holl, employs a load-bearing exterior wall made of precast concrete panels perforated with small, regularly spaced windows.

[20] The nine square grid of square windows creates a distinct pattern in the dorm room.

[21] The dense grid of windows extends to public areas like the lobby.

SANAA's design process could not conceive of the abstraction of the architecture appearing like a model without the deep understanding of the various materials systems and trades (plumbers, carpenters, formwork, rebar, etc.) necessary to execute the seemingly simple objective of a frameless window.

SCALING EFFECTS

Simmons Hall (2002) at MIT employs a repetitive pattern of windows to a startling perceptual effect. [19] The undergraduate residence hall occupies a narrow site on Vassar Street, between a railroad track and the MIT athletic fields. Architect Steven Holl conceived the building as a rectilinear block with the "porosity of a sponge." A dense grid of square cells making up the majority of the facade and a handful of large irregular openings that correspond to communal spaces. Each room behind the facade has nine windows in a three-by-three grid. [20] The windows are each 2' x 2' (600mm x 600mm). The precast concrete grid that makes up the facade was dubbed "perforated concrete," or "perfcon" by the designers. The dimension of the waffle grid allows for cast-in-place concrete slabs to tie the prefabricated panels together for a nine-and-half foot (2.9m) planning module and a nine and half foot (2.9m) floor-to-floor height.

The effect of the dense grid from the outside is waffle-like. The window-to-wall ratio has perforated the surface to the point where there is no plane — just a series of thick frames. The dense array of windows in a grid also plays with our perception of the scale of the building. The surprisingly dense array of windows gives the impression that the building is three times larger than its actual 10 stories. Seen across the athletic fields, Simmons Hall appears like a massive linear megastructure run aground.

The residence hall houses 350 beds in single and double rooms and contains several apartments, a cafeteria, a multipurpose room, and several multi-story student lounges. Large cutouts and indentations in the massing are designed to create some relief from its 385-foot (117m) length. As a result of the notches at the top of the building, the single building must behave like three structures, which necessitates three elevators and six egress stairs. The designers liken Simmons Hall to a hive or a fragment of a city. As such, it has multiple paths through the building, creating many opportunities for students to meet and interact.

Construction photos of Simmons Hall show the large waffle panels being lifted into position. [19]

22 Le Corbusier's sketch of a full-height glass wall spanning from floor to ceiling and wall to wall.

Because the precast has poor insulating value, the grid of the load-bearing facade was then clad in an aluminum rainscreen over several inches of rigid insulation. The deep recesses for all the windows create a surplus surface area of cladding.

From the interior, the grid of windows spans from floor to ceiling and from partition to partition, creating transparency in atypical places. Instead of a single large panoramic window, the "sponge" facade subdivides the room view into nine smaller windows. The window's depth provides shade from the high summer sun and invites the low-angle winter light. Multiple windows in each room allow students to open those along the top and bottom rows for convection ventilation. It also means that each window has its own mini curtain; nine windows and nine curtains.

WINDOW AS WALL

Le Corbusier's sketch shows a human figure standing in front of a full-height glass wall. The furniture and curtains denote the foreground as interior. 22 The trees and diagrammatic sun make it clear that the background is exterior. The low horizon and the cropped view of the tree trunks suggest that we are in an elevated condition, perhaps in an apartment overlooking a park. The thin lines that define a plane are understood as the building envelope, rendered in the most minimal way. The double line at the floor reveals that Le Corbusier imagined the frame of the window to be sitting on top of the slab. Le Corbusier has scaled up the window until it spans from floor to ceiling and wall to wall. It is the most minimal separation of inside and outside, a glass membrane stretched across an opening. This sketch captures much of the ambition of Le Corbusier's architecture to provide shelter while reconnecting occupants with the natural world. He sketched what today we would refer to as a window wall.

FOUR CHICAGO WALLS

Questions of windows and building enclosure and structure are all amplified in the tall building. Four tall buildings in Chicago's Loop illustrate the transformation of the window to the window wall to the curtain wall: the Monadnock Building (1891), the Reliance Building (1894), Lake Point Tower (1968), and Aqua Tower (2009).

The Monadnock Building, built in 1891, and the Reliance Building, built in 1894, represent two distinct paradigms of structural systems and

[23] The Monadnock Building (1891), designed by Burnham & Root, was the last major building constructed with masonry load bearing walls.

[24] The Reliance Building (1894), designed by Charles Atwood for D. H. Burnham Company, was built only three years after the Monadnock Building, but its steel frame and terracotta facade represented a paradigm shift in architecture.

enclosure. The two structures are proximate in time, three years apart; in their siting, only four blocks apart; and in authorship, Monadnock was designed by Daniel Burnham and John Wellborn Root of Burnham & Root, and Reliance was designed by Charles Atwood for D.H. Burnham and Company. However, the two buildings represent entirely different world views and mark a turning point that would signal a paradigm shift in architecture — namely, the shift from load-bearing walls and windows to structural frames and envelopes.

Standing on the corner of South Dearborn Street and West Jackson Boulevard in Chicago, the Monadnock Building has a powerful presence. [23] Few buildings meet the ground with the same heft as the Monadnock. Its load-bearing brick walls are six feet (1.8m) thick, and the recessed storefront windows reveal the depth of the masonry structure. Above the first floor, the walls taper back with a gentle curve that continues up the building's 16 stories. The distinctive flare at the base of the Monadnock emphasizes the variable thickness of the perimeter walls that decreases along its height. The building is remarkable for its size and proportions but also its unadorned use of brick. The smooth flare and the filleted corners are devoid of the ornamentation that was typical at the time of its construction. The bay windows projecting outward in alternating bays and the smooth transitions from the plane to the window projections give the building the appearance of a rippling fabric, anticipating in an almost subconscious way the future evolution of the curtain wall.

Four blocks to the north, at North State Street and West Washington Street, the Reliance Building marks the corner with a completely different disposition. [26] In contrast to the dark brown masonry monolith down the street, the Reliance is a delicate assembly of white glazed terracotta and plate glass. [24] Where the window-to-wall ratio of the Monadnock is approximately 50 percent glass, the Reliance is far more glass at approximately 80 percent. The lightness of the Reliance is a function of its steel skeleton structure. [25] Architectural historian Joanna Merwood-Salisbury notes the Reliance Building's large plate glass windows and terracotta tile appear to form a single surface, appearing as an "undulating veil rather than a sculpted mass."[17] The horizontal articulation of the ornamental spandrel panels and the discontinuity of the vertical members reinforce the expression of a weave — a horizontal warp to the vertical weft.

It is difficult to overstate the significance of this shift from the load-bearing paradigm represented

[17] Joanna Merwood, "The Mechanization of Cladding: The Reliance Building and Narratives of Modern Architecture," in *Grey Room* 04, (Cambridge, MA: The MIT Press, 2001), p. 54.

25 The steel frame of the Reliance Building (1894) under construction highlights the lightness of the revolutionary structural system.

26 The glass and terracotta cladding of the Reliance Building (1894) showcases the non-load-bearing nature of the exterior wall as it shifts from structure to envelope.

by the Monadnock, to the new paradigm of frame represented by the Reliance. A diagrammatic section shows the relatively similar height, but the sharp contrast in the conceptualization of the wall. 27 In the language of evolutionary biology, the crustacean exoskeleton had evolved into a vertebrate in three short years.

A few blocks east of the Reliance, a third tower illustrates the expressive potential of the curtain wall. Lake Point Tower was designed by George Schipporeit and John Heinrich, two former students of Mies van der Rohe. 28 The tower occupies a prominent site east of Michigan Avenue, making it an in-the-round tower unlike any other in the city. The design consists of a Y-shaped plan with filleted corners. A bronze curtain wall wraps the entire structure, giving it a distinctly monolithic appearance.

Lake Point Tower was completed in 1968, several years after the pioneering curtain wall projects in New York and Chicago by Mies van der Rohe and Skidmore, Owings & Merrill. The curtain wall represents the complete evolution from window to envelope, with the vertical mullions extending past the edge of the slab to form a continuous membrane. The gravity load of the curtain wall is supported on each level, so it does not carry any accumulated load. The curtain wall modules allows the building facade to act like a panelized membrane stretching over the structural frame. The facade of Lake Point Tower learns from the facades of Mies' Lake Shore Drive apartments (1951), Seagram Building (1958), and JFK Federal Building (1966). The Lake Point Tower facade minimizes the spandrel panel to give the appearance of an all-glass envelope. Unlike a commercial office building, the tower's residential use allows the floor packages to be minimized to the shallowest possible spandrel zone.

The filleted corners of Lake Point Tower give the building its unique sculptural qualities. The Y-shaped plan and its curved corners ensure that residents' views do not look into the other units. Lake Point Tower's curved geometry is unlike any other tower in the tightly gridded city of Chicago. The curtain wall module produces a faceted curved corner with 12 flat modules to round the 90-degree corner at the prow of each of the wings of the tower. 29 The flat modules can sufficiently approximate the radius of the filleted corners due to the scale of the floor plate and thus produce the effect of an undulating fabric as the facade wraps the structure. The curtain wall here appears distinctly curtain-like.

The last Chicago tower in this series of cladding case studies is the Aqua Tower (2009) on North

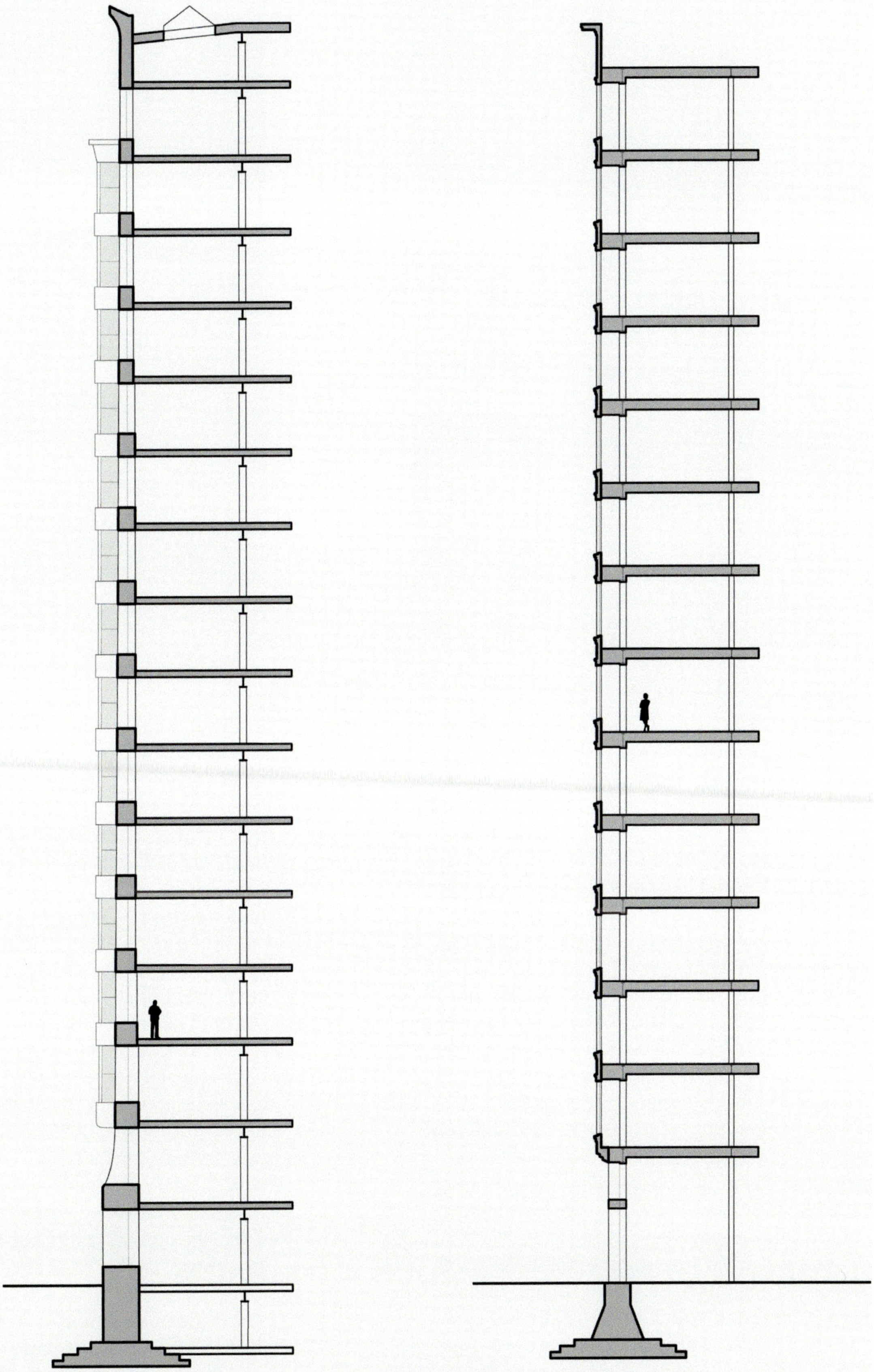

27 Diagrammatic sections of the Monadnock Building (1891) and the Reliance Building (1894) compare the thickness of the load-bearing wall with the lightness of the curtain wall envelope.

28 Lake Point Tower (1968), designed by Schipporeit and Heinrich, employed the curtain wall to great effect, creating a taut glass membrane around the sculptural building.

29 A close-up photo of the Lake Point Tower curtain wall highlights the narrow spandrel panel and the faceting at the filleted corners.

Columbus Street. 31 Aqua Tower, designed by Jeanne Gang of Studio Gang, marks a milestone in the expressive potential of the tall building, but also a return to a simpler envelope strategy, the window wall. Its curvilinear slab edges, which vary from floor to floor, produce the appearance of a rippling mass, even though the building envelope is a straightforward, cost-effective, orthogonal glazed box.

Aqua Tower combines hotel and residential programs within the single extruded shaft. As an all-concrete structure, the slabs are ideal for damping sound transmission between floors, which is essential for residential uses, but also presents the opportunity for the expressive slab edge. Gang, working with the developer Jim Loewenberg, an architect himself, proposed to make each slab edge a unique combination of undulating curves. The visual effect of the varying slab edges is a stratified undulating mass, one that Gang associates with a geological formation. 30

Essential to the success of Aqua Tower is the cost-benefit analysis of the unique concrete slab edges and the standard building envelope. For Gang, the concrete formwork is where the design efforts — and added costs — are focused. The building envelope is economical and pragmatic. Gang selected an off-the-shelf window wall system to span floor to ceiling, allowing the concrete slab to cantilever out to the undulating edge. The dark tinted glass and dark mullion paint selection allow the glass to recede visually. The balcony railings, consisting of black-painted steel pickets, blend into the dark tones of the window wall system beyond, allowing the white-painted concrete slabs to create the primary reading of the building.

The design of the unique slab edges was conceived with its construction in mind. Concrete slabs are typically formed on a reusable formwork platform larger than the building floorplate. The formwork for slab edges is normally constructed outboard of the building footprint. For the unique profiles of the Aqua Tower, the architects produced 82 unique edge-of-slab footprints. Each profile was produced in CAD as a "contour" operation, which takes a modeled volume and slices contours at regular intervals. To construct the curvilinear slabs, the contractor utilized a flexible steel plate "pour stop" anchored at key points to the horizontal formwork platform. 32 Once the rebar was laid out within the boundaries and the concrete was placed, the formwork was dismantled and reused.

When it was completed in 2009, Aqua Tower caused a sensation. The curving edges of the balconies made the almost 2,000,000-square-foot

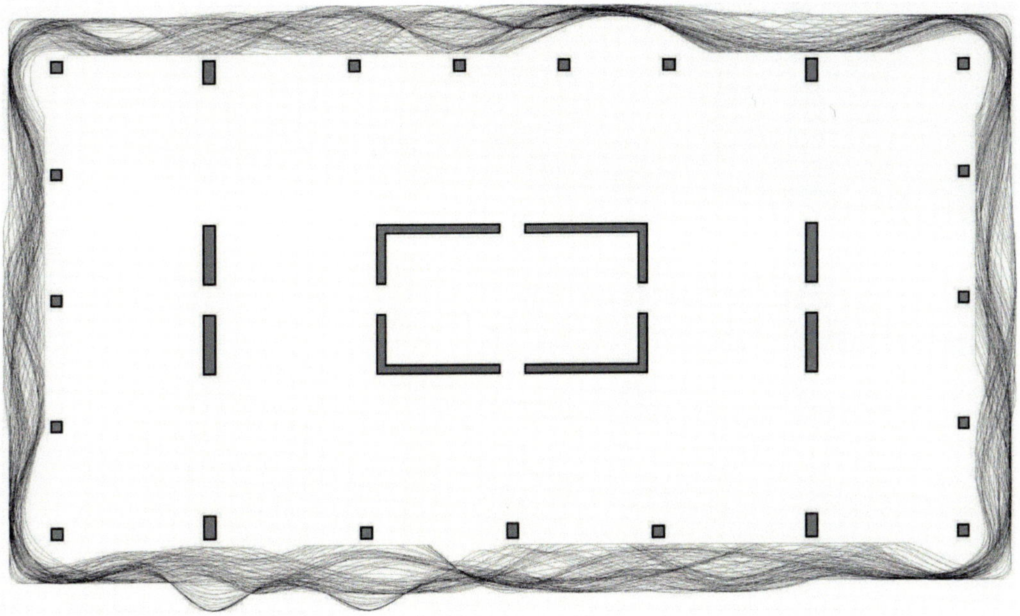

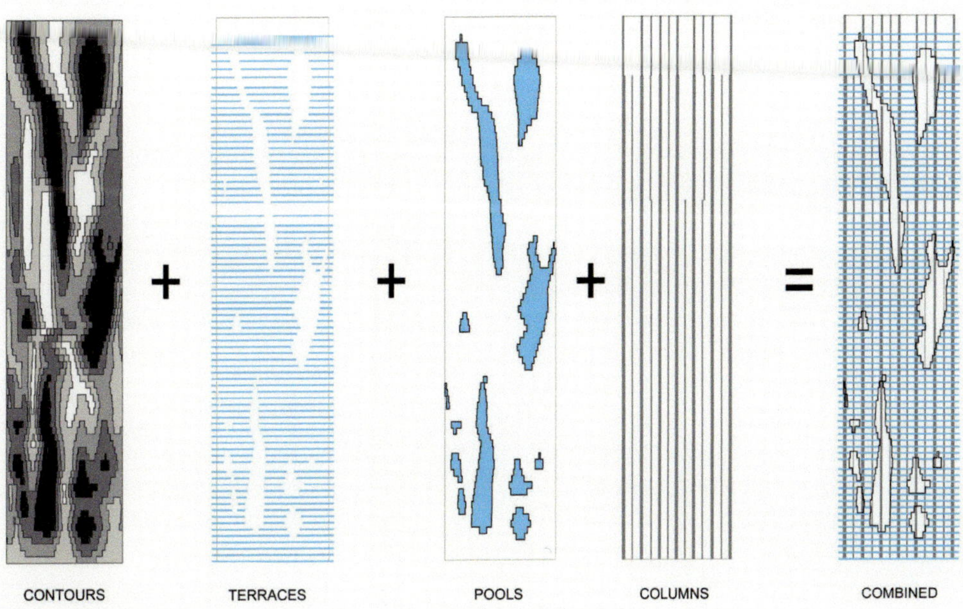

30 Diagrams of Aqua Tower (2009), designed by Jeanne Gang of Studio Gang, show the composite of all the irregular slab edges, and conceptual diagrams of the facade.

3|1 Views looking up at Aqua Tower (2009) highlight the white-painted slabs and soffits and deemphasize the dark glass window wall system.

3|2 Jeanne Gang with steel plate "pour stop" demonstrating how the Aqua Tower curved slab edges were achieved.

building seem curvy and organic. Views looking up at the tower amplified the slab edges and minimized its enclosure, giving the impression of a wave pattern, with the whole building appearing to ripple, thus earning the name Aqua Tower. The use of a glass window wall played a critical role in the effect of undulation, namely, to efficiently enclose the building and recede from view.

The architects describe the geometry of Aqua Tower and its balcony layout as responsive to available views and optimized to create self-shading of the glass facades.18 Aqua Tower was undoubtedly one of the first large-scale manifestations of a computationally derived formal expression. Digital modeling software commands, like the "contour" command, are clearly legible in Aqua Tower. To what extent are our design tools inscribed in the forms and geometries of our built environment? Along these lines, we might also ask: Is the straight edge of the T-square legible in the taut edges of the elevation of Villa Savoye? Is the 45-degree angle of the drafting triangle evident in the geometry of I.M. Pei's Bank of China project? Does the computer's copy-paste function correspond to the repetition of floor plates in the 432 Park Avenue skyscraper? Ultimately, we also must consider how the materials and construction methods incorporated into the design processes produce a feedback loop of possibility and imagination.

18 Despite the arguments for contextualism and energy savings from solar shading, the building has been criticized for not addressing the thermal bridging and heat loss through the continuous cantilevered concrete slab.

19 Rem Koolhaas, *Delirious New York: A Retroactive Manifesto for Manhattan*, New ed. (New York, NY: Monacelli Press, 1994).

TOO SMALL

Koolhaas argued in *Delirious New York* (1978) that while Europeans wrote the manifestos for modernism, it was the Americans who first built the modern metropolis — but without a manifesto. For Koolhaas, New York is a masterpiece without a master, or a practice without a theory. He sets out to write a "retroactive manifesto" for New York, calling its premise, "Manhattanism," thus giving the city the theoretical argumentation it never had.19 For Koolhaas, the feverish prose of Le Corbusier's manifestos and conceptual sketches of La Ville Radieuse and its Cartesian skyscrapers found their corresponding counterparts already constructed in the North American metropolis. On Le Corbusier's first visit to New York in 1935, he declared the New York skyscrapers "too small" and proposed replacing them with his own Cartesian skyscrapers. Koolhaas describes Le Corbusier's "all-consuming ambition to invent and build the New City commensurate with the demands and potential glories of the machine civilization. It is his tragic bad luck that such a city already exists when he develops this ambition,

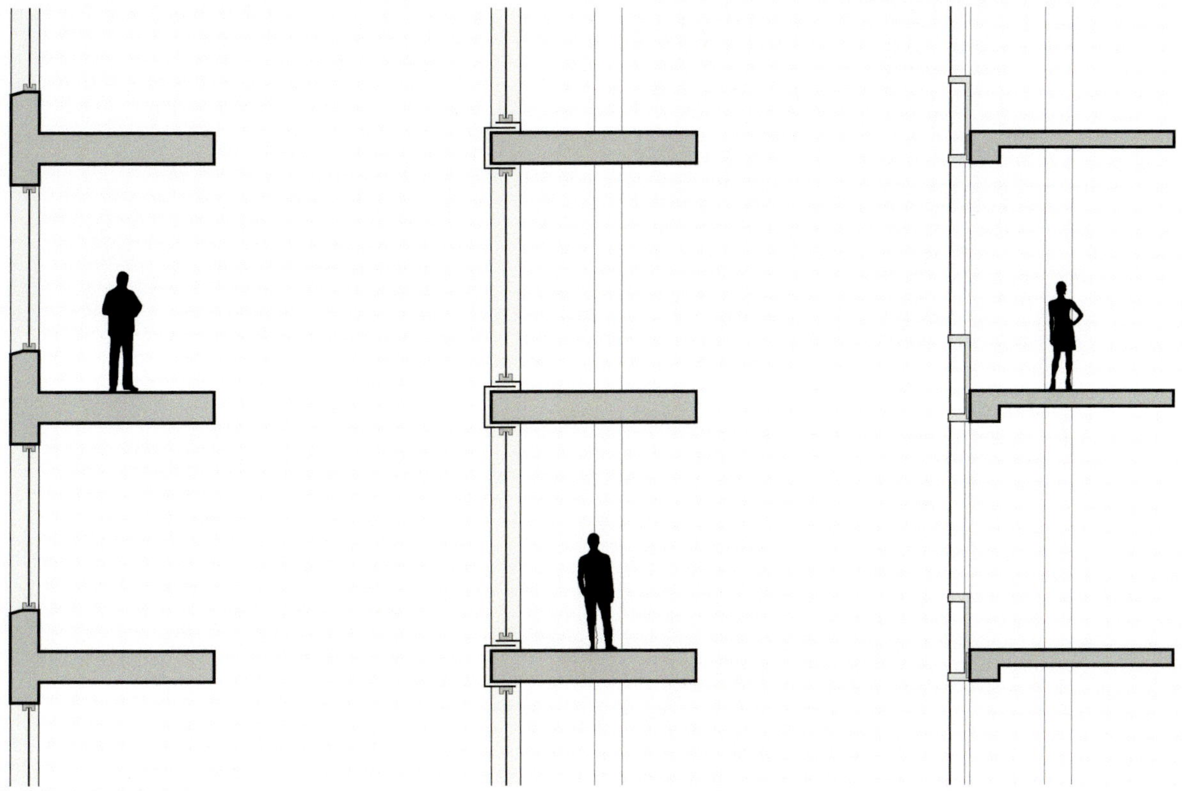

33 Diagram comparing the window (left) in a load bearing wall, the window wall (middle) that sits on the slab, and the curtain wall (right) that runs past the slab edge to form a continuous building envelope.

namely Manhattan."[20] Le Corbusier needed to discredit the constructed New York and replace it with a conceptual New York that he has yet to invent. In Koolhaas's *Delirious New York*, the order of operations is inverted: the construct precedes the concept.

The opportunity to build a version of Le Corbusier's Cartesian Skyscraper would present itself after the Second World War, in 1947, as part of the international advisors to the United Nations for the design of their new headquarters.[21] New York architect Wallace Harrison was appointed Director of Planning for the project, and he steered the process from planning and site selection through design and construction. 35 Convinced that the new secretariat should follow his version of modern architecture, Le Corbusier produced volumes of design iterations, dominated the workshops, challenged Harrison's leadership and claimed authorship of the design even though it departed from his recommendations.

The 39-story slab of the United Nations Secretariat, completed in 1951, is generally regarded as the first large-scale, all-glass curtain wall building in the United States. 34 The building runs perpendicular to 42nd Street, with its primarily east and west facades clad in glass and the north and south facades clad in white marble. The curtain wall consists of aluminum mullions at four-foot (1.2m) centers, spanning 12 feet (3.7m) from floor to floor, and attached to the edge of the concrete slab. The vision glass above the sill is divided into two equal lites in a double-hung configuration, allowing the windows to be operable. The spandrel zone under the sill consists of blue-green tinted glass in front of a masonry knee wall, providing the necessary fire separation.

The east and west facades of the secretariat building include 300,000 square feet (28,000sm) of glass, almost seven acres (2.7 hectares). The all-glass facade was pioneering and not without controversy. Le Corbusier argued for a *brise-soleil* to be added to the west facade, similar to the design of the Ministry of Education and Health in Rio de Janeiro, but the exterior shades were omitted due to cost and maintenance concerns. Soon after the building opened, the users complained about overheating, and the air conditioning system could not keep up with the cooling demand.[22]

When the UN building was completed in 1951, it was referred to as a "glass cliff," associating the abstract slab with a geological scale. The UN Secretariat was soon followed by other significant curtain wall buildings, including the Lever House

20 Rem Koolhaas, *Delirious New York*, pp. 249–251.

21 The 10 advisors to the United Nations: Sven Markelius, (Sweden), Le Corbusier (France), Ssu-ch'eng Liang (China), G.A. "Guy" Soilleux (Australia), Nikolai Bassov (Soviet Union), Ernest Cormier (Canada), Gaston Brunfaut (Belgium), Howard Robertson (United Kingdom), Julio Vilamajó (Uruguay), and Oscar Niemeyer (Brazil).

22 Le Corbusier was an advisor to the architects of The Ministry of Education and Health in Rio de Janeiro, completed in 1943.

34 The United Nations Secretariat Building is regarded as the first large-scale, all-glass commercial curtain wall buildings constructed in the United States. Its east and west elevations are entirely glazed, proving hard to heat in the winter and hard to cool in summer.

35 The United Nations Headquarters (1952), designed by an international team of architects, including Wallace Harrison (USA), Le Corbusier (France), and Sven Markelius (Sweden).

(1952) by Skidmore, Owings & Merrill, the Seagram Building (1958) by Mies van der Rohe and Philip Johnson, and the Pepsi-Cola Building (1960) also by Skidmore, Owings & Merrill. Each new curtain wall introduced significant advancements, including modular assemblies, integrated window washing, and prefabrication.[23]

GLASS MYTHOLOGIES

With its origins in the early 20th century, the all-glass curtain wall realizes an ambition to construct crystalline buildings where the unique properties of transparency and its associations with openness would have the capacity to transform society. Writing in 1914, author Paul Scheerbart imagined a utopian society inhabiting kaleidoscopic cities made up of buildings of colorful glass in *Glasarchitektur*.[24] The material properties of glass would produce spiritual and emotional effects on the citizens. Architect Bruno Taut, who was a close friend of Scheerbart's, produced a series of illustrations of faceted glass structures within mountainous settings that he referred to as "Alpine Architecture." The utopian imaginary is grounded in a critique of the current context, social, political, and material, and speculates on possible alternative worlds[25]

Scheerbart's writings elaborate on a materially-enabled utopia in which glass architecture dispenses with social ills like tyranny and corruption. The associated qualities of glass, transparency, honesty, and integrity are transferred to an ideal society, and the inhabitants of the glass world are free to bask in the radiant light of a new material-symbolic paradigm.

Scheerbart and Taut write: "If we want our culture to rise to a higher level, we are obliged, for better or worse, to change our architecture. And this only becomes possible if we take away the closed character from the rooms in which we live. We can only do that by introducing glass architecture, which lets in the light of the sun, the moon, and the stars, not merely through a few windows, but through every possible wall, which will be made entirely of glass — of coloured glass. The new environment, which we thus create, must bring us a new culture."[26]

The future worlds imagined by Scheerbart and Taut in 1914 eventually materialized in post-war projects such as the UN Secretariat Building and Lever House, in New York. Subsequent post-war American buildings adopted curtain wall technologies and radical new building systems, quickly going from the exception to the norm. The curtain wall today has become the *lingua franca* of commercial

23 For a history of the curtain wall see, Scott Murray, *Contemporary Curtain Wall Architecture*, 1st ed. (New York, NY: Princeton Architectural Press, 2009), and Alejandro Zaera-Polo, *The Ecologies of the Building Envelope: A Material History and Theory of Architectural Surfaces*, Print edition. (New York, NY: Actar Publishers, 2019).

24 Paul Scheerbart and Bruno Taut, *Glass Architecture and Alpine Architecture* (London: November 1972).

25 For more about Scheerbart and Taut's glass-inspired utopian visions and their climatological implications, see Chapter 8, "Climate and Conditioning."

36 John Hancock Tower (1976) was built in Boston's Back Bay neighborhood, across the street from H.H. Richardson's Trinty Church. The taut, reflective glass curtain wall extends from sidewalk to parapet without interruption.

37 One United Nations Plaza (1976), designed by Kevin Roche of KRJDA, wraps the hotel and office programs in a taut, reflective, glass skin.

architecture due to its combination of performance, speed of construction, and economies of scale.

26 Paul Scheerbart and Bruno Taut, *Glass Architecture and Alpine Architecture*, p. 41.

MIRROR MIRROR

It was not until 1976 that the curtain wall reached its apotheosis of tautness and abstraction with the completion of two buildings in the same year: the John Hancock Tower in Boston, 36 designed by Henry Cobb of Pei Cobb Freed & Partners, and the United Nations Plaza in New York, 37 designed by Kevin Roche of KRJDA. Both towers are prismatic volumes clad in highly reflective all-glass curtain walls, and both are remarkable for their abstraction and manipulation of perceptual scale.

The Hancock Tower (1976) immediately attracted significant negative attention largely due to its adjacency to H.H. Richardson's Trinity Church in Copley Square. At 790 feet (240m), the Hancock Tower dwarfs Trinity Church in a way that many found inappropriate. Located in Boston's Back Bay, a neighborhood known for its architectural cohesion and mid-rise masonry architecture, the mirrored glass cladding of the Hancock Tower stands in sharp contrast.

The Hancock Tower's massing acknowledges its significant neighbor by presenting a narrow face toward Trinity Church along Saint James Avenue. The tower's footprint is a parallelogram that creates a triangular plaza east of the building along Clarendon Street to provide some spatial relief to Trinity Church. The proportions of the tower and its parallelogram geometry make it appear extremely slender at times, and rather broad at others. The narrow faces are rendered even narrower by a V-shaped notch that bisects the narrow facades into even slenderer coplanar faces.

The Hancock Tower is distinctive for its uninterrupted, taut, mirrored glass curtain wall that extends from the ground to its full height, forming a prismatic solid. The typical curtain wall module is 4' x 11' (1.2m x 3.4m), spanning floor to floor, without a spandrel panel. The reflective coating on the glass masks the slab edge in the upper portion of the vision zone. The combined effect of the abstract and angular geometry and the material reflectivity gives the Hancock an otherworldly, monolithic appearance. A sense of the scale is hard to grasp because the typical scaling devices, such as the horizontal striping of the spandrel panels, are absent.

Rather than sit on the ground or transition to a stone base, the 60-story tower meets the ground with a narrow drainage trench, giving the impression that it continues below the surface of the surrounding

38 The TRUTEC Building (2006) in Seoul, South Korea, was designed by Barkow Leibinger with a faceted glass curtain wall that breaks up the images of the context while producing kaleidoscopic effects

39 The irregular geometry of the TRUTEC Building facade modules.
40 Installation of the custom unitized TRUTEC curtain wall modules.

sidewalks. These subtle details contribute to Hancock's enigmatic quality and reveal a level of abstraction and nuance unmatched for a building of its scale.

The UN Plaza (1976) in New York City employs many of the same techniques as Hancock Tower. Consisting of two towers arranged perpendicular to each other, the UN Plaza towers are also clad in a highly reflective blue-green glass, which masks the activities within as well as the slabs behind spandrel panels.[27] The tower massing is also modified with faceted cuts that transform the volume as it rises. One United Nations Plaza runs parallel to 44th Street and contains offices on the lower levels and a hotel on its upper levels. The hotel requires a narrower floorplate than the office portion, and the sloped setback corresponds to the transition in program.

Unlike the Hancock Tower, where the curtain wall modules are maximized to have only one module per floor, the UN Plaza buildings subdivide the floor-to-floor height into equal subdivisions, creating a horizontally proportioned module. The glass module is 4'-7" x 2'-7 ½" (140cm x 80cm) high. Because the floor-to-floor heights of an office are taller than those of a hotel, the office floors are subdivided into four equal modules, while the hotel is subdivided into just three equal modules. The two-foot-seven-and-a-half-inch module is a common denominator for both sections of the curtain wall, allowing the same module to extend from the office to the hotel without changing proportions, transitioning from one program to the other, while masking the floor-to-floor height changes and reinforcing the monolithic quality of the building.

27 See Eeva-Liisa Pelkonen et al., *Kevin Roche: Architecture as Environment* (New Haven, CT: Yale University Press: In association with Yale School of Architecture, 2011), pp. 50–58.

KALEIDOSCOPE

When Berlin-based architects Barkow Leibinger were asked to design the TRUTEC office building in Seoul, they saw it as an opportunity to explore patterns and effects of the unitized curtain wall. 38 Their preliminary studies show repeating modules like tiling patterns with inlaid diagonal subdivisions. 42 Their material research explored the effects of reflective surfaces folding out of the picture plane and capturing reflections of the sky and the surrounding context.

The architects worked with local facade contractors to develop a custom curtain wall of standard extruded profiles and relied on CNC fabrication to customize the angles to allow for a unique out-of-plane faceting of the glass. 39 Their design consists of a series of repeatable units with a fixed set of variations, working within the constraints of the building

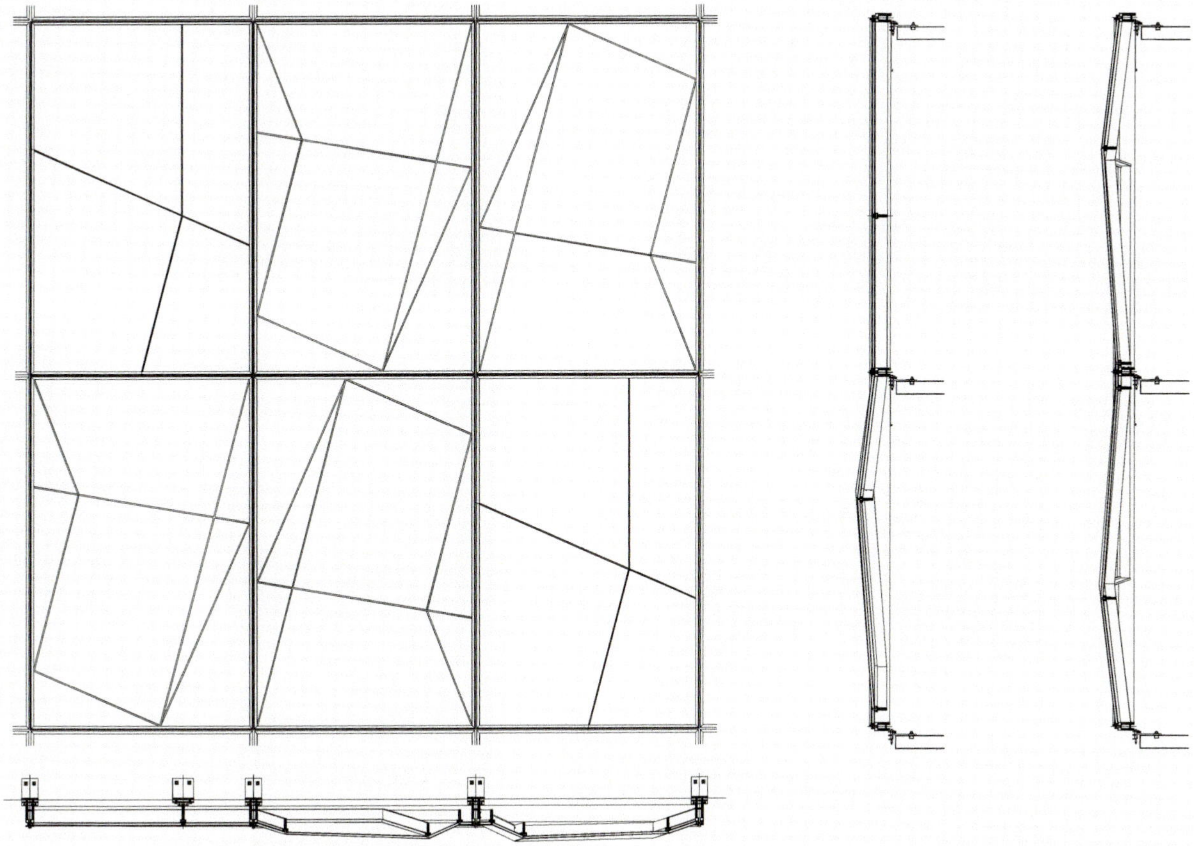

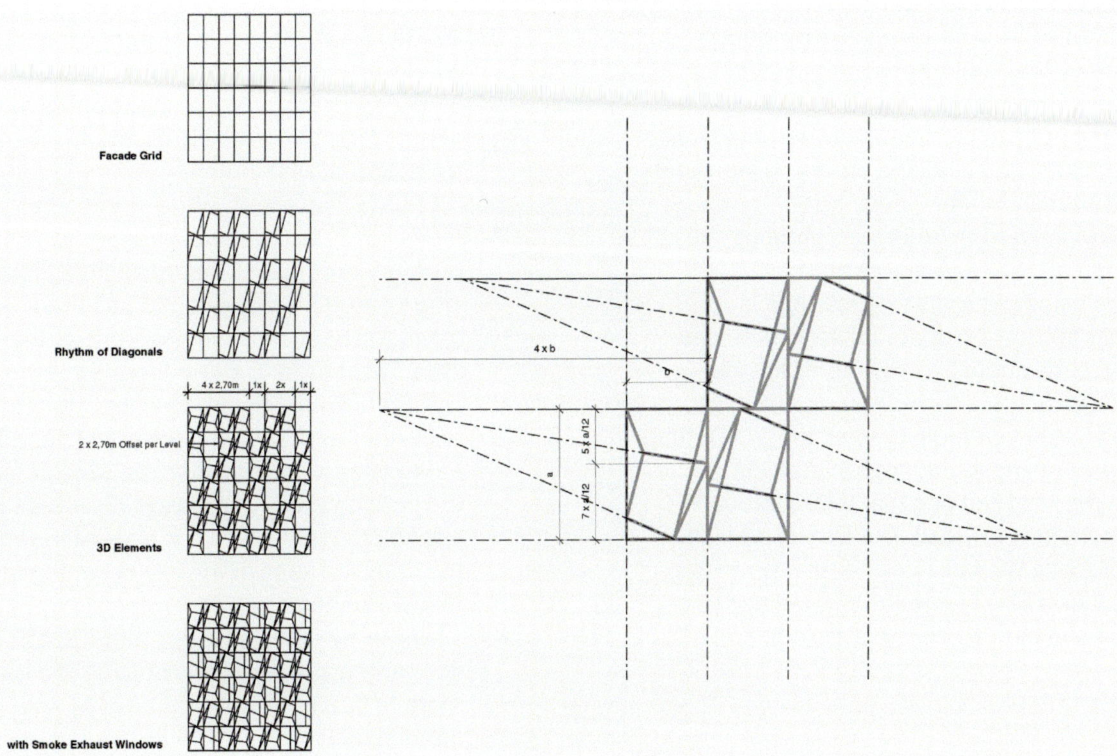

4.1 Curtain wall drawings for the TRUTEC Building, including partial elevation, partial section and partial plan.
4.2 Concept diagrams of the TRUTEC Building facade strategies.

43 Elevation diagram of Jean Nouvel's 100 11th Avenue project (2010) documents the glass types by their rotation from the plane of the building: Right, left, bottom, and top, as degrees of rotation, from 2 to 5 degrees.

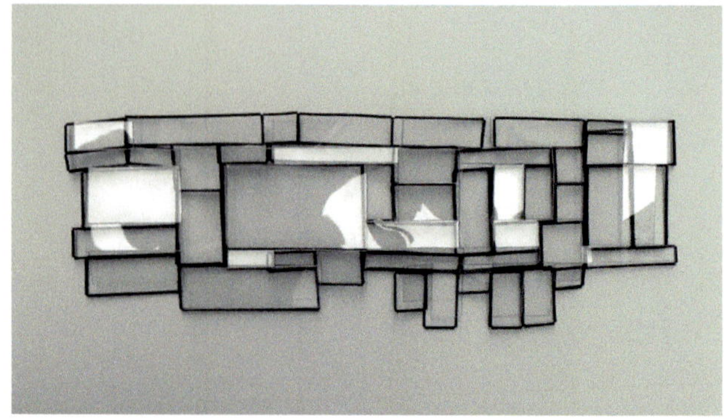

44 A concept model for the facade of 100 11th Avenue (2010) shows the individual windows, each with a slightly different tilt, assembled into a larger panel.

45 Akin to the early "divided-lite" window in Vermeer's painting, the 100 11th Avenue curtain wall megapanel is subdivided into smaller windows. It is preassembled in the factory and brought to site in the largest possible unit size.

system and available technologies to produce a facade strategy that explores themes of variation and repetition. They explain: "Building through reflection and optical fragmentation, we could react while still not knowing what the emerging surroundings would ultimately offer. The polygonal surfaces fragment any given contextual surroundings into a kaleidoscope of infinite pixels."[28]

Barkow Leibinger's working method incorporates fabrication operations into their design process. An early publication by the architects, *An Atlas of Fabrication*, is organized by operations: 2D-Cutting, Casting, Bending, Punching, 3D-Cutting (Revolving).[29] Tying techniques to projects suggests a mode of working that thinks through material operations in detail. The form of a project does not precede their working out the "how" of a project. Rather, a set of techniques is researched and studied and then applied to a project.

VISION MACHINE

The intentional legibility of assemblies and components, as discussed in Chapter 4, "Part to Whole," often runs counter to the aspirations of tautness or smoothness. Jean Nouvel's residential project at 100 11th Avenue in New York City subdivides the building envelope into discreet window frames and rotates each of the 1,650 lites of glass, slightly out of plane to achieve the unusual effect of a mosaic of tiles of varying sizes, each reflecting the surroundings in a slightly different color. 46 The shimmering effect of Nouvel's tower occurs within the orthogonal frame of a prefabricated panel, where all the glass is rectilinear when viewed frontally, but small triangular returns close off the panels between out-of-plane glass lites. The west and south facades of the 23-story residential building are clad in a tiled mosaic pattern of curtain wall frames. 44 Nouvel describes the building as a "vision machine," designed to capture the striking views across the West Side Highway and the Hudson River. It is also intended to reflect the different hues of the sky, resulting in dramatic, polychromatic optical effects.

The curtain wall mega panel spans floor to floor and wall to wall, maximizing the transportable module size and minimizing field joints. The largest mega panel is 12' x 37' (3.6m x 11m). 45 Within the mega panel, a composition of "aggregated picture windows" is fit tightly into a rectilinear unit. The composition utilizes 36 different window frame types with three different types of glass and four different angles of rotation.[30] The varying rotations

28 See https://barkowleibinger.com/archive/view/trutec_building, accessed 6.2.2023.

29 "How do tools drive architecture? Tools shape materials that make forms, not the other way around. Tools are inclusive: for us the handcrafted and the digitally controlled are equally legitimate… It is a revolution of choice. We believe that the most compelling shifts in design occur when new technologies and fabrication capabilities are made available and interpreted in new ways," Frank Barkow et al., eds., Barkow Leibinger: *An Atlas of Fabrication* (London: Architectural Association, 2009), preface.

30 See www.frontinc.com/project/100-11th-avenue, accessed 4.4.2022.

4 6 Jean Nouvel's striking 100 11th Avenue residential tower (2010) reimagines the curtain wall as an aggregation of discrete windows, tiled to form a pixelated glass wrapper.

4 7 The International Commerce Centre (2010), designed by Kohn Pedersen Fox, is clad in a curtain wall with a shingled profile, breaking up the mass of the expansive facade.

4 8 The ICC tower facade seems to peel away from the building to form entrance canopies on three sides and a tapered atrium on the north.

of the windows created challenges in terms of fabrication and water tightness, resulting in the recommendation to perform all the glazing in a shop environment.

Within the mega panel, individual windows are cataloged as belonging to one of five types: leaning out, leaning in, tipping right, tipping left, and neutral. The stack joints are designed to withstand the larger building movements, while the interior joints are designed to fit tightly within the frame. Even though their glass is the same type, the south- and west-facing facades pick up a range of different lighting conditions, creating a spectacular shimmering effect. The facade of 100 11th Avenue makes it a singular building that produces a proliferation of optical effects. Its design incorporates modularity and multiplicity, calibrating the mosaic concept with a rigorous and systematic approach to construction.

TEXTURED SKIN

Curtain walls that reveal their modularity and their logics of aggregation often challenge the singular picture plane of the taut facade. The International Commerce Centre (ICC) in Hong Kong highlights the use of a modular curtain wall at a very large scale. 4 7 At 108 stories, rising to 1,588 feet (484m), the tower provides nearly 3 million square feet (275,000sm) of office and hotel program. The entire building is clad in a reflective silver glass, totaling 1.5 million square feet (142,000sm), or 11.5 acres (4.6 hectares) of curtain wall. For architect William Pedersen of Kohn Pedersen Fox (KPF), the tall building represents a unique kind of urban-scaled form, often conceptualized through volumetric carving operations. The ICC tower is a square floor plate with recessed corners that vary in depth along the height of the shaft to allow the upper half of the building to taper inward and the lower segment to flare outward. The varying depths of the reentrant corners give the tower a distinct silhouette on the skyline.

The broad faces of the tower are articulated with a slightly inclined and overlapping profile, giving the entire facade a shingled look. 4 9 The lapped geometry was conceptualized as a modular assembly with the stepped profile aligning with the sill condition at each floor. By rotating the curtain wall panel six degrees out of plane, the surface area of the building increases, with additional material needed to close the gap between the rotated planes. From the interior point of view, the sloped exterior wall is a subtle change from the typical plumb wall,

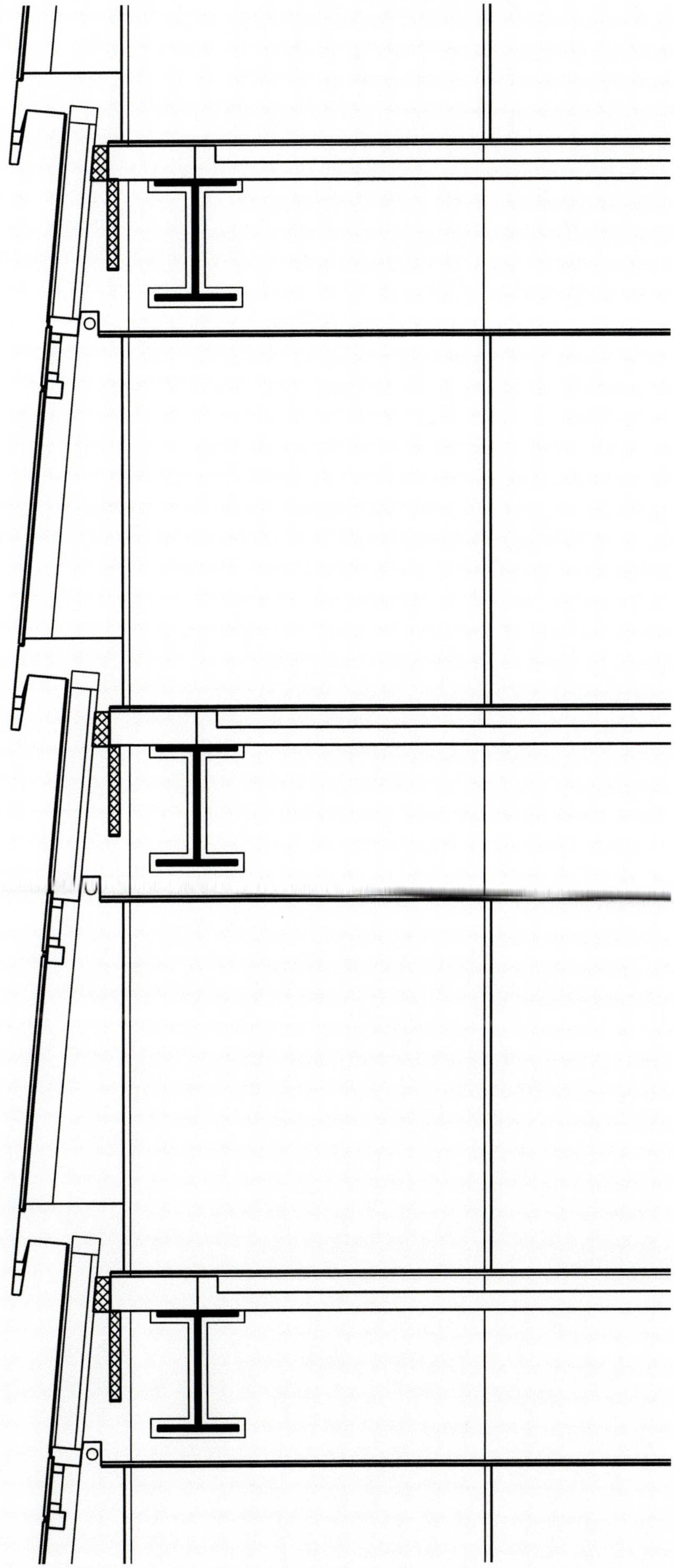

4 9 A partial wall section of the typical ICC curtain wall shows the offset between curtain wall panels with an overhanging profiled edge.

5|0 2050 M Street (2020) in Washington, DC, was designed by Joshua Ramus of REX as a fluted glass monolith with a unitized curtain wall with curved lites of glass, producing subtle and surprising optical effects.

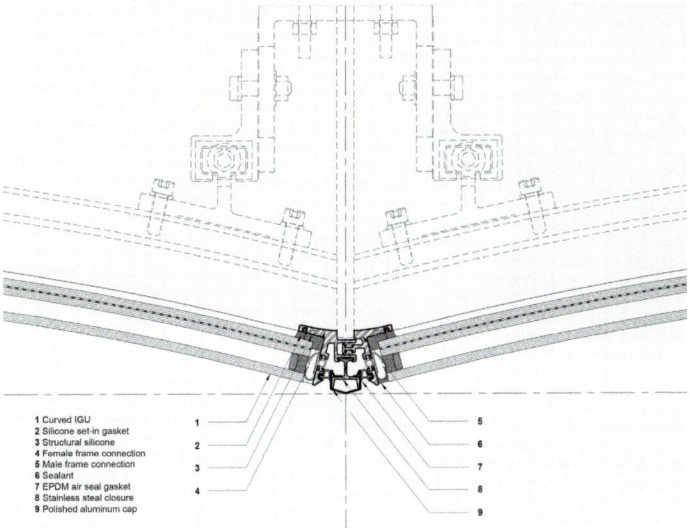

51 Plan detail of the "mullionless" curtain wall assembly containing the movement joint and an extruded aluminum profile, in the dimension of the insulated glass unit (IGU).

52 The curtain wall stack joint and anchor, mounted to the face of slab conceals the horizontal joint in the dimension of the floor package.

but the expanded sill condition feels generous. In fact, the outside corners of the tower are designed to project with a slight angle to give the occupants the effect of a bay window and articulate an extended plane of the four primary faces. From a regulatory point of view, the building's gross floor area was calculated at the edge of the slab, while the building lease line was calculated from the inside face of the glass. Because of the shingled facade and the expanded sill, there is a significant difference between the two perimeters, resulting in more leasable area than a typical flat facade. A slight offset on one floor would not be significant, but each condition is multiplied a hundred-fold on a hundred-story building. Even a gain of a few square feet per floor is quickly compounded by the height of the tower. The logic of the tall building is one of repetition and multiplication, where every condition is amplified by its seriality.

The subtle effect of the overlapping panels evokes the texture of traditional roof tiles, or dragon scales. Rather than meet the ground, the four faces of the tower peel up to form canopies on three sides and an enclosed atrium on the north side. 48 Varying the degree of rotation between panels gives the base of the tower its unique flaring profile, while the articulation of each floor with a lapped profile breaks up what would otherwise be a monumental flat surface. The subtly textured facade produces a shimmering play of light and reflection while dematerializing the mass of the building. The articulation of the floors in the glazing geometry gives the building visual scaling cues while also registering the scale of the curtain wall module.

FLUTED FACADE

Introducing curvature to the taut glass facade, 2050 M Street (2020) in Washington, DC, employs curved glass curtain wall modules to achieve a unique fluted effect. 50 Designed by Joshua Ramus of REX, 2050 M Street is at the extreme limits of a unitized curtain wall. With it's painstakingly procuring curved insulated glass lites and minimized aluminum frames, it achieves subtle but spectacular optical effects from both the street and the interior.

Referred to by Ramus as a "trophy" office building, he proposed a fluted glass facade of concave glass panels. The glass flutes conform to a typical curtain wall module 5 feet (1.5m) wide by 11 feet 3 inches (3.4m) tall but lack the vertical mullions common to a unitized facade, relying instead on the curvature of the glass to produce stiffness in section to resist the wind load on the floor-to-floor

5 3 The all-glass attraction called "Levitation" at One Vanderbilt, designed by Kohn Pedersen Fox, lets visitors stand in an all-glass room 1,063 feet (324m) above street level.

span.31 Aluminum mullions typically perform three functions: they support the glass, span between floors, and create weather-tight seals at the joints between prefabricated modules. The typical seal consists of an aluminum extrusion with gasketed interlocking profiles. For M Street, the curved glass was deemed stiff enough to resist the local wind-loading conditions, thus eliminating the need for traditional aluminum mullions. Ramus' design for a mullion-less curtain wall reminds us that the all-glass facade is only as transparent as its entire system, including the aluminum mullions that typically support the glass.

31 The curvature induced stiffness is also the aspect of the serpentine walls at UVA as discussed in Chapter 2.

The interlocking seal between panels is integrated into a minimal aluminum profile at the joint and is largely hidden in the one-inch (25mm) insulated glass unit (IGU). Less a mullion than a glazing adapter, the custom extruded joint conceals its functions in the dimension of the IGU. 5 1 The curved and laminated glass unit presents its own challenges in terms of fabrication, as the glass lites need to be curved perfectly so that the inside curvature of one lite perfectly matches the outside curvature of the other to ensure a tight lamination. The monolithic outer light also has a reflective coating applied to its second surface to improve solar performance.

The experience from the inside of the building is a remarkable glass curtain that ripples across the facade without the typical interruption of vertical mullions. The joint between curved lites of glass is an aluminum extrusion only two inches (50mm) wide. The glass spans from floor to ceiling, and the finished floor is flush with the sill plate, which conceals the gap between the undulating profile of the back of the glass and the straight line of the edge of the slab. The horizontal stack joint is integrated into the glazing frame and aligns with the concrete structural slab. 5 2 Although the curved glass introduces a distortion of the view — this is not a picture window — the overall effect of the facade is astonishing, with the building appearing like a shimmering prism of glass, reflecting and refracting the city around it.

ALL-GLASS

The unimpeded panoramic view still has its own currency in architecture, particularly in supertall high-rise buildings like the pencil towers in Manhattan and observation decks on new buildings like One Vanderbilt and 10 Hudson yards, both by KPF. The Summit One viewing deck at the top of One Vanderbilt creates a series of attractions that include an outdoor viewing terrace, exhibition spaces

5|4 The full-scale performance mockup tests the five-sided glass room and each of its connection details.

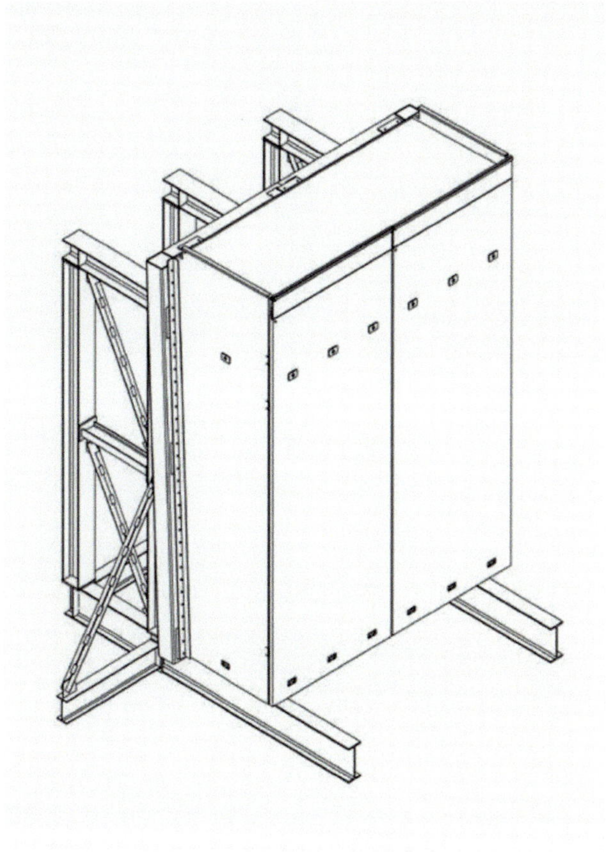

5|5 To achieve the all-glass experience, the architects specified laminated glass for safety and redundancy, as well as a mortise and tenon connection between the glass floor and the glass walls.

with abundant mirrored finishes on the floors and ceilings, and an attraction called Levitation, which allows the viewer to occupy an all-glass bay window, outside the plane of the building face, 1,063 feet (324m) above street level. 5|3 The projecting room has glass walls, floors, and roof. The enclosure employs multiple laminations to achieve the strength and redundancy required for the glass floor application. The glass-to-glass connections at the corners minimize the visual obstruction with the use of a point support at the head and a unique mortise and tenon connection at the floor. 5|4 5|5 Experiences like the glass panorama of Summit One seek to dissolve architecture to "almost nothing," including the roof and floor, to heighten the sense of vertigo. The design of Summit One is for bodily experience as well as picture taking — a space for portraits and selfies. The projecting room of the Levitation attraction is like an inhabitable fisheye lens through which to survey the city, but also a scene designed to be photographed through the 21st-century optical instrument, the smartphone camera.

From the translucent glass lites in Vermeer's 17th-century interior scenes, to the windshield-like windows of Le Corbusier's Villa Savoye, to the mixed media panorama of Diller + Scofidio's Slow House, the story of evolution and transformation of the window to window wall to curtain wall highlights technical innovations, cultural associations, and behavioral patterns. Just as large plate glass storefront windows are said to have produced the window shopper and the *flâneur*, the all-glass observation deck might also produce the selfie-taking vertigo seeker. The design and construction of glass windows and envelopes are never neutral. The waterfront house is not simply viewing the surroundings; it is determined by them. The view is often the determinant for the house — its primary parameter. The glass window is a technology — a primary interface between a house and its context. It frames our view of landscapes and contexts, natural and man-made. It is also a powerful cultural marker: the window frames and frameless windows define our interior worlds, just as they define our views of the world beyond and how we see ourselves in it.

1 The gravity-defying cantilevered loop design of the CCTV Headquarters (2012), designed by OMA with Arup, showcases the structural performance of the tower while gesturing, through its facade, about how it does it.

Chapter 7 Force, Form, and Gesture

The principle of *entasis* introduces an anomaly within the dominant narrative of structural rationalism that has largely defined our understanding of classical Greek architecture, and that of western architecture derived from it. The slight swelling of the midsection of many ancient Greek columns is at the limit of perceptibility. [2] These deviations, long overlooked by architects, archeologists, historians, and surveyors, cannot be explained by structural or functional rationale alone, and aesthetic and optical arguments for these modifications remain difficult to prove. Prevailing theories suggest that it is a corrective measure to compensate for the perceptual and optical effects of human vision. In the 1840s, British architect, archeologist, and scholar Francis Penrose defined entasis as the "swelling given to a column in the middle parts of the shaft for the purpose of correcting a disagreeable optical illusion, which is found to give an attenuated appearance to columns formed with straight sides, and to cause their outlines to seem concave instead of straight."[1]

The architecture of ancient Greece has served as a model for much of Western architecture, codifying the classical columnar orders, establishing a vocabulary of structural and ornamental components, and defining cultural notions of architectural beauty. Greek architecture appears as a material manifestation of a philosophical tradition that articulates principles of reason. Entasis, an inexplicable deviation at the center of the aesthetic ordering system, points to the complex relationship between our expectations of pure, geometric expression and our perceptions and embodied experiences of our constructed environments.

Entasis remains an enigma at the heart of architecture's origin story. It is a geometrical deviation, that is, not a function of error in fabrication or execution, but rather an intrinsic inflection so subtle that its existence was initially debated, and once established, its motives remained ambiguous. Some argue that the swelling at the midpoint of a column is a visual expression of the weight columns carry.[2] The representation of force through form — of structural principles expressed through architecture — is the subject of this chapter. The expression of structure, like the principle of entasis, can be read as a representational, didactic gesture. Specifically, the question of how architecture is made legible shifts design away from a one-to-one relationship between form and its self-evident function, and toward principles that define architecture as a discursive object.

1 Francis Penrose, *An Investigation of the Principles of Athenian Architecture; or, The Results of a Survey Conducted Chiefly with Reference to the Optical Refinements Exhibited in the Construction of the Ancient Buildings at Athens* (London, UK: Macmillan, 1888), p. 39.

2 Architectural historian Vincent Scully argues that the columns of the Greek temple are analogous to "divine persons," and a "celebration of bodily force." Describing the swelling of columns he states, "One straight axis runs down the center, expressive of the imposition of a point load upon the figure, around which the body swells symmetrically as it takes the compression from above." Vincent Joseph Scully, *Architecture: The Natural and the Manmade*, 1. paperback ed. (New York, NY: St. Martin's Press, 1993), p. 50.

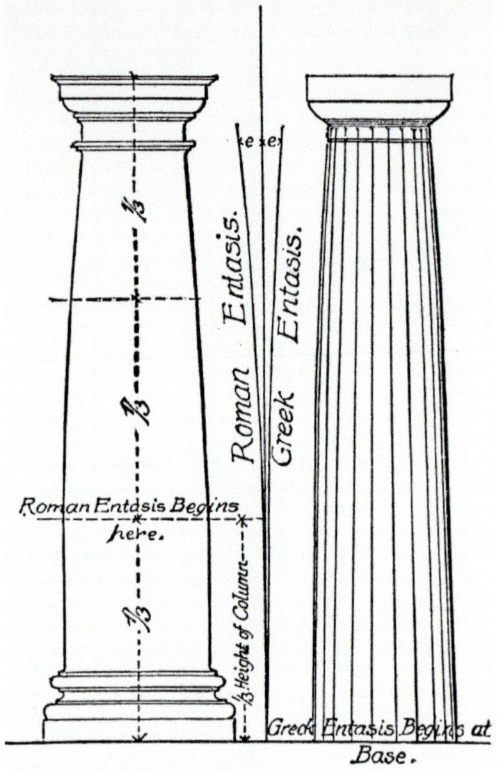

2 The principle of *entasis*, from classical Greek architecture, introduces a subtle deformation to the straight column, making it appear to swell at its mid-section, but serves no structural or functional purpose. It is purely an optical effect, that translates force into the gesture of form.

FINDING FORM

German architect, Frei Otto began his career observing structural and material behaviors in natural phenomena: The intricate networks of spiderwebs, patterns of cracks in arid landscapes, surface tensions in soap bubbles, geometries of stalactites and stalagmites, and the curvatures of draped materials. His close observations of naturally occurring behaviors led him to a theory of self-formation processes in both natural and man-made conditions. In each precedent he studied, Otto discerned the interactions of materials with the forces acting on them.[3] He observed, for example, how spiderwebs exhibit tremendous tensile strength as well as a highly evolved model for material optimization and efficiency, how cracking patterns formed in drying mud demonstrate the result of evaporation and material shrinkage, and, that the natural curvature formed by draped materials, such as fabric, string, and cables approaches an optimum known as a catenary curve.

Otto developed form-finding, both a design process and a research methodology, through his work at the Institute of Lightweight Structures (ILEK), which he founded at the University of Stuttgart in 1964. His earliest experiments were conducted with physical models for observing and analyzing the behavior of materials subjected to external forces. Physical experiments with minimal surface films, pneumatically and hydraulically stressed membranes, angle of repose studies, branching structures, and inverted catenary chain models were also developed at the institute. 5 Otto experimented with soap film stretched between rigid frames to better understand minimal surfaces and test the equal distribution of tension across them, which informed the shape and performance of larger constructions. In many ways, Otto's material form-finding could be understood as an early mode of material computation.

Frei Otto's winning entry for the German Pavilion at the 1967 Expo in Montreal was his first large-scale project utilizing his tensile structures research. 3 Otto and his team used physical models made with stretchable tulle fabric as well as weighted wire models for performing physical calculations of form and measurement. The team developed a cable net system suspended from eight rigid masts. The intersections in the netting were held together with a custom cross-clamp fitting. The 20-inch (50cm) grid acted as a ladder, making the entire surface of the roof accessible and allowing for the installation of a PVC-coated polyester fabric suspended below

[3] Werkbund Bayern and Museum Villa Stuck, eds., Frei Otto, Bodo Rasch: *Finding Form: Towards an Architecture of the Minimal*: The Werkbund Shows Frei Otto, Frei Otto Shows Bodo Rasch: Exhibition in the Villa Stuck, Munich, on the Occasion of the Award of the 1992 Deutscher Werkbund Bayern Prize to Frei Otto and Bodo Rasch (Stuttgart: Axel Menges, 1995).

3 The German Pavilion from Montreal's Expo 67, was designed by Frei Otto as an experiment in large-scale tensile structures employing masts and cable nets.
4 Frei Otto applied the principles of tensile structures to the Munich Olympic Stadium (1972).

5 Frei Otto developed the technique of "form finding" by allowing material to deflect under its own self-weight. In this photo with a catenary model and mirror, Otto is demonstrating the principle of the invertibility; A cable structure in pure tension, inverted, is an optimal form in pure compression.

the cable net like a tent. Otto's successful temporary pavilion in Montreal popularized the modern tensile structure, and his form-finding process of using models as analog analytical tools would become a methodology for subsequent projects.

A few years later, Otto applied these principles of tensile structures to the Munich Olympic Stadium (1972), a project designed in collaboration with architect Günter Behnisch, and structural engineer Jörg Schlaich. 4 Largely defined by a large, interconnected network of tent-like canopies, Otto and Behnisch designed the stadium in Munich to contrast with state architect Werner March's 1936 Olympic stadium in Berlin that showcased Nazi Germany in the lead up to World War II. March's monumental structure, complete with repeating columns and piers, was intended to recall the rational tradition of ancient Greek architecture. The 1972 Olympic Park was conceived of as an extension of the surrounding landscape, with bowl-shaped amphitheater spaces integrated into an artificial topography. Using building rubble from the aftermath of World War II as part of the fill for the grounds, the stadium and surrounding park literally reshaped the ruins of the past to create an Olympic complex.

In keeping with the relatively informal composition that complemented the smooth contours of the park, the largest of the tent structures covered only the western half of the main stadium and wound through the park to connect to two other pavilions. Otto and his team used physical models to simulate the surface geometries of the peaks and valleys of the structure. In construction, steel masts working in compression were complemented by a cable net surface working in tension. The peaks and saddle geometries of the membrane surfaces that swept over the iconic stadium in Munich were determined by the downward pull of the tensile forces, an aspect of the project only made possible by the rigorous tests that ultimately saw Otto's form-finding principles realized at an enormous public scale.

SO, INVERTED

English polymath, scientist, philosopher, and architect, Robert Hooke, was the first to theorize the inverse relationship between tension cables and compression arches. In 1675, he wrote: *"Ut pendet continuum flexile, sic stabit contiguum rigidum inversum,"* which translates to: "As hangs a flexible cable so, inverted, stand the touching pieces of an arch." 6 7 A catenary (from the Latin word for chain) is the curve that a cable or string assumes under its own weight when supported at its ends. Catenary

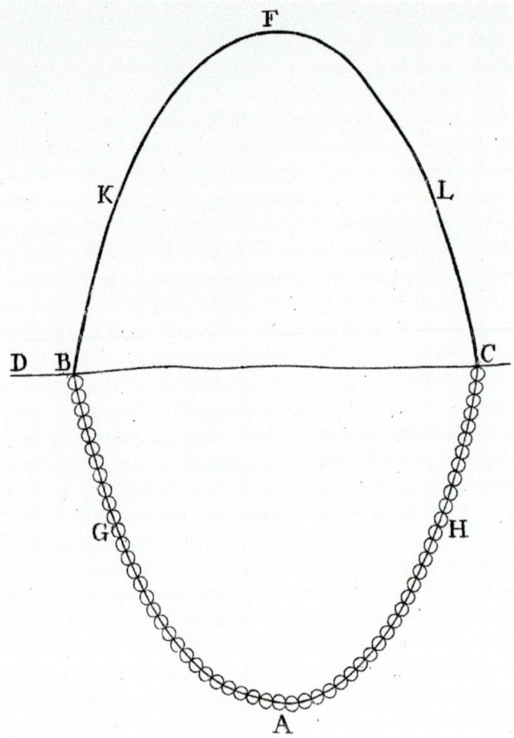

6 Diagrams by Giovanni Poleni, illustrating Robert Hooke's principle of invertibility, whereby a cable in tension, inverted forms a pure arch in compression.

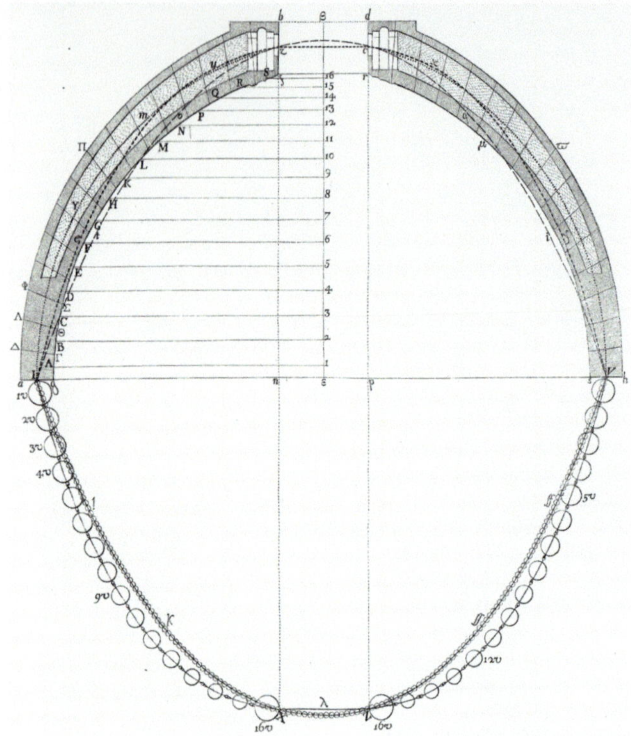

7 Giovanni Poleni's graphic calculation of the of the Dome of St. Peter's in Rome (1748) employed a loaded chain in tension to determine its funicular shape in compression.

curves, also known as funicular, are geometries governed by the lines of the flow of forces.[4] A catenary curve is an idealized geometry for a hanging structure where the self-weight of the chain finds its equilibrium and the structure holds a curve in pure tension. The inversion of the catenary curve also forms an idealized arch, where all elements are held in pure compression.

Catalan architect Antoni Gaudí understood the invertibility of catenary geometry well, thanks in part to his hanging chain models that informed the design of many of his structures, including the Sagrada Família church in Barcelona. 8 The construction of Sagrada Família began in 1882 as a Gothic Revival design by Francisco de Paula del Villar. When del Villar resigned in 1883, Gaudí was appointed chief architect, and dedicated his life to the project until his untimely death in 1926. Progress on the church was complicated by the Spanish Civil War and incomplete drawings and records. For nearly 100 years, designers and builders have relied primarily on photographs of plaster models and limited drawings and sketches to continue construction on what is widely accepted as the largest incomplete Catholic church in the world. Ongoing work toward Sagrada Família's completion has continued with the aid of interpretive designs by New Zealand architect, Mark Burry. Burry deploys computational techniques to produce digital models of the design and uses robotic technologies to carve the stone blocks used in its construction. [5]

Gaudí's physical study models share a process of form-finding with Otto's experiments that demonstrates how physical models remain an indispensable means for observing and calculating force and form.[6] Gaudí used elaborate upside-down chain models to derive catenary geometries that he found to be highly efficient due to material alignment with the thrust lines. 9 The chain models for the Güell Colony Crypt, for example, included a mirror that he would slide beneath to visually invert the catenary and transform it into an arch. In some cases, Gaudí's models included small weights hung from chains to simulate loading across the arches and buttresses. These experiments in form and force translate otherwise invisible thrust lines into highly expressive form, manifesting possibilities by both material and computational means. Over the last century and a half, paradigms for design and construction have evolved dramatically — from load-bearing stone to the introduction of steel-reinforced concrete, to computer-controlled stone carving and post-tensioned masonry construction. Inverted catenary geometries remain ideal for a full-compression stone

4 Edward Allen, Wacław Zalewski, and Joseph Iano, *Form and Forces: Designing Efficient, Expressive Structures* (Hoboken, NJ: John Wiley & Sons, Inc, 2010), p. 230.

5 Antoni Gaudí, Mark Burry, and Deutsches Architekturmuseum, eds., *Gaudí Unseen: Completing the Sagrada Família* (Berlin: Jovis, 2007).

6 Frei Otto included photos of Gaudí's suspended models in his 1995 book, Frei Otto, Bodo Rasch: *Finding Form: Towards an Architecture of the Minimal*: The Werkbund Shows Frei Otto, Frei Otto Shows Bodo Rasch: Exhibition in the Villa Stuck, Munich, on the Occasion of the Award of the 1992 Deutscher Werkbund Bayern Prize to Frei Otto and Bodo Rasch.

8 The Sagrada Família in Barcelona, Spain, is the unfinished life's work of Antoni Gaudí, who worked on its design from 1883 until his death in 1926.

9 Antoni Gaudí developed the design of the Sagrada Família with weighted hanging chain models to arrive at highly efficient and expressive catenary forms. Gaudi transformed hanging chain models in tension into stone arches in compression.

structure. Consistently, the balanced calibration of the self-weight of the material and the attenuated geometry of the arch optimizes the structure for expected load cases.

Emerging from an eclectic Neo-Gothic, Art Nouveau style, Gaudí developed a unique formal language that materialized the geometric principles of ruled surfaces, hyperboloids, and boolean operations through physical models and analog computations. Scholars and researchers still refer to Gaudí's analog parametric models, which demonstrate how a change in one parameter, such as the support point of a weighted cable, results in a cascade of effects as the system adjusts.

That the Sagrada Família is so well suited to the sophisticated digital scripting and fabrication tools developed nearly a century after Gaudí's death attests to his material and mathematical imagination. Far ahead of his time, his work and methods continue to inspire and inform the design processes and architectural possibilities afforded by our current moment.

STANDS THE ARCH

The 1947 Jefferson National Expansion Memorial competition, won by Finnish American architect Eero Saarinen for his proposal of the Gateway Arch for St. Louis, Missouri (completed in 1965), resulted in a freestanding structure derived from an idealized diagram of a weighted catenary curve. 11 The competition celebrated US westward expansion and American frontierism, which would mark St. Louis as the "Gateway to the West." Saarinen's 630-foot tall design was envisioned as analogous to the 555-foot (170m) Washington Monument, or the 305-foot (93m) Statue of Liberty.[7]

[7] As with many monuments, the intended symbolism and its public reception change over time. The Saint Louis Arch has become a symbol of Saint Louis the way the Eiffel Tower has become a symbol of Paris. More recent interpretations of the Gateway Arch include a broader historical context that includes acknowledgement of settler colonialism and the displacement of Native Americans.

The Gateway Arch is both a cultural and technical artifact. The symbolism of its iconic profile was a primary impetus for the project, and the meaning it conveys to certain audiences makes it a product of its time. As a structure, the arch represents a significant engineering achievement, pushing the boundaries of what was technically achievable when it was built.

The 630-foot height (192m) of the Gateway Arch is matched by its 630-foot (192m) base — dimensions that made the arch too tall and too wide for scaffolding to be used in its construction. The span of the arch is more than 10 times that of the Gothic arches spanning the nave of Chartres Cathedral (54 feet, 16.4m). Given these challenging dimensions, the builders at MacDonald Construction Company devised an ingenious method by which they mounted

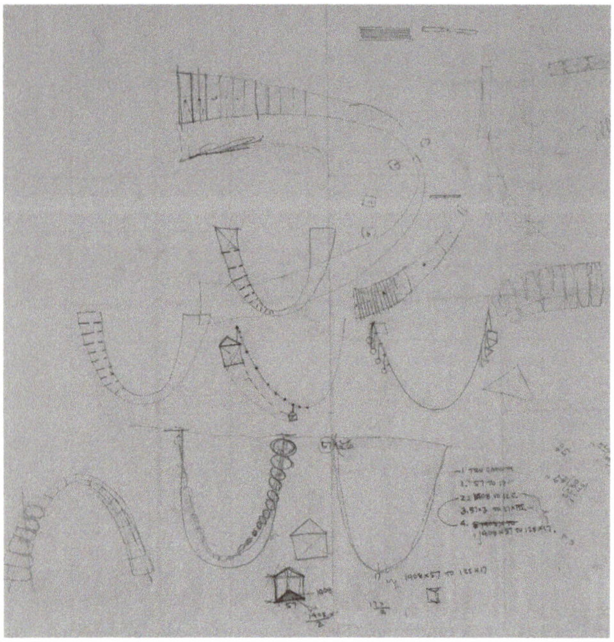

10 Eero Saarinen's conceptual sketches of the Gateway Arch competition in 1947 show Saarinen applying the principles of inverted catenary forms in tension and compression. The middle right hand sketch appears to show a weighted chain, reminiscent of Gaudi.

11 Saarinen photographed gesturing with a model of the Gateway Arch.

a platform crane on steel rails to the outside of the arch's legs, allowing the legs to be constructed independently. Each leg was constructed toward the center, with the climbing crane hoisting segments into place. This approach required not only imagination and ingenuity, but tremendous precision in fabrication and coordination in the field. A temporary steel truss bridge was installed at the height of 530 feet (161m) to stabilize the legs as they leaned toward one another. 12 This allowed builders to limit the differential movement as they inserted the final keystone module. The symbolism of two independent structures leaning toward each other and joining in the middle might symbolize unity or co-dependence, depending on the context and framework of interpretation.[8]

The logic of the Gateway Arch follows the sequence of a traditional arch. Individual wedge-shaped pieces are set along a curve until a final keystone can be inserted to join two halves. Parts along the curve form a new whole that, in equilibrium, spans the void between supports. Each individual part is arrested in midair by its adjacent components. The keystone at the center transfers its weight to adjacent modules, which then transfer their loads through the arch until they reach their base or abutment, which resists the thrust.

[8] The act of construction as an act of community building is well documented. The barn raising of rural Amish communities enlists the help of a community to build a barn quickly. Similarly, the construction of a church or cathedral in medieval Europe often took generations to complete and drew on the resources of an entire community.

[9] The documentary movie "Monument to the Dream" features closeup shots of the drafting of the arch geometry in elevation, as the tracing of a French curve. A French curve is a template made up of multiple segments of curves that follow the Euler spiral, or a curve that changes continuously with its length.

The geometry of the Gateway Arch consists of a near catenary curve, with a wide base and a profile that tapers as it rises.[9] The cross section of the arch reveals an equilateral triangle with a face dimension of 54 feet (16m) at its base and 17 feet (5.2m) at its keystone. It is constructed as a segmented triangular tube, made up of 142 tapered segments. 13 Each segment was prefabricated as a steel plate sandwich with a quarter-inch (6mm) stainless steel outer skin and a carbon steel interior skin. The largest panels measured 12 feet x 54 feet (3.6m x 16.4m) and were shipped by rail from Pittsburgh, Pennsylvania. On site, each panel was assembled and hoisted into place, joined and welded together. The lower portions of the arch were also filled with rebar, concrete, and post-tensioning cables. A precise eighth-inch (3mm) joint was maintained between the steel segments.

The Gateway Arch's stainless steel skin glistens in the light, and its sharp corners appear to alternate between thick and thin depending on the angle of the viewer. Seen in direct sunlight, the triangular cross section of the arch highlights the contrast between its lit and shadowed faces. Standing at one foot of its base, the arch appears to launch into space, arcing like a surreal projectile that recedes from view as it reaches a tremendous height, only to fall along a measured path mirroring that from which it

12 The installation of the "keystone" during the construction of the Gateway Arch (1965, eighteen years after the design competition). The Saarinen design necessitated several construction innovations, including the climbing cranes and the trussed brace to minimize the differential movement of the two half arches.

13 Prefabricated segments were assembled on site by piecing together triangular modules to form the tapered sculptural form of the Gateway Arch.

came, landing 600 feet away. The structure's face is subdivided into three panels, but otherwise there are few scale references, and it rises without articulation or differentiation. The tapered geometries of the joints appear to converge in perspective, at once attenuating the perceptual height of the arch and rooting it firmly to the ground.

The Gateway Arch gives expression to gravitational loads and translates the flow of forces into architectural form. A sketch from Saarinen shows the arch both upright and inverted as a catenary. 10 The radial geometry of the keystone-shaped modules of the arch are perpendicular to the force vectors of the thrust lines, and we can understand the joints as expressions of gravity and wind acting on the form. Saarinen's immaculately constructed monument is, among many things, a perceptual device, teaching us to see the formal expression of force in architecture.

Saarinen's Kresge Auditorium (1955) on the Massachusetts Institute of Technology (MIT) campus in Cambridge, Massachusetts, though typologically different in nearly every way, is as instructive a demonstration of force acting on form. 15 The building's primary feature is a thin-shell concrete dome derived from the geometry of a perfect sphere — yet, due to subtractions of sections from its shell, the building meets the ground at only three points and is largely triangular in plan. 16 With a thickness of only three-and-a-half-inches (89mm) across its surface, it is extremely light and thin. The building's edges are articulated with upturned concrete beams that transfer loads to the three primary footings and fortify the edges of the thin shell. 17 During construction and following the removal of the formwork, the edge beams experienced unanticipated deflections, resulting in the installation of additional steel columns immediately behind the curtain wall for extra support.[10] The spherical geometry of the dome was mathematically well-defined, but it fell short of optimal structural form, despite its thin-shell conception and construction.

[10] An account of the construction and structural remediations of Kresge Auditorium is in included in the paper by Plunkett, J & Mueller, Caitlin. (2015). "Thin Concrete Shells at MIT: Kresge Auditorium and the 1954 Conference," Conference: The Fifth International Congress on Construction History, Chicago, 2015.

Notwithstanding its mixed results and reception, Kresge Auditorium is an exemplary case study in the tensions that can arise between architecture, engineering, and construction — three disciplines that necessarily intersect around structurally ambitious buildings. Saarinen's architectural vision for a truncated sphere that would rest on three points overrode structural and practical concerns. Charles Whitney, the structural engineer, likely would have pointed out that the spherical geometry with unsupported edges was not an optimal structural

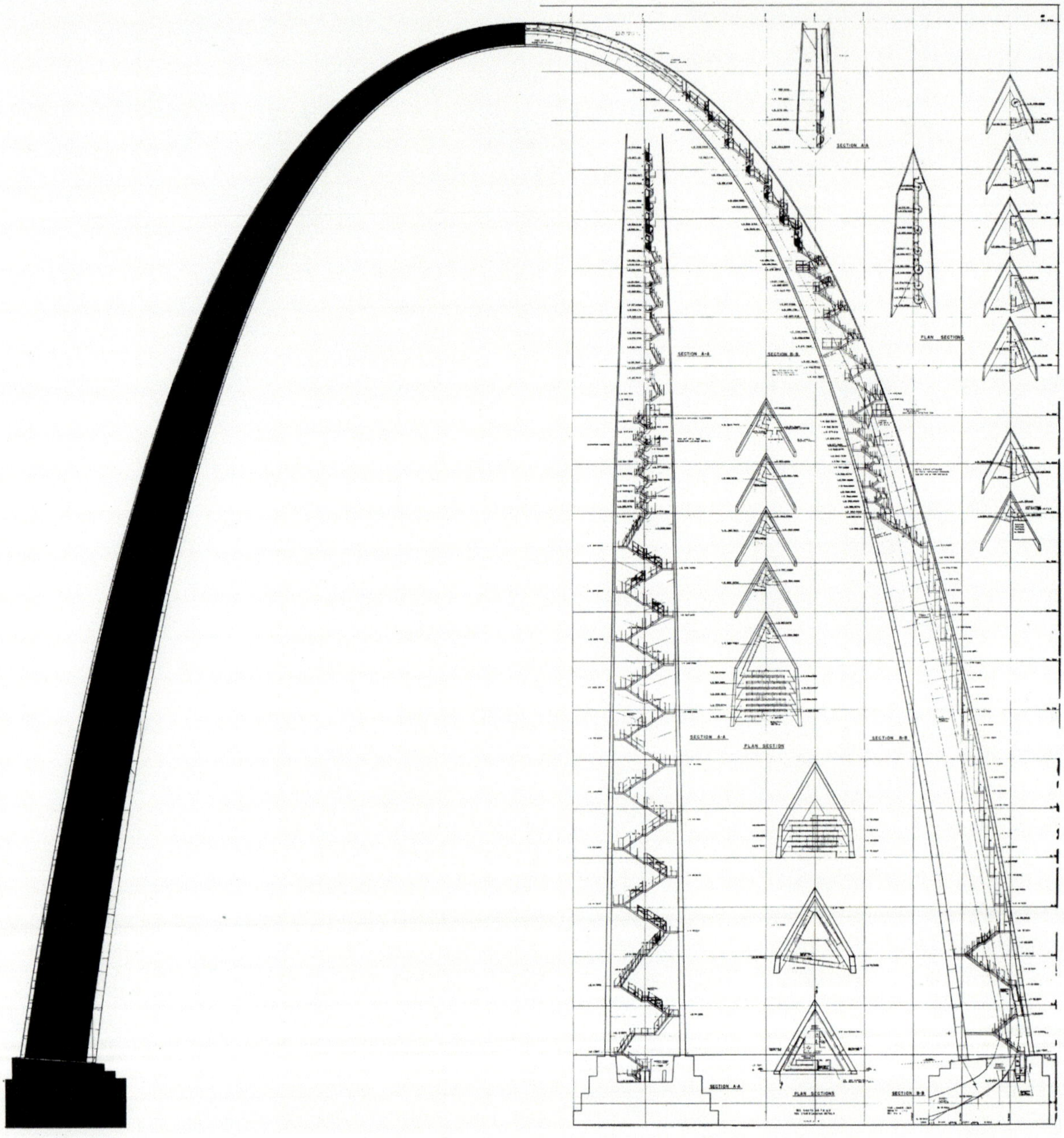

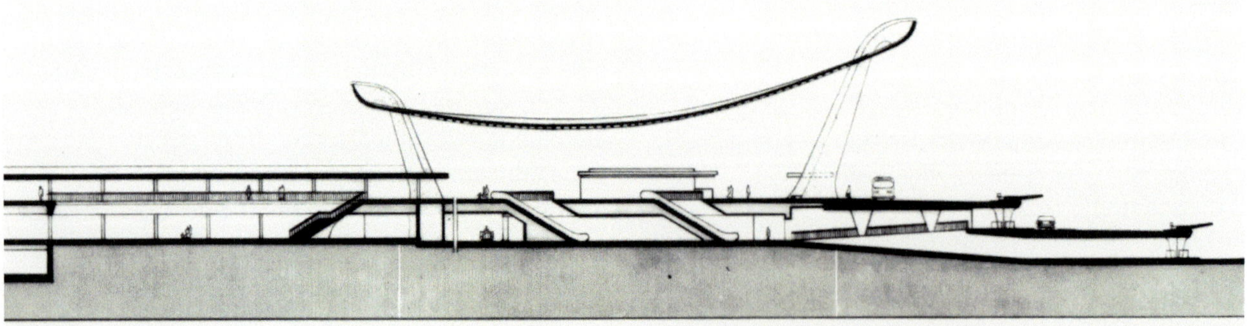

1
4
The Gateway Arch and Dulles Airport, by Eero Saarinen, are both conceived as catenary structures. The Gateway Arch is a catenary curve of a compression structure. Dulles Airport is a curved surface under its own self weight in tension.

15 Eero Saarinen's Kresge Auditorium (1955) at MIT is a thin-shell concrete structure derived from a segment of a pure sphere.

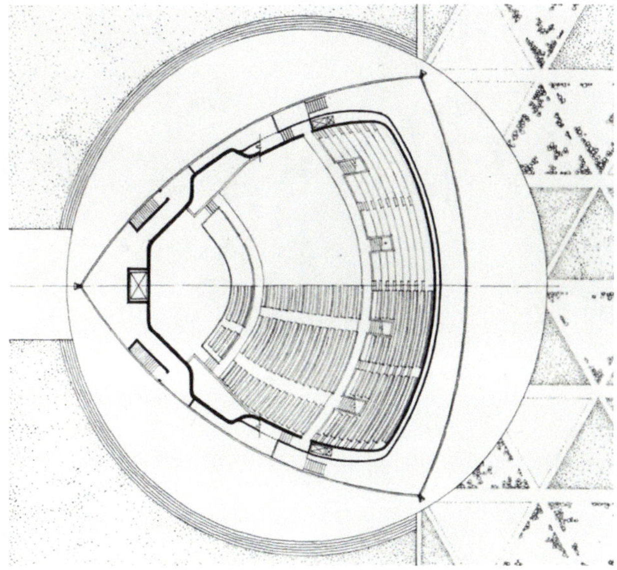

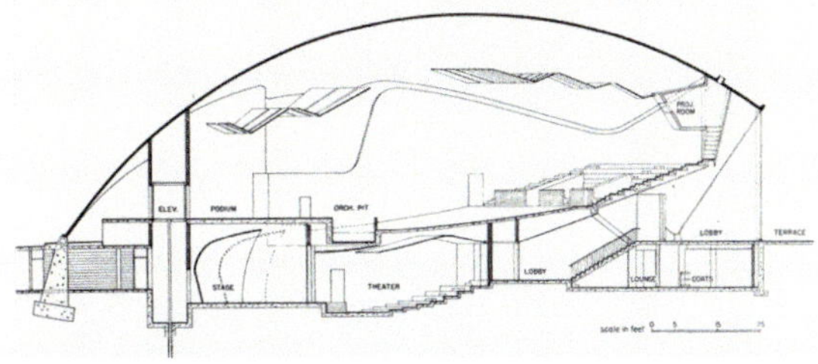

16 Plan of Kresge Auditorium (1955) is a cropped segment of a sphere.
17 Section of Kresge Auditorium (1955) showing the footing on the left where one third of the loads are concentrated. The free edge on the right shows an upturned and thickened slab to bring loads to the other two footings.

form, and therefore unable to take full advantage of the thin-shell construction. The builder, George A. Fuller Company, surely would have expressed reservations around implementing the unusual design. Kresge Auditorium demonstrates the tenuous, and necessarily delicate interaction of geometry, material, and performance. As such, it is a cautionary tale that points to the importance of full appreciation for multiple disciplinary perspectives in architectural design, engineering, and construction.

Saarinen's work is rich with examples of structurally expressive form that falls precisely at the intersection of pure geometry, structural performance, and constructability. The Gateway Arch derived its form from a structurally optimized diagram of a catenary compression arch. Kresge Auditorium tested the section of a sphere as a thin-shell domed structure. Saarinen's Dulles Airport is the third in a series of structurally expressive projects — again deploying the catenary curve in tension for the primary roof form. 14

Completed in 1962, Dulles Airport was the first airport built to serve jet aircraft. As such, the main terminal's design expresses the streamlined geometries of a new era in transportation. The main terminal consists of a 600-foot long (183m) pavilion, with parallel edges that curve upward — delineating "land side" and "air side." The structure itself evokes a draped fabric tent, suspended between two rows of inclined piers. 14 The gentle curvature of the canopy appears to be the result of its own weight evenly distributed and settled into a state of repose.

To achieve the appearance of a monumentally scaled yet lightweight pavilion hovering over the landscape, Saarinen designed the building as a catenary curve — a stressed ribbon structure of steel cables and precast concrete panels supported by rows of inclined concrete pylons, 150 feet (45m) at their base, and nearly 200 feet wide at their highest point. The building's pylons are made of site-cast concrete shaped into piers with filleted curves at their base and a tapered, swans-neck geometry that reaches through an opening in the canopy to support it from above. The edges of the canopy are also made of site-cast concrete formed to align with the curvature of the cables that support this portion of the structure. The site-cast edges act as rigid beams, transferring the cable loads back to the concrete piers spaced 40 feet (12m) apart.

Supported by pairs of cables strung between the roof's rigid frame, the precast panels are distributed uniformly across its span, and the catenary curve results from the weight of the entire roof system. 18 19 As with many precast systems, the roof sections

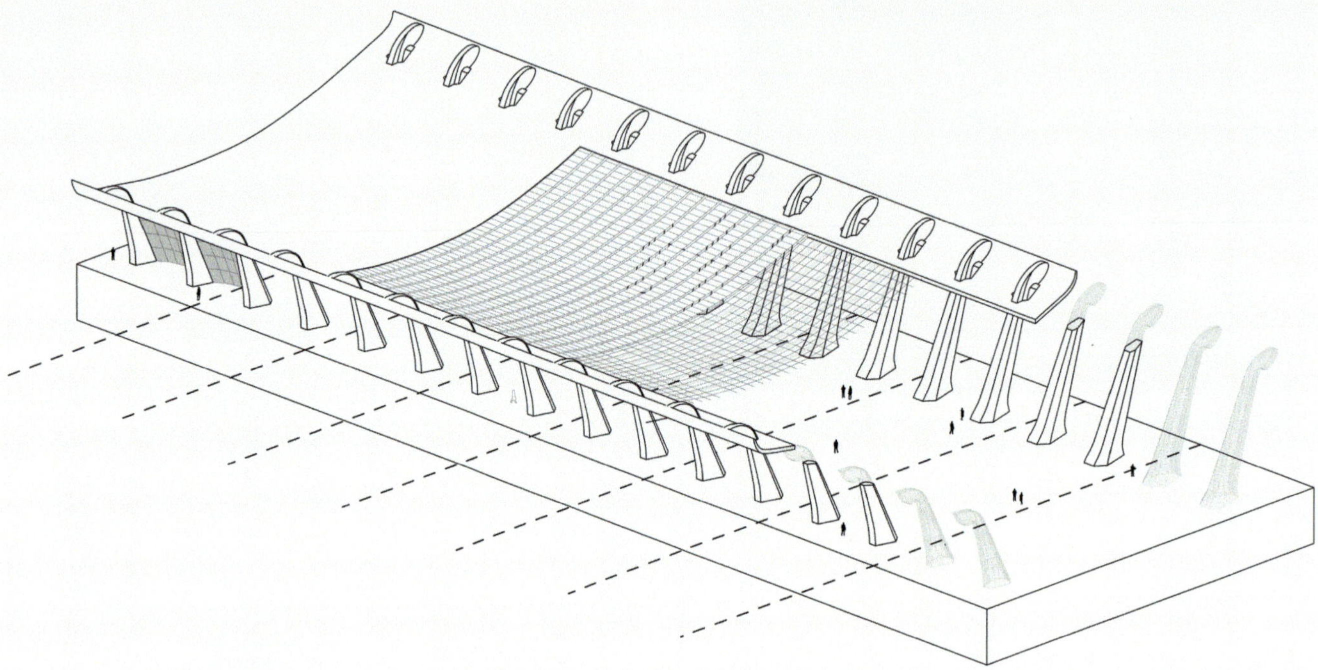

18 Isometric "process drawing" of Dulles Airport showing the construction sequencing of the concrete piers, on the right, the cable spanning between piers, and the completed and enclosed terminal, on the left.

at Dulles Airport are tied together with the rebar that extends from them, forming a series of site-cast ribs. A smooth stucco soffit was later added to the underside of the roof to give it a monolithic appearance.

The expression of structural principles governs the geometry of Dulles Airport. The gentle curve of the roof expresses its self-weight. The incline of the piers counteracts the horizontal pull of the cables. The concrete piers taper on all four sides and flare at their bases, expressing at once their flexibility and connection to the ground. The pier is essentially an upright cantilevered structure leaning outward, with the two levels of the slab and the foundations acting as backspan. Tracing the loads from the midspan of the roof, we can follow them along the curvature of the suspended underbelly to the plane of its rigid edge. Loads along the edge are picked up by the curved neck of the pylons that then plunge downward through the opening in the curved plane of the canopy, continuing until they disappear and pivot at the fulcrum of the structural slab. The piers continue into the lower level where they engage with the building foundations below.

Saarinen died of a brain tumor in 1961 at the age of 51. He never saw the Gateway Arch completed. Several of Saarinen's designs were completed after his death including the Gateway Arch, Dulles Airport, the TWA Flight Center at JFK Airport, and the CBS Building in Midtown Manhattan. Saarinen's formal explorations search for a structurally expressive vocabulary rendered in a wide range of material systems. The Gateway Arch's steel-plate triangular tube structure achieves a graceful catenary-like curve by aggregating discrete panelized modules. Kresge Auditorium transforms the geometry of an arch into a spherical concrete dome. And the convex, arcuate geometries of the arch and the dome find their complement in the concave geometries of the catenary tensile structure that so defines Dulles Airport. In each case, structural force is translated into form and material. While these projects are not a direct translation of force into form, they are the results of Saarinen's bold yet careful processes of experimentation, interpretation, and translation to produce new structural gestures and sculptural forms.

CATENARY CLERESTORIES

In 1998, Sergey Brin and Larry Page founded Google, a startup that revolutionized how information is accessed via a command line search function. For the first 20 years of their growth and development, the company primarily occupied office spaces that were renovated to suit their needs, including the

19 The construction of Dulles Airport's catenary roof structure involved the suspension of steel cables between concrete edge beams, and the placement of precast concrete panels across the cables.

20 Google Bay View "squomes," designed by Bjarke Ingels Group (BIG) and Heatherwick Studio, seen in an aerial photo. Note the catenary geometry of the roof segments and the smile-shaped clerestory windows between them. The large floor plate buildings benefit from daylight introduced by the clerestory windows.
21 One of the "smile trusses" being hoisted into position during construction.
22 Building section showing the catenary geometry of the roof and the one-story office floorplate within.

Googleplex in Mountain View, California. Their real estate strategy has been likened to a hermit crab, occupying shells built by other companies. In 2015, Google launched a design competition to reimagine its first fully commissioned architectural headquarters.

Ultimately, the project was awarded to two firms, Bjarke Ingels Group (BIG) and Heatherwick Studio, that worked together in a 50-50 collaboration. From the outset, the brief was interested in process as much as outcome, and so the two teams began by speculating on the future of work to come up with a "moonshot," defined by Google as an idea so far-fetched that striving for it inevitably catalyzes other innovations.

Jointly imagined, the Google flagship design is informed by ideas about user experience and access to daylight and views, as well as the anticipated need for flexibility to accommodate programmatic instability. A long span structure, "like a hangar rather than an office building"[11] that "draped" over a series of structural supports with clerestory windows dividing sections, emerged as the typology capable of meeting Google's aspirations. 20 Thomas Heatherwick has said he found inspiration in Bedouin tents and the nearby zeppelin hangar at Moffett Field. The designers also studied the visionary proposals for a dome over Manhattan put forth by Buckminster Fuller (1959), and Otto's cable net structures (1974). The idea of a long span shed structure that would serve as a permanent infrastructure overhead while also allowing for continuous transformation of the interior arrangements, including the configuration and elevation of the structure's floor slabs, appealed to Brin and Page. Early renderings depicted robotic gantries attached to the roof masts that could be used in the construction, and continual reconstruction, of a dynamic interior landscape. The terraced interior floor slabs, or "tables," break up the massive interior floor plate into smaller sections, or neighborhoods.

Working within the traditions of long span tensile structures, Google Bay View combines the convex global profile of a shallow dome with the multiple concave profiles of tensile structures. The subdivision of the extensive roof into 125-foot (38m) bays, connected with what the design team called "smile trusses," allows for the clerestory windows that separate opaque roof sections. 21 23 The smile trusses are essentially bowstring trusses that thicken at the center and correspond to moments produced by external loads. The concave geometry of the trusses combines the two curved arc segments,

11 Bjarke Ingels, ed., *Formgiving* (Köln: TASCHEN, 2020), p. 550.

23 Google Bay View under construction, with the catenary roof segments being installed between smile trusses.
24 Roof installation with compound curved roof sections supported by a tube-net structure.
25 Building cross-section showing the office floor plate perforated with openings to bring daylight to the lower level.

joining upper and lower roof sections while introducing daylight into the plan. 22 25

Spanning between major trusses, Google Bay View's roof sections are infilled with a triangulated steel-tube network (or tube-net), capable of tension and compression to resist seismic loads, that support a perforated metal deck. 24 The building's roof is clad with customized "dragon scale" photovoltaic panels that tessellate its curvature and contribute to the shared goal of producing all needed energy on site. Rainwater harvesting from the massive roof and the extensive installation of geothermal wells also contribute to its sustainability goals and LEED platinum rating.

The interiors of Bay View are vast and light-filled, unlike any workplace in the world. 27 The tallest columns at the center of the plan are 130 feet (40m) tall and the office partitions and conference rooms are dwarfed by the height of the space. The various "tabletop" floor plates are set at different elevations, stepping slightly from the center to the perimeter and connected with ramps. The footprint of one of Bay View's two buildings is approximately 350,000sf (32,000sm), or 8 acres (3.25 hectares) making it large enough to house six football fields. Even from the center of the deep floorplate plan there is abundant natural light and views out to the landscape beyond.

Seen from a distance, the "squomes" (square domes) of the Bay View campus appear mirage-like, with crests and ridges appearing like a new, artificial landscape. 26 The glazed gaps between roof panels and perimeter glazing that infill the steps in the canopy, convey a sense of continuity and porosity. Like Otto's Munich Stadium, the landscape seems to move under and through the lightweight tensile structure, while multiple curves rising and falling at once suggest the supple material and fabric forms of a suspended membrane in repose.

The catenary curves of both Dulles Airport and the Google Bay View buildings, are governed primarily by gravity loads. The self-weight of the structure and the length of its span is the main determinant of form for these low-rise buildings. For tall structures, gravity loads remain significant, but the lateral loads, of wind and earthquakes, become the the primary drivers for structural design.

VERTICAL AND LATERAL

At the origins of buildings that test the limits of engineering, the Eiffel Tower and the structural design of the Statue of Liberty, both the work of French structural engineer Gustave Eiffel, achieve unprecedented heights by taking full advantage

26 The Google Bay View buildings create a landform structure, appearing like the contours of an artificial landscape. The individual roof segments are joined by the clerestory geometry of the smile trusses.

27 The daylit interior landscape of the Google Bay View building with the tube-net structure overhead and the stepped floorplate below.

28 The structure of the Statue of Liberty consists of a symmetrical, tapered, cross-braced primary frame, and a secondary, irregular, and asymmetrical, frame to support the cladding.

29 The Statue of Liberty pre-assembled in France before being disassembled and shipped to New York.

of both geometry and industrially-produced iron components to counteract the powerful effects of the wind's lateral force. Eiffel's two towers — one draped in a thin copper skin to emulate the fabric folds of Lady Liberty's flowing gown, 29 the other a bare iron skeleton so offensive to the aesthetic sensibilities of Parisians that they petitioned to halt its construction — together mark a watershed moment for both architecture and engineering.[12]

Trained as a chemist in the École Central des Arts et Manufactures, one of Europe's first engineering schools, Eiffel worked in a machine and tool factory producing ironworks before he was hired to serve as project manager for the construction of a railway bridge in Bordeaux. The development of the railroad in France, and the necessity of creating near-level tracks for train locomotives, meant the construction of hundreds of tunnels and bridges. Eiffel developed new techniques for erecting long span bridges made of latticed iron girders supported by masonry piers, many of which would translate to the conditions and structural ambitions of both the Statue of Liberty and the Eiffel Tower.

The concept and design of the Statue of Liberty was the work of the French sculptor Frédéric Auguste Bartholdi. Well connected, Bartholdi traveled widely and witnessed the construction of the Suez Canal, one of the great feats of engineering of his time. Inspired by the colossal stone statues of the Egyptian pharaohs, he proposed an enormous statue that would stand at the entrance of the canal. Bartholdi's initial sketches for what he called "Progress Carrying the Light to Asia," depict a robed female figure with one arm raised to hold a torch that would act as a lighthouse. After the design was rejected, he reworked the idea for "Liberty Enlightening the World," that would be realized in a different harbor, New York, as a symbol of Franco-American friendship.

Gustave Eiffel was commissioned to design the interior structure for the Statue of Liberty after the original engineer, Eugène Viollet-le-Duc passed away in 1879. Eiffel's design, informed by his experience with lightweight trusses supporting railroad bridges, consisted of a tapered wrought iron pylon, 100 feet (30m) tall, subdivided into eight cross braced bays. 28 A series of secondary triangulated frames attach to the eight sections, each negotiating the varying gap between the central orthogonal core and the curvilinear contours of the statue's fabric folds. The structure for Lady Liberty's right arm consists of an asymmetrical truss of diminishing dimensions that attaches itself to the top three frames of the central core. The diagonal members of her outstretched arm

[12] Ulrich Pfammatter traces the co-evolution of the parallel disciplines of architecture and engineering at the moment when scientific, technical, and industrial practices were radically transforming both disciplines. Ulrich Pfammatter, *The Making of the Modern Architect and Engineer: The Origins and Development of a Scientific and Industrially Oriented Education* (Basel; Boston: Birkhäuser-Publishers for Architecture, 2000).

30 The colossal scale of the hand of the Statue of Liberty is evident in the photo taken in the foundry of Gaget and Gauthier.

31 A view of the interior of the Statue of Liberty, showing the undulating surface of copper sheet, supported by the contoured straps, and the intermediate steel members connecting the surface of the statue back to the center frame.

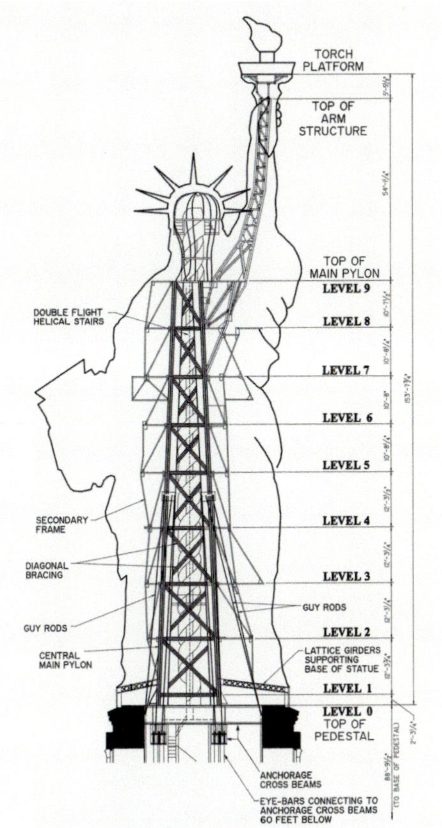

32 Section through the Statue of Liberty showing the central cross-braced frame and the truss supporting the outstretched arm.

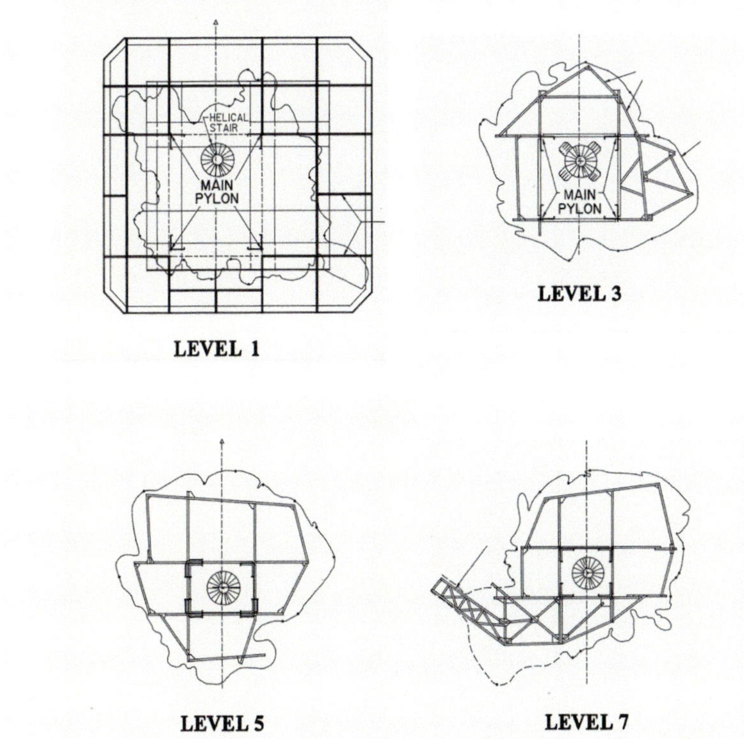

33 Plans cut through the Statue of Liberty showing the central core and the irregular structural framing to support the surface of the statue.

continue through her core as diagonal cross bracing members and extend to the opposite side, providing points to develop a backspan that counterbalances the loads from the cantilevered arm.[13] 32

The exterior skin of the Statue of Liberty is made up of hammered copper sheets 3/32 inch (2.5mm) thick. An inner webbing of iron strapwork follows the contours of the exterior material and provides connection points for the tertiary structure of flat iron bars to connect the copper skin back to the secondary trusswork. 31 33 The lightness of the skin and the intermediate structure also provides the copper some flexibility, key for deflection to absorb the impact of the wind in New York harbor.

Eiffel's drawings for the structure of the Statue of Liberty show the silhouette of the statue that helps to calculate the tributary area of wind load that the internal structure needs to resist. 34 The Statue of Liberty's profile diminishes as it rises, creating a smaller effective area at the very top of her outstretched arm. Meanwhile, wind speed increases with height as the wind encounters less friction with the ground. Eiffel's drawings include structural calculations based on graphical methods including a force polygon to project the anticipated loads. The profile of the Statue of Liberty and the curve of the assumed distribution of wind forces are plotted along an arc that approaches an ever-steeper vertical at the top of the statue and flares out toward the base. The line is labeled "*courbe des moments fléchissants dus au vent*," or "curve of bending moments due to wind." The drawn profile of the bending moments, mirrored, resembles the iconic profile of the future Eiffel Tower.

Soon after the designs for Eiffel's "300 Meter Tower" were presented to the public, a group of Parisian artists, writers, and intellectuals, penned a letter protesting its construction, claiming that it would "dishonor and disfigure" the city of Paris, calling it a "useless and monstrous...odious column of bolted metal."[14]

Eiffel, trained not as an architect but an engineer responded to his critics, "I believe that the tower will have its own beauty. The first principle of architectural beauty is that the essential lines of a construction be determined by a perfect appropriateness to its use. What was the main obstacle I had to overcome in designing the tower? Its resistance to wind. And I submit that the curves of its four piers as produced by our calculations, rising from an enormous base and narrowing toward the top, will give a great impression of strength and beauty."[15]

13 Joseph Harriss, *The Tallest Tower: Eiffel and the Belle Epoque* (Boston: Houghton Mifflin, 1975), p. 50.

14 Joseph Harriss, *The Tallest Tower*, pp. 20–21.

15 Joseph Harriss, *The Tallest Tower*, pp. 24–25.

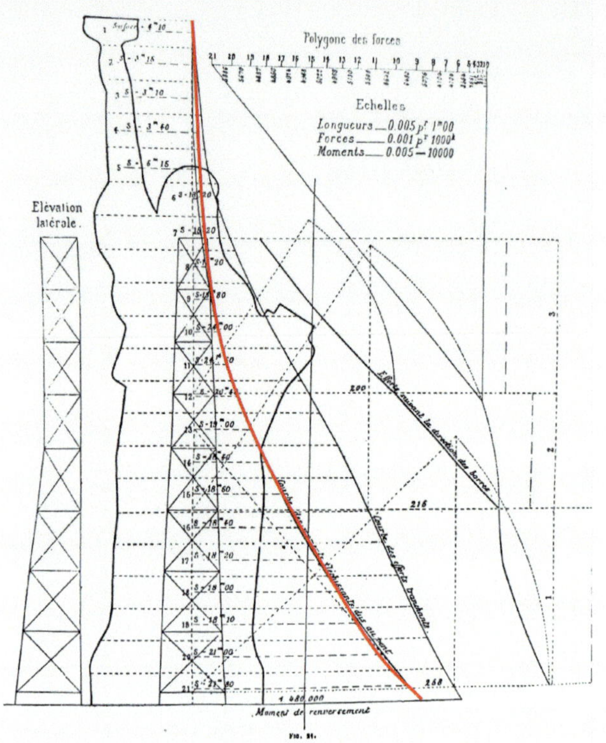

34 The structural diagram of the Statue of Liberty shows the central frame as a tapered and cross-braced trapezoid and the profile of the statue. The drawing, prepared by Gustave Eiffel, shows the structural forces acting on the Statue of Liberty. The curved line labeled, "*courbe des moments fléchissants dus au vent,*" or "curve of bending moments due to wind," visualizes the forces acting on the surface area of the statue. It also looks like the profile of the Eiffel Tower.

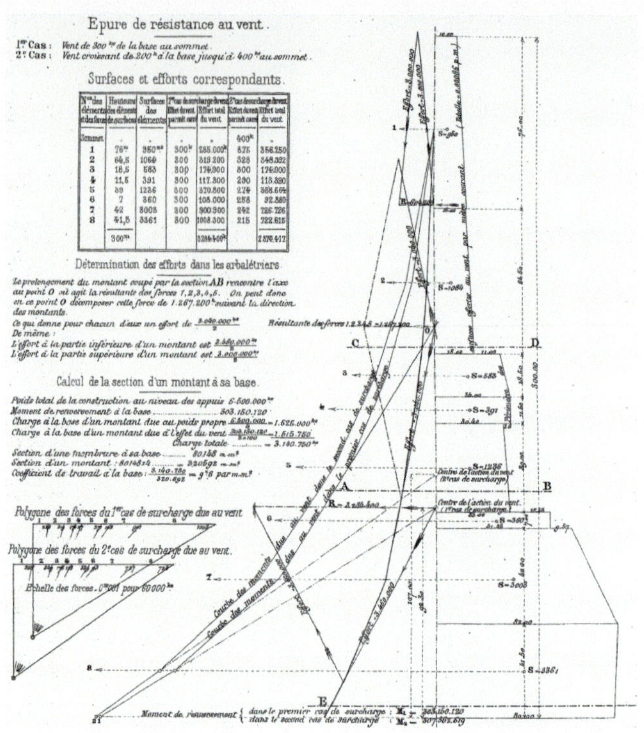

35 The structural diagram of the Eiffel Tower shows a similar graphic static calculation of wind forces. The curved geometry of the Eiffel Tower, with its splayed base and tapering profile, is based on the optimal form as determined by the resistance to wind forces.

The preliminary design for the Eiffel Tower was designed by two engineers working at Eiffel and Company, Maurice Koechlin, and Émile Nouguier. They were assisted by in-house architect Stephen Sauvestre. An early drawing by Koechlin shows the splayed geometry of the lattice-trussed piers with a concave profile. 37 Bridges distributed at roughly sixths along its height tie the four piers together horizontally. In their drawing, alongside the 300-meter pylon, are other monuments, drawn to scale on top of one another: the Cathedral of Notre-Dame, the Statue of Liberty, the Vendome Column, and the Arc de Triomphe. Combined, they are dwarfed by the height of the proposed tower.

On January 8, 1887, Eiffel received a signed contract to design, engineer, and build the Eiffel Tower on the Champs de Mars in the center of Paris for the 1889 Exposition Universelle. He had 24 months to realize the construction of the tallest structure ever conceived. Engineering challenges were not limited to the structure of the tower. Eiffel helped procure the raw materials, fabricate the structural components, and hire the labor force. He also helped develop innovative means for dewatering the foundations using a pressurized chamber to keep water out as workers built the footings. The legs of the tower were constructed separately before they could be connected in the air to form the first platform.[16] The cranes used to build the tower would also need to climb the tower as it rose.

A caricature of Eiffel published in *Punch* at the time of the exposition depicts him as a colossal tower, standing with his legs splayed apart. 36 The hem of his overcoat aligns with the first platform. His shoulders are smoothed out and his neck is attenuated so that his head is shown atop the tower's crown. A 1000-foot tower visualized as a human figure with legs splayed apart, is a useful analogy for understanding the forces acting on the tower. Resisting the force of the wind, we intuitively spread our feet apart to lower our center of gravity and spread the distance between supports.

Reaching well beyond new heights and lengths, the towering iron structures of the late 19th-century exceeded known precedents constructed of stone and wood. It is worth noting that much of the attraction for visitors to the Eiffel Tower remains not beholding its sight from the ground, but the aerial view from the panoramic observation deck. Standing at the base of the tower, visitors are captivated by its soaring height. On the observation deck, they become giants. The Eiffel tower was a phenomenal achievement of engineering and construction,

16 Like the Gateway Arch in Saint Louis, the construction cranes used to build the tower were also climbing the tower as it went up.

17 Joseph Harriss, *The Tallest Tower*, p. 63.

18 Lewis Mumford, "The Brooklyn Bridge," *American Mercury*, 23 (August 1931), p. 449.

36 A caricature of Eiffel, published in *Punch* magazine, shows the engineer assuming the stance of his eponymous tower.

37 The original sketch of the "300 Meter Tower" produced by Maurice Koechlin and Émile Nouguier in 1884, shows the tapering profile of the tower, and its height relative to other Parisian landmarks, stacked up on the right.

offering a new paradigm of engineering beauty. It was also a machine for altering perception, changing how Parisians saw their city, and gesturing toward other engineered futures.

The Eiffel Tower nearly doubled the height of the Washington Monument (1888) that had formerly held the title of world's tallest structure. The Washington Monument, constructed of masonry, stands 555 feet (169m) tall and is 55 feet (16.7) wide at its base. The Eiffel tower stands at 1000 feet (300m) and is 410 feet (125m) wide at its base. Eiffel's use of iron trusswork and splayed geometries leveraged the strength and ductility of the new building materials, while simultaneously shifting the form-finding paradigm of his time as it related to wind. The Washington Monument is an example of a monolithic structure. The Eiffel Tower breaks with monumental lineages and introduces a structural lattice made of assembled parts and designed to resist the forces of wind.

For Eiffel, form follows force. "The tower would stand because of Eiffel's grasp of a paradox: victory over wind is achieved not through brute force, but by reducing the supporting elements of an openwork structure until the wind has virtually nothing to resist it. Otherwise stated, the real strength of the Eiffel Tower is in its voids as much as its iron."[17]

[19] The Arc de Triomphe on the Champs Elysee has a span of 48 feet (15m) between piers. The arch between the legs of the Eiffel Tower span approximately 265 feet (81m). Joseph Harriss writes that Eiffel understood the psychological effects of the semi-circular arches on the public. "Those magnificent circular arches between its legs, for example, that most visitors assume support the tower, actually bear no load at all. With their stylized floral motif, their role is purely decorative – and perhaps psychological, Eiffel knowing that the public would "understand" a structure supported by arches but feel less at ease with one resting only on curving pylons." Joseph Harriss, *The Tallest Tower*, p. 92

In describing the Brooklyn Bridge (1883), historian and theorist Lewis Mumford admired the contrast of masonry piers and a delicate steel cable suspension structure, describing what he called the architecture of the future as "light, aerial, open to sunlight, an architecture of void rather than solids." [18] A description equally applicable to the Eiffel Tower. The triangulated open web trusses span distances far greater than the arcuated vaults of masonry construction. Eiffel arranged the iron members and planned for the voids between them, which allowed the trusses to be extremely thin and diaphanous.

Comparing Koechlin's initial sketches to the final constructed tower, the structural concept is the same, but the final design differs in its articulation. Koechlin's platforms span the piers in shallow arcuated profiles, while the built tower has a semi-circular arch. The addition of the arch is a gesture attributed to architect Stephen Sauvestre, who was tasked with making the tower more palatable to the jury and the public. Eiffel understood that the public's frame of reference for monuments and grand civic structures included arches and ornamental motifs.[19]

The addition of the non-load bearing arch to the profile of the Eiffel Tower is significant for several

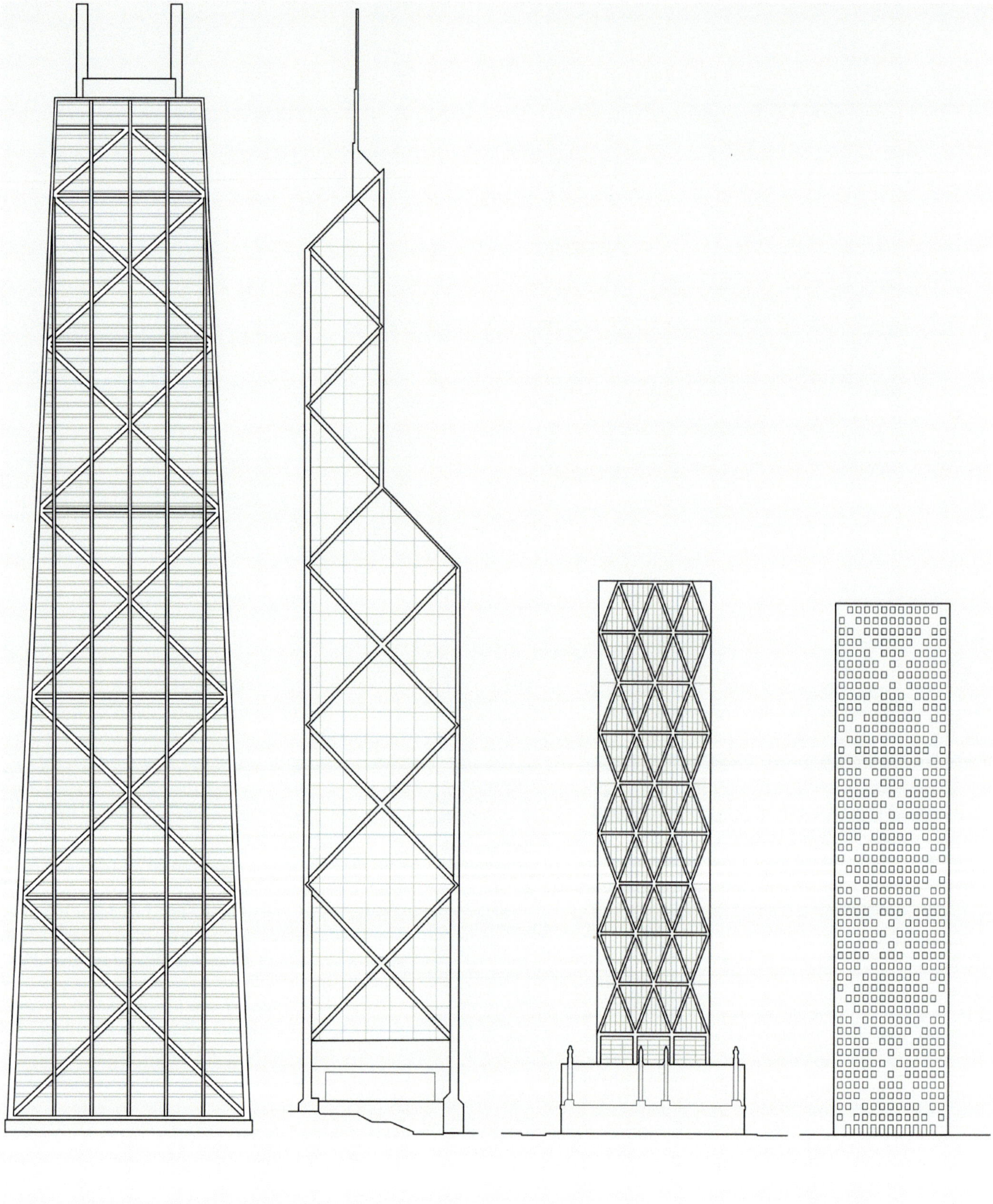

3.8 Elevations of significant exoskeletal buildings including John Hancock Center, Bank of China, Hearst Tower, and 780 3rd Avenue.

39 John Hancock Center (1970), designed by Bruce Graham and Fazlur Khan of Skidmore, Owings & Merrill (SOM). The distinctive tapered shaft and external cross-bracing give the building its signature look.

reasons. The tower was not simply an engineering feat, but also a psychological artifact. Its scale and materiality (or dematerialization) upset conventions and unsettled established norms. The additive arch was a familiarizing gesture, an acknowledgement of the strangeness of the Eiffel Tower as a new landmark in the city, its cultural life, and identity. The fact that the arch was added by an architect to a form derived by engineers speaks volumes about the disciplinary tensions between the two professions, as well as the relationship between large-scale civic projects and their public reception.

[20] In fact, the tapered steel frame core of the Statue of Liberty has an uncanny resemblance to the exoskeletal form of the John Hancock Center.

EXOSKELETON

The John Hancock Center in Chicago employs similar strategies to that of the Eiffel Tower and the interior structure of the Statue of Liberty.[20] 32 38 Skidmore, Owings & Merrill's (SOM) 100-story tower, completed in 1970, briefly captured the title of world's tallest building owing to its innovative trussed tube structural frame. The original design consisted of two towers, one residential and one office, with a shared podium occupying the entire block. The SOM team, led by architect Bruce Graham and structural engineer Fazlur Khan, proposed combining the two uses in a single tower with offices in the bottom half and residences in the top. The tapered form of the tower, like a truncated obelisk, accommodates the small residential floor plates above and the wider office floor plates below, it also corresponds to a form optimized to resist wind loads. The diminishing profile at the top limits the surface area subject to wind loading, while the splaying of the base efficiently distributes its loads.

Like Eiffel, Khan was a structural engineer, and he developed the unique trussed tube concept for the John Hancock Center only after testing several versions of perimeter framed-tube structural designs. The Hancock's perimeter columns are combined with diagonal members in the same plane to form a trussed tube. 40 The tower consists of five and a half X-braced tiers that make up the tapered form. The use of diagonal bracing members in the exterior plane of the building mass locates the structural members furthest apart, maximizing their efficiency, optimizing space, and producing the signature image of the building.

While the structure is placed at the perimeter of the building, the members are covered in fireproofing and clad with insulation and anodized aluminum panels that allow for differential deflections between the structure and envelope, limit thermal bridging, and thermally induced deflections. 41 The curtain

40 The John Hancock Center under construction, with two and a half of its five and half structural tiers completed.

41 Close up view of the cladding for the John Hancock Center during construction. The cross-braced structure is already clad, and the curtain wall infill is being installed.

wall system of bronze mullions and tinted glass infills the primary structural members yet remains independent of the structure. Initial concerns about the presence of the diagonals disturbing the views of residents seems to have been overcome by the acknowledgement that, today, residential units with the iconic diagonals are now considered more valuable than those without.

At the time, architecture critics decried the expression of the diagonalized facade of the John Hancock Center as "utilitarian," describing its overall mass as "grossly out of scale," and "inharmonious."[21] The conspicuous precence of the trussed structure on the facade classified the building as more a product of engineering than architecture.

David Billington, in *The Tower and the Bridge: The New Art of Structural Engineering* writes, "For critics, the Hancock is mere technology, the result of a purely utilitarian urge, whereas the works of the designer from the Bauhaus are legend to those that arose in the past in society that were not democratic. What critics sensed about the Eiffel Tower was that it was not architecture in any traditional sense. But the Hancock goes beyond the Eiffel Tower in that it directly symbolizes a new city life — a life that escapes the image of control."[22]

21 Carl W. Condit, *Chicago 1930–1970, Building, Planning, and Urban Technology* (Chicago, IL: The University of Chicago Press, 1974), p. 110.

22 David P. Billington, *The Tower and the Bridge: The New Art of Structural Engineering*, Princeton paperbacks, science/architecture (Princeton, NJ: Princeton Univ. Press, 1985), p. 262.

These examples highlight various moments where the necessity of structure and its articulation in built form challenged conventional understandings of traditional categories of building and infrastructure, or architecture and engineering. The idea that an articulated structure, like the cross bracing of the Hancock Tower, renders it utilitarian, relegating it to the realm of engineering, in many ways the counterpoint to the impetus for the non-structural arch added to the Eiffel Tower to make it appear more architectural. The decorative arch aims to domesticate the work of pure engineering, while the necessary X-bracing on the Hancock expands the definition of what had previously been understood as architecture.

DIAMONDS NOT CROSSES

The exoskeletal superstructure concept was further developed by I.M. Pei in his design for the Bank of China Tower in Hong Kong, completed in 1989. 38 42 The 70-story office tower diminishes in mass as it rises with triangular extruded quadrants terminated with oblique faces that slope away from the center, giving the composition the effect of a set of ascending crystalline forms. The building mass is clad in a taut glass skin with expressed metal panel

4.2 The Bank of China (1989), designed by I.M. Pei with Les Robertson, derives its form from the triangulated cross-braced exoskeleton of the perimeter truss design, resulting in a distinctive building profile as an expression of its structural demands.

43 The Bank of China under construction with two of its five and a half structural tiers completed.

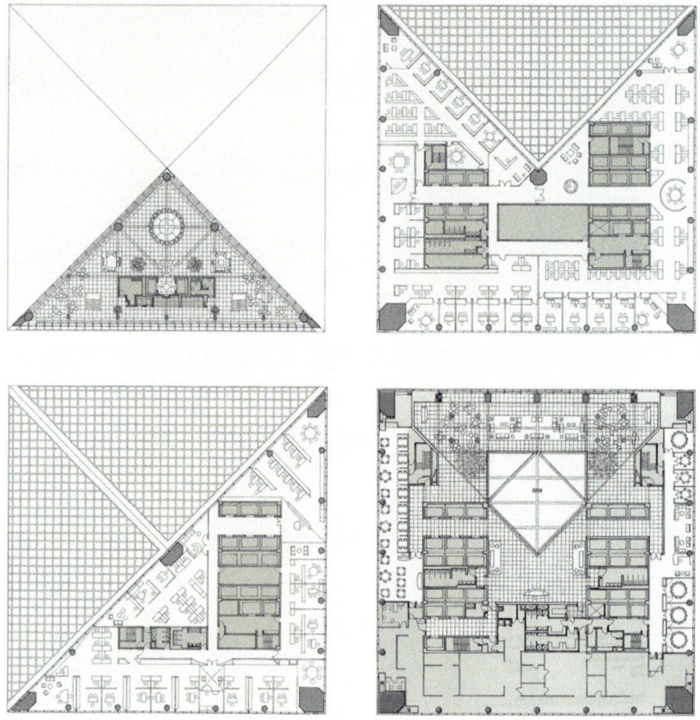

44 Plans for the Bank of China show the strategy of subdivision and setback as the tower rises. The lower levels are the full square floorplate. Mid-levels are half a square triangle. The upper levels are a one quarter triangle.

diagonals that translate the interior structure to the exterior.

The structural engineer for the Bank of China was Leslie Robertson, who had previously helped design the structure of the World Trade Center's Twin Towers in New York City. Robertson applied his experience with supertall towers and tube-in-tube structures to develop a new structural type tailored to Pei's sloped, stepped massing. Hong Kong is in a typhoon region where wind loading requirements are double those in New York and are four times the code-mandated earthquake loads in Los Angeles.[23] Robertson developed a large-scale, space-frame concept with diagonally trussed frames that sit inboard of the cladding and run parallel to the prismatic exterior volume. The offset centerlines of the space frames of the various extruded masses converge at the corners with composite steel and concrete "megacolumns." 43 44

The Bank of China client objected to the expressed X-bracing on superstitious grounds. The "x" mark was understood as the sign of a condemned man, and the diagonal expression on the facade was deemed inauspicious by the local Feng-shui master. Pei responded by removing the horizontals from the facade, transforming the "x" into a diamond shape, a symbol of prosperity.[24]

[23] Leslie E. Robertson, *The Structure of Design: An Engineer's Extraordinary Life in Architecture*, First American edition. (New York: The Monacelli Press, 2017), pp. 263–264.

[24] This story is recounted by Leslie Robertson in his book, *The Structure of Design*, p. 268.

The diagonal exoskeletal frame, also called a diagrid, has been applied to dozens of buildings following the Bank of China. Its efficiency in resisting lateral loads makes it particularly effective for tall buildings where the primary drivers of structure and cost are the effects of wind on the steel tonnage of the superstructure. Norman Foster utilized a multi-story diagrid for his 2006 Hearst Tower design in New York City. 38 45 The eight-story tiers of attenuated diamond structure enclose the small footprint of the rectangular tower. The building's "bird beak" corner articulation follows the triangulated trussed tube configuration, seemingly taking a cue from Pei's Bank of China where the massing runs parallel to the triangulated geometries of the mega trussed super structure. Foster used a similar exoskeletal diagrid for his Gherkin Tower in London, completed in 2004. For the Gherkin, officially known as 30 St Mary Axe, the diagrid is suppressed within the rotational figure of the bullet-shaped massing, with each diamond module containing four floors.

Rem Koolhaas's Seattle Central Library (2004), and the Prada store in Aoyama, Tokyo, by Herzog & de Meuron (2003), also employs the diagrid, though at a much smaller scale. The Seattle Central Library is wrapped in a steel diagrid corresponding to the scale of a facade module, thereby eliminating the

45 The Hearst Tower (2008), designed by Foster + Partners, employs an exoskeleton in an attenuated diagrid configuration. The fine grain cross bracings stack four and a half structural modules high, while each package of large-scale triangulation is eight stories high.

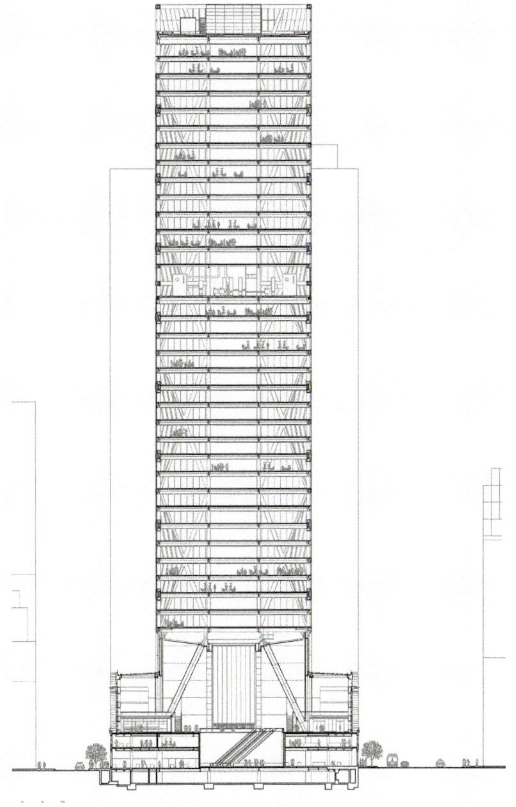

46 Section through the Hearst Tower, showing the eight story packages and the tower in relation to its Art Deco podium base.

aluminum curtain wall system by mounting the glass directly to the steel frame. The Prada store's diagrid aligns with the slope of the stairs on the interior, and extends to form diamond-shaped fitting rooms, all while remaining within the sky exposure plane of the Tokyo zoning envelope.

In contrast to the highly expressive diagrid projects of the early 2000s, SOM's 780 3rd Ave in New York presents an enclosed or subsumed diagrid tower. 38 47 Completed in 1983, the 49-story office building appears to be something of an exercise in restraint as an extruded rectangular prism with a punched window, red granite facade. The regular grid of punched windows is marked by the infilled windows corresponding to a four-tier perimeter, trussed tube exoskeleton. The 12-story tiers are legible in the solidity of the exterior wall, a subtle way of expressing the diagonal members within.

The punched window facade articulation of 780 3rd Ave came at a moment when the principles of modern architecture were called into question by the historicist discourses of postmodernism. A desire to recuperate premodern elements of architecture, including scaling devices like punched windows, led architects to reconsider the all-glass skins popularized by buildings like SOM's Lever House and Mies van der Rohe's Seagram Building.

The paradox in 780 3rd Ave is that it employs a highly efficient frame and skin approach, with its tiered and X-braced perimeter tube structure, yet its expression is suppressed behind a stone-clad curtain wall. The tension between the human-scaled punched windows and the multi-story braced frame is cleverly negotiated with the infilled window modules, so the truss is legible in the omission of windows within the grid.

These projects reveal an inherent incongruity between the multi-part structural principle of triangulation of the multi-story truss and tendencies toward orthogonal building components. The basic difference between a solid wide flange beam and a multi-component truss is greater efficiency. Individual members spread further apart, creating increased stiffness, while saving material. The introduction of the exoskeletal trussed tube prompted the articulation of oblique structural elements with orthogonal panels in their facade articulation and allowed for different degrees of expression (John Hancock Center) or suppression (780 3rd Ave) of structural systems.

Interestingly, the structural tube projects that preceded the John Hancock Center relied on non-triangulated perimeter tube structures, with dense orthogonal grids of columns and beams

47 780 3rd Avenue (1984), designed by SOM, employs an exoskeletal structure, but it does so discreetly, with the cross-bracing hidden in the pattern of punched windows.

48 One Shell Plaza (1971) in Houston was designed by Bruce Graham and Fazlur Khan of SOM, with a concrete tube-in-tube design. The slight ripple in the exterior wall hints at the increased stiffness where the perimeter tube connects to the tube of the center core.

that essentially formed a perforated tube. SOM structural engineer Fazlur Khan's early tube-in-tube designs involved orthogonal concrete framed perimeter structures like the 38-story Brunswick Building and the 43-story DeWitt-Chestnut Building in Chicago. In each case, the dense grid of the exterior tube works together with the shear walls of the service core to resist the lateral forces of the wind loads. The modular regularity of the gridded facades obscures the concentrations of load that vary across the facades. Khan's design for the 50-story One Shell Plaza (1971) in Houston, Texas, for example, negotiates the orthogonal tube-in-tube structure and the punched windows, while also acknowledging load concentrations where the tubes connect, namely at the outer corners of the service core.[25] 48 50 The facade of One Shell Plaza swells very subtly at the quarter points of its broad faces and roughly at third points on its short faces. The exterior wall piers in these areas project further than is typical, creating the effect of a bulge or ripple in plan. The line of glass remains in plane, allowing the variable depth of the structure to read against the datum of the glazing.

One Shell Plaza is clad in Roman travertine, and the tinted glass windows recede graphically against the light-colored frame. From many angles, the tower appears like any other boxy office building in the Houston skyline, and viewed frontally, the effect of the varied depth of the facade members is almost imperceptible. Approaching the tower from Louisiana Street or Smith Street, however, the slight flare of the base is apparent, and the oblique viewing angles across the long faces of the tower reveal the ripple in the arrangement of orthogonal members. 49 The dense grid of columns transitions to larger stone-clad piers at the building core's corners and where its deeper columns are most pronounced. 50

Architect Bruce Graham likens the swell of the facade of One Shell Plaza to a "splayed skirt," akin to the flared base of the Monadnock Building in Chicago. The "undulation effect" translates the variability of internal forces to the exterior face of the building.[26] It telegraphs the demand for greater stiffness associated with the inner core to the exterior facade, creating an implied curvature or swelling in the areas needing most stiffness.

The subtle bulging of the piers at One Shell Plaza, an expression of the internal loads translating into a slight convexity on the facade, appearing almost unnoticed to the casual observer, might recall the phenomenon of *entasis*. Theories of classical architecture associate subtle deviations like entasis as a surplus of form, intended not only as

25 Fazlur Khan tested the subtle expression of structural forces with the thickened structural members in Two Shell Plaza, in Houston, and the Marine Midland Bank, in Rochester, New York.

26 Mir M. Ali and Fazlur Rahman Khan, *Art of the Skyscraper: The Genius of Fazlur Khan*, 1. publ. (New York, NY: Rizzoli, 2001), pp. 88–90.

50 The plan of One Shell Plaza (1971) shows the tube-in-tube design with the shear walls of the center core bracketing elevator shafts and the perimeter structure of densely spaced columns. The swelling in the building facade corresponds to the connections between inner tube and outer tube.

49 The bulge in the stone-clad facade is visible when viewed on the oblique. The deeper members hint at the forces acting on the tower, roughly aligned with the corners of the center core.

expressions of force, but also as giving architecture the qualities of human flesh that tenses or relaxes its musculature in the presence of force. Reading the undulations of One Shell Plaza as a kind of contemporary entasis situates the building not only in the realm of reason and structural rationalism, but also in the domain of gesture and embodiment.

From its introduction in the late 1960s, the exoskeletal perimeter tube has enabled the construction of extreme structures by responding to the demands of lateral forces with increasingly larger triangulated structures, each with varying degrees of visible expression. The turning point, with its superlative height, was the John Hancock Center, where its articulated perimeter structure led architectural historian Robert Mark to claim that "because of its giant scale, its unusually light structure, and the openness of its base, this landmark building might well be characterized as the Bourges Cathedral of our time."[27]

When completed in 1973, the Twin Towers of the World Trade Center displaced the John Hancock Center as world's tallest building. 51 Standing at a height of 1,368 feet (417 m) for the north tower and 1,362 feet (415.1 m) for the south tower. The square prisms of the Twin Towers extruded 110 stories from their street-level plaza without setbacks. Looking up from the base of the towers was a sublime experience, where architecture appeared at the scale of landscape. The vertical grain of the towers' dense exterior walls was undifferentiated for most of its height, except for the mechanical penthouse at the top and the triple-height lobby at its base, where bundles of three piers converged into a single pier with branched one-to-three column transfer panels dubbed "tree columns."

Minoru Yamasaki, the architect, working with structural engineer Leslie Robertson, of the firm Worthington, Skilling, Helle & Jackson, designed the dense facades of the towers as perimeter tubes, consisting of closely spaced columns, 3-foot 4-inches (1 m) on center with narrow windows infilling the spaces between them, creating a perimeter tube structure. The structure resembled Fazlur Khan's design for the John Hancock Center; however Robertson did not employ diagonal bracing, choosing instead to express all the members with a strict vertical grain connected by substantial horizontal spandrel beams. The structural design for the twin towers achieved their extreme height by remaining extremely light. The tube organization placed all of the structure at the building's perimeter, for greatest efficiency, while floor slabs consisting of

[27] Robert Mark, *Light, Wind, and Structure: The Mystery of the Master Builders*, McGraw-Hill Edition, New liberal arts series (New York, NY: McGraw-Hill Pub. Co, 1990), p. 177.

5 1 The Twin Towers of the World Trade Center (1973), photographed during construction, reveal their lightness in the apparent transparency of the shafts.

5 2 Simmons Hall (2004), designed by Steven Holl with Guy Nordernson, appears like a massive, gridded structure of indeterminate scale. Seen from the oblique, along Vassar Street, the colored metal panels hint at the structural performance encoded in the colors.

open web trusses connected the steel frame core to the outer structure. Notably, Robertson pioneered the use of viscoelastic dampers at the connection between the truss and the column to limit the amount of sway in the upper stories.[28]

Robertson's structural design for the towers was largely developed using early computer software and punch cards to calculate the stresses in the steel and specify the members. With Yamasaki's insistence that the towers extend straight up from top to bottom, Robertson designed the columns as box-beam cross sections uniform in dimension, despite the different structural demands of their location. The buildings appear to adhere to a strict dimensional regularity thanks to the employment of 14 different strengths of steel and plate thickness of the perimeter tube columns to resist wind loads and distribute stresses across the facades.

Following the destruction of the towers on September 11, 2001, structural engineer and Princeton University professor Guy Nordenson curated an exhibition at the Museum of Modern Art called *Tall Buildings*. Nordenson worked with students to produce color-coded elevations of the two towers, revealing the varied types of steel columns used. 5 6

28 Even with the dampers, the top floors of the twin towers were known for their uncomfortable movements and at the time of their destruction on September 11, 2001, the topmost floors were unleased.

29 Guy Nordenson, *Guy Nordenson: Reading Structures: 39 Projects and Built Works, 1983–2011* (Zurich: Lars Müller Publishers, 2016), p. 26.

The distribution of 14 variations of steel strength displays not only Robertson's design choices, but also the effect of the variation of wind pressure around the building due to the site-specific wind directionality, and the wind's passage through the city relative to the environment, and each other. These different conditions are effectively suppressed and sublimated in the relentless visual uniformity of the actual structure and architecture.[29]

In 2004, Nordenson applied a similar logic to the Simmons Hall dormitory at MIT, designed in collaboration with architect Steven Holl. Holl's concept was to express the porosity of the building with a uniform-gridded facade of aluminum-clad precast concrete. 5 3 Like the World Trade Center facades, the building's face dimensions of both vertical columns and horizontal beams are held constant. At Simmons, the face dimension of the members is 12 inches (30cm) in both directions, in all conditions. The depth of the dormitory's concrete panels varies slightly from 12 inches (30cm) deep at its base to 10 inches (25cm) at its top. Deep recesses in the massing create challenges for the perimeter walls, as they need to transfer loads, much like vierendeel trusses, around the openings in the structures. Like Robertson's subtle steel

53 Simmons Hall (2004), seen frontally from across the sports fields, looks monochromatic and silvery gray.

variations at the World Trade Center, Nordenson varied the building's steel reinforcing bars to make necessary adjustments in rigidity. 54 Despite the variable rebar strategy, panels around substantial openings required additional reinforcement. Nordenson and his team infilled the cells to increase their stiffness, creating what they called "freckles" that scattered across the facade, clustering around large openings and cantilevers. Their approach recalls the solid cells that correspond to the diagonal members of SOM's 780 3rd Ave, making legible areas of higher stress and shear loads.

When viewed from across the MIT athletic fields, Simmons Hall appears like a silvery Cartesian megastructure, notched selectively to form a giant three-dimensional puzzle. Approaching the building along Vassar Street, however, the building flashes red, orange, and yellow in its window returns. 52 The different colors represent the size of the rebar in the precast panels within, turning the building facade into an abstract diagram of forces. Nordenson recalls a moment where he saw a color-coded elevation diagram produced by engineer, Amy Schreiber. 57 Nordenson showed the diagram to Holl, who found it fascinating, and adopted it for the facade colors.[30]

[30] Guy Nordenson, *Guy Nordenson: Reading Structures*, p. 309.

Both in reference to history, and the emergence of data visualization in architecture, Nordenson and Holl's Simmons Hall designed a color-coded facade that broadcasts information about the structural performance of the concrete facade panels, allowing — or perhaps, inviting — viewers to read the forces acting on the structure.

DESIGNED INSTABILITY

OMA's competition-winning scheme for the China Central Television (CCTV) Headquarters in Beijing was a radical rethinking of the skyscraper. 1 Instead of a single extruded shaft, the architects proposed a contorted monumental arch made up of six sections in a continuous and bent rectangular tube. The building's two vertical sections tilt toward each other and both rise approximately 50 stories. The two horizontal portions create L-shaped, orthogonal tubes. One rests on the ground, while the other connects the two vertical shafts in the air, forming a massive, bent lintel beam. The building mass implies an enormous cubic void formed atop a landscaped plinth. Unlike the Gateway Arch in Saint Louis, CCTV does not ultimately provide an image of balance or stability. Rather, the bent, trabeated structure appears particularly unstable, straining to maintain

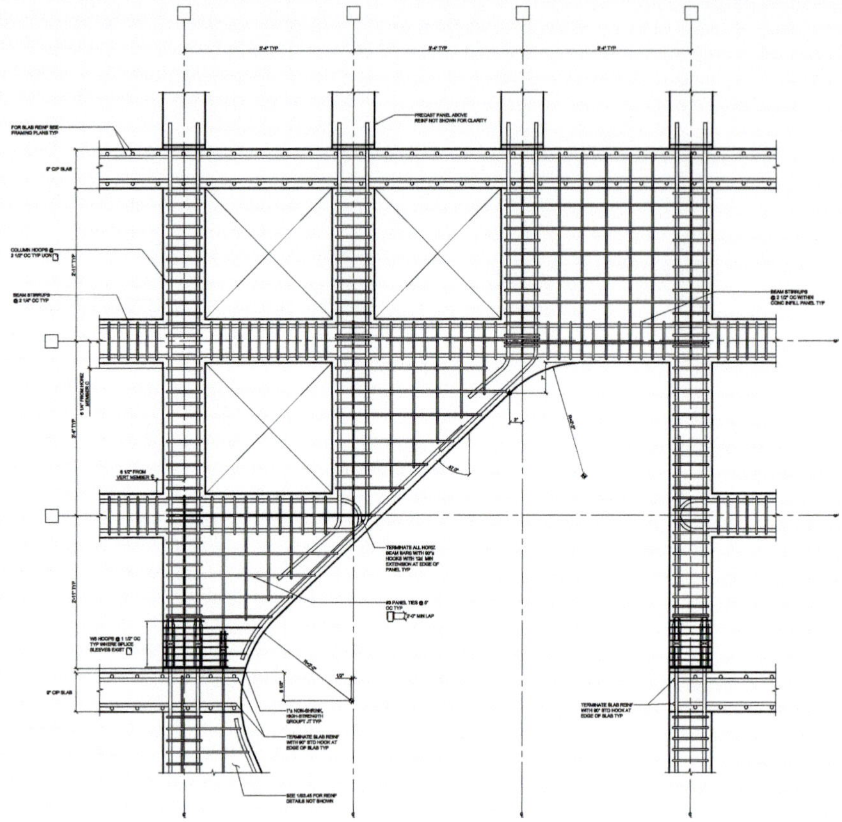

5 4 Rebar drawing of an atypical precast panel for Simmons Hall with the exceptional "void" window and the filled in cells adjacent to the opening to manage the stress concentrations.

its upright posture through the counterbalancing of its lower back span wings and the stiff "moment of connection" of its cantilevers. 5 9 Architectural writer William Millard describes CCTV's unstable stance as its most salient feature — a quality he refers to as "visually intuitive instability."[31]

Defying gravity, the bent bar loop challenges common expectations that monumentally scaled arches are meant to provide an image of stability and continuity. Koolhaas partnered with Arup structural engineer Cecil Balmond to design the seemingly precarious structure. Balmond, a long time OMA collaborator, describes the public image of the engineer as "a hard person of science — who makes the impossible work." Balmond's design for CCTV builds on the triangulated perimeter tube structure and modifies it for the irregular cantilever and its resulting torsion, creating a visual diagram of forces through the arrangement and density of structural members. 5 8

The structural design of CCTV's diagrid mesh outlines the algorithmic process that drove the pattern: give all diagonals the same properties, analyze the mesh for seismic loads, double the mesh density where members are overstressed, delete mesh members where they are underutilized, reanalyze the structure, and repeat until the mesh pattern meets the design criteria. Unfolded elevations of the building's multiple faces show the color-coded mesh patterns derived through this analysis with the axial forces in the perimeter members. 6 1 OMA principal Ole Scheeren describes the process of deriving the facade design: "What you see is the actual steel structure, derived analytically. The beauty of it is that the seemingly irrational image is completely rational. We wanted this highly unusual structure to be discernible on the exterior."[32]

From the exterior, the CCTV building appears to be a fully glazed unitized curtain wall between the diagonal members of an irregular diagrid frame. The building's steel perimeter trusses are inboard of the curtain wall, and so the diagrid expressed on the exterior is a two-and-a-half-foot wide (800mm), inset profile that mirrors the structural member behind it. 6 0 Much like the Bank of China building and Hearst Tower, CCTV's continuity creates a structure entirely enclosed by its facade. Critically, the gap between the real structure and the facade profile keeps base building movement from translating to the facade members. Is the exterior profile decorative? The steel section of the profile serves a double function as blast protection for the primary structure, so technically, the profile plays a structural role that shifts its valence from decorative to performative.

31 William B. Millard, "Dissecting the Iconic Exosymbiont: the CCTV Headquarters, Beijing, as Built Organism," in Rem Koolhaas and Brendan McGetrick, eds., *Content: Triumph of Realization* (Köln: Taschen, 2004), p. 490.

32 Sandra Hofmeister and OMA — Office for Metropolitan Architecture, eds., *OMA — Architektur Und Baudetails: = OMA — Architecture and Construction Details*, 1. Auflage., Edition DETAIL (München: Edition DETAIL, 2019), p. 131.

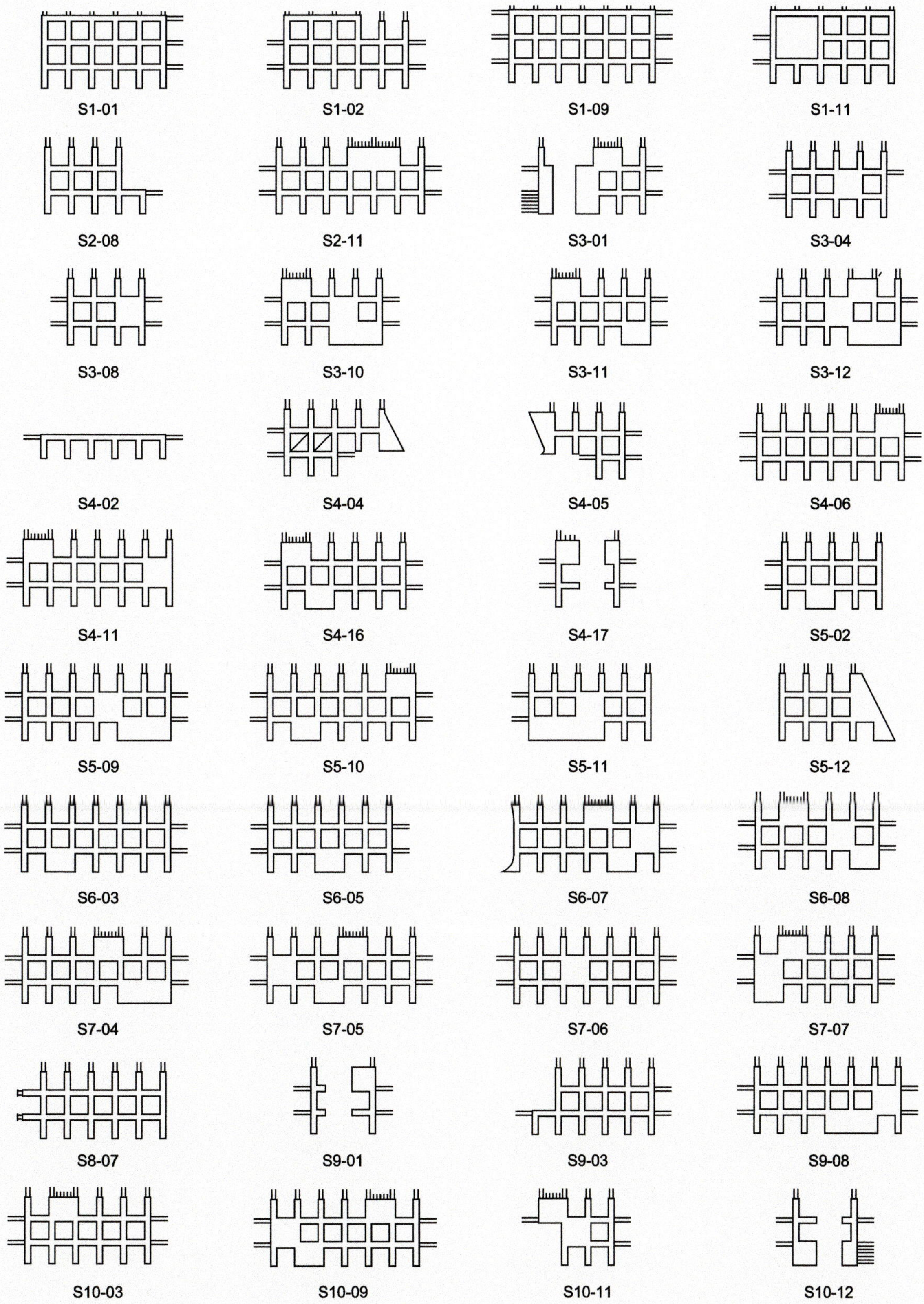

5.5 An array of precast concrete panel types from Simmons Hall including typical and atypical modules that respond to local conditions and load concentrations.

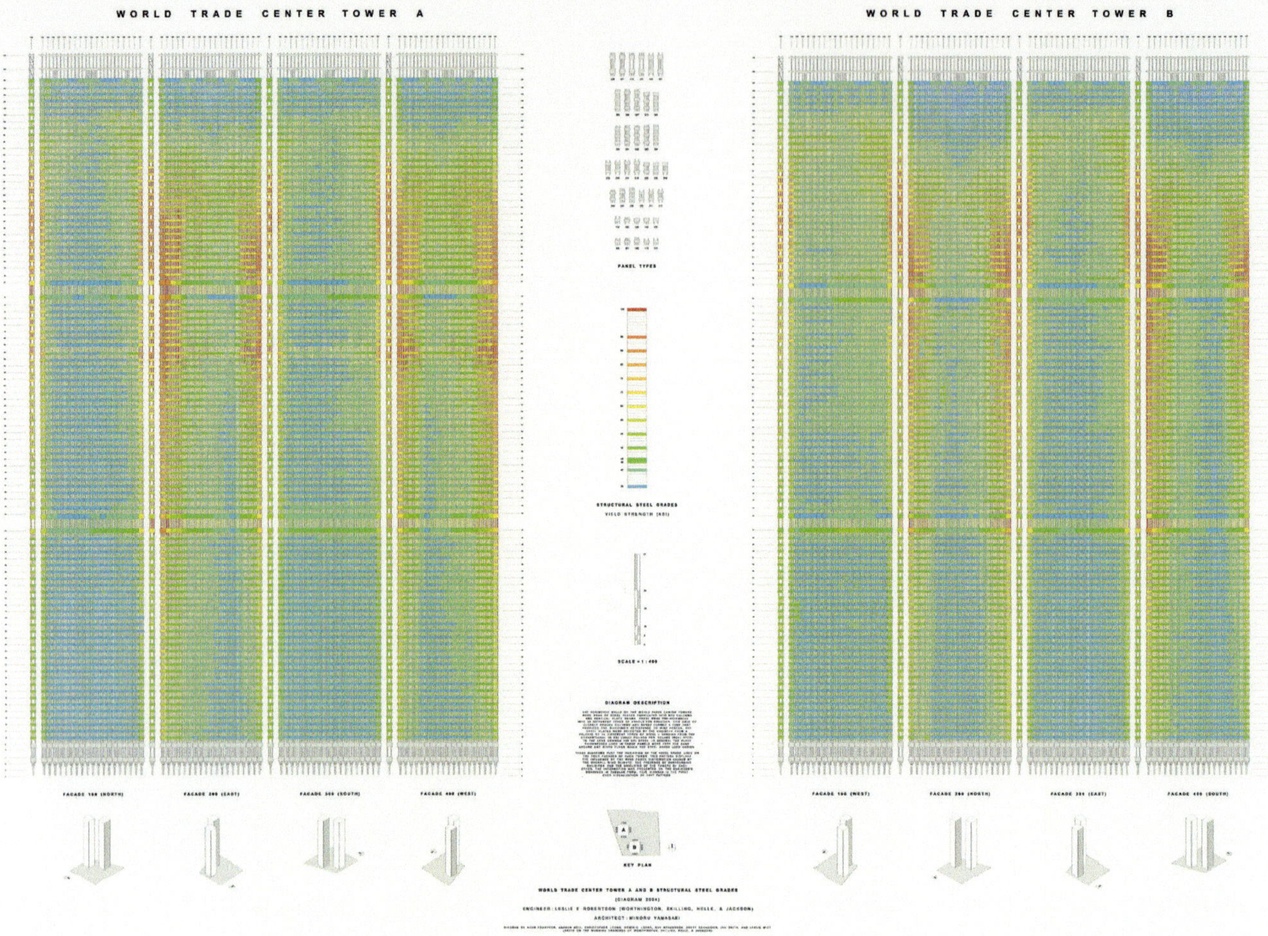

56 Diagram of the World Trade Center structural steel grades produced by Guy Nordenson and Brett Schneider. While the structural design of the Twin Towers relied on the perimeter box columns to be of equal dimension, the structural engineer, Leslie Robertson, varied the steel strength to respond to necessary wind loads. The color-coded diagram shows the grades of steel across the unrolled elevations of the two towers.

57 Diagram of the structural reinforcement in the precast concrete panels on Simmons Hall. The color spectrum corresponds to smaller #5 bars to larger #9 bars. The color-coded diagram served as the inspiration for the color-coded facade panels of the building envelope.

58 The structural design of the CCTV Headquarters, designed by OMA with Cecil Balmond of Arup, consists of a diagrid mega-tube that is varied to respond to increased stresses, creating areas of higher density of members where demand is higher, and sparser members where less are needed. The translation of structural loads into a visible pattern of members on the facade makes the invisible internal forces legible.

The uniformity of structural expression in in the World Trade Center speaks to the architectural attitudes of its time, where a legible ordering principle masks subtle differences in structural demands, but also speaks to the low resolution of the data utilized in the earliest days of computer aided structural analysis. The color- and density-coded patterns of Simmons Hall and CCTV reveal shifting attitudes about complexity, and an interest in translating information into patterns — making data decorative. 56 57 58

Regardless of the impetus, or the era, the articulation of a facade as a pattern or texture distinct from its load-bearing frame is inherited from the principles of the modern movement. If the facade of a building transfers no loads, then the fenestration is freed of load path continuities. This was an important revelation for modernists, and has since remained a primary consideration for architects, engineers, and builders. As we saw in Chapter 6, Le Corbusier's ribbon window, for example, is an early departure from the vertically aligned punched windows of the past, and critically, an emphatic gesture of structural discontinuity. Fazlur Khan's perimeter tube structure at John Hancock Center is a compelling and efficient resolution of large lateral forces in extreme buildings, but also represents a return of structure to the surface of architecture, and with it, the expressive capacity of architectural form to become the transmitter of otherwise invisible interior forces. With the perimeter tube, form again follows force.

33 Shohei Shigematsu et al., eds., *OMA NY: Search Term* (New York, NY: Rizzoli, 2021), between images 2199 and 2200. The Search Term book is indexed to images, and not pages.

STRUCTURAL PATTERNS

OMA's Faena Cultural Forum in Miami Beach, Florida, a performance and event venue completed in 2016, highlights some of the problematics of the contemporary architectural facade as an expressive structural surface. 62 The four-story, white-painted concrete structure was designed by Shohei Shigematsu of OMA New York. The building's two-part composition consists of a conjoined cube and cylinder. The cube houses a black box theater, and the cylinder a domed event space. The exterior walls are marked by an irregular pattern of windows produced as an overlay of curves and vertical piers, resulting in 350 unique punched windows. The unusual pattern, described by the architects as "resembling an organic form, like a palm tree or a seashell,"[33] is the result of their iterative design process that tested dozens of options before arriving at this particular pattern.

59 Construction photo of the CCTV Headquarters shows the density of steel diagrid elements concentrating at the "elbow" near the cantilevered wing. The curtain wall cladding at the base of the tower highlights the internal structure with a recessed channel on the facade.

The Faena forum's cylindrical massing is a response to its context, producing a convex form in dialog with the concave curve of the Saxony Hotel across Collins Avenue. Its curvilinear geometry finds many precedents in the city's art deco district, well known for its filleted structures found up and down Miami Beach. OMA New York's book *Search Term*, filled with model documentation and diagrams, provides a glimpse into the design process. Two thumbnail diagrams, images 2314 and 2315,❸❹ underscore the importance of a memorable arrival experience in Miami Beach. Rather than add a canopy to a building, Shigematsu proposed a subtraction, removing a wedge-shaped slice from the cylindrical volume facing Collins Avenue, and creating a dramatic cantilever that invites visitors under the dramatically sloped soffit.

The undercut is complemented by a suspended grand stair and a shallow reflecting pool. The building's 45-foot (13.7m) cantilever dramatizes the approach and entry of its patrons, creating the appearance that the cylinder above is balancing precariously on its edge. A series of facade studies conducted by OMA explored strategies for cladding the cylinder in different patterns and textures, including dot screens and compositions of punched windows. The breakthrough occurred when the design strategy for the facade responded to its structural needs. Two thumbnail images are appended with a simple note: "We decided to show a different facade approach expressing the tension lines needed to support the cantilevered cylindrical volume."❸❺ The first image shows a series of thin bands reaching diagonally across the cylinder to form a mesh of straps tying its teetering volume back to the stable cube. The second image shows a series of thicker curves springing from the ground and wrapping around the cylinder like a series of supporting arches.

❸❹ Shohei Shigematsu, *OMA NY: Search Term*, images 2314 and 2315.

❸❺ Shohei Shigematsu, *OMA NY: Search Term*, image 2346.

The final design of the Faena Forum's facade is a combination of three patterns: A delicate splay of thin catenary straps drawn upward as if in tension, a series of thicker fanned arches springing from the support points at its base as if in compression, and a thin overlay of vertical members described as a "hurricane grid," presumably to further subdivide glazed areas. 67 The decision to express the facade structure's lines of force, in tension and compression, made the unique overlaid pattern of load-transferring surfaces and perforated voids possible. The result is a graphic overlay of stripes and curves, some of which converge at the support points, some of which form segmented eye-shaped patterns on the prow of the building.

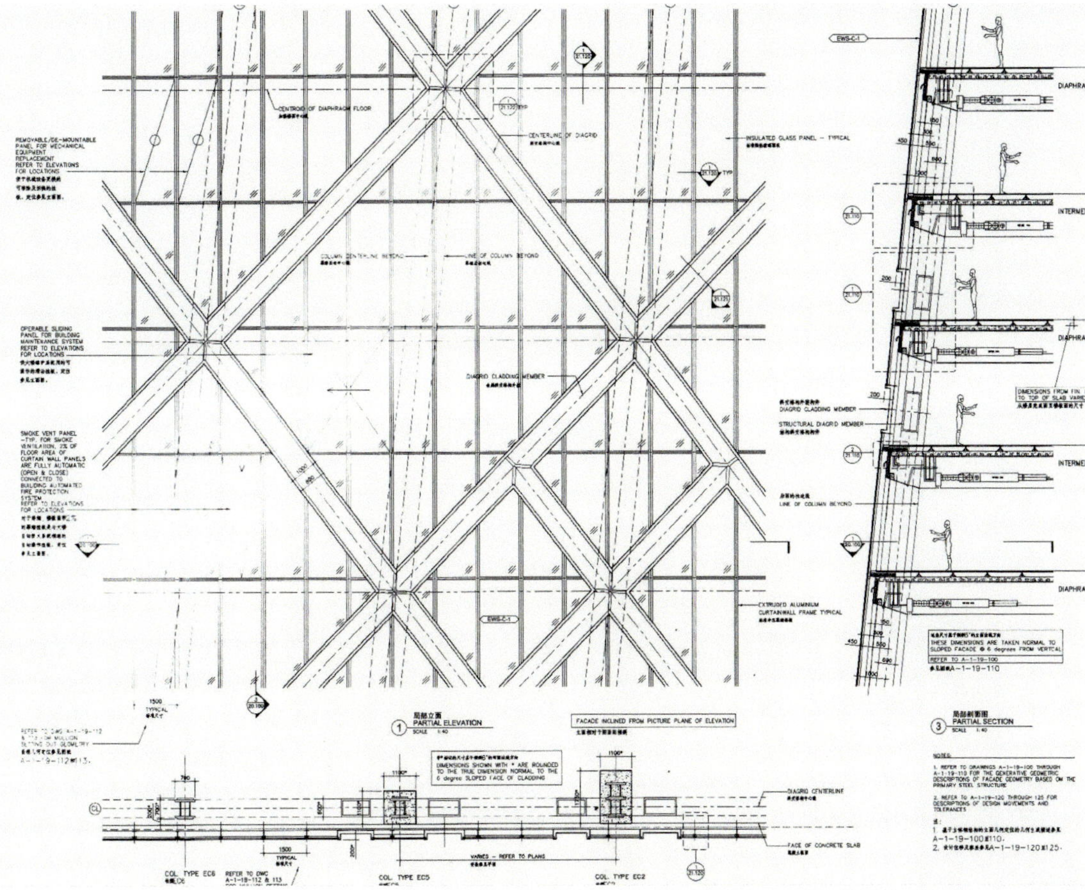

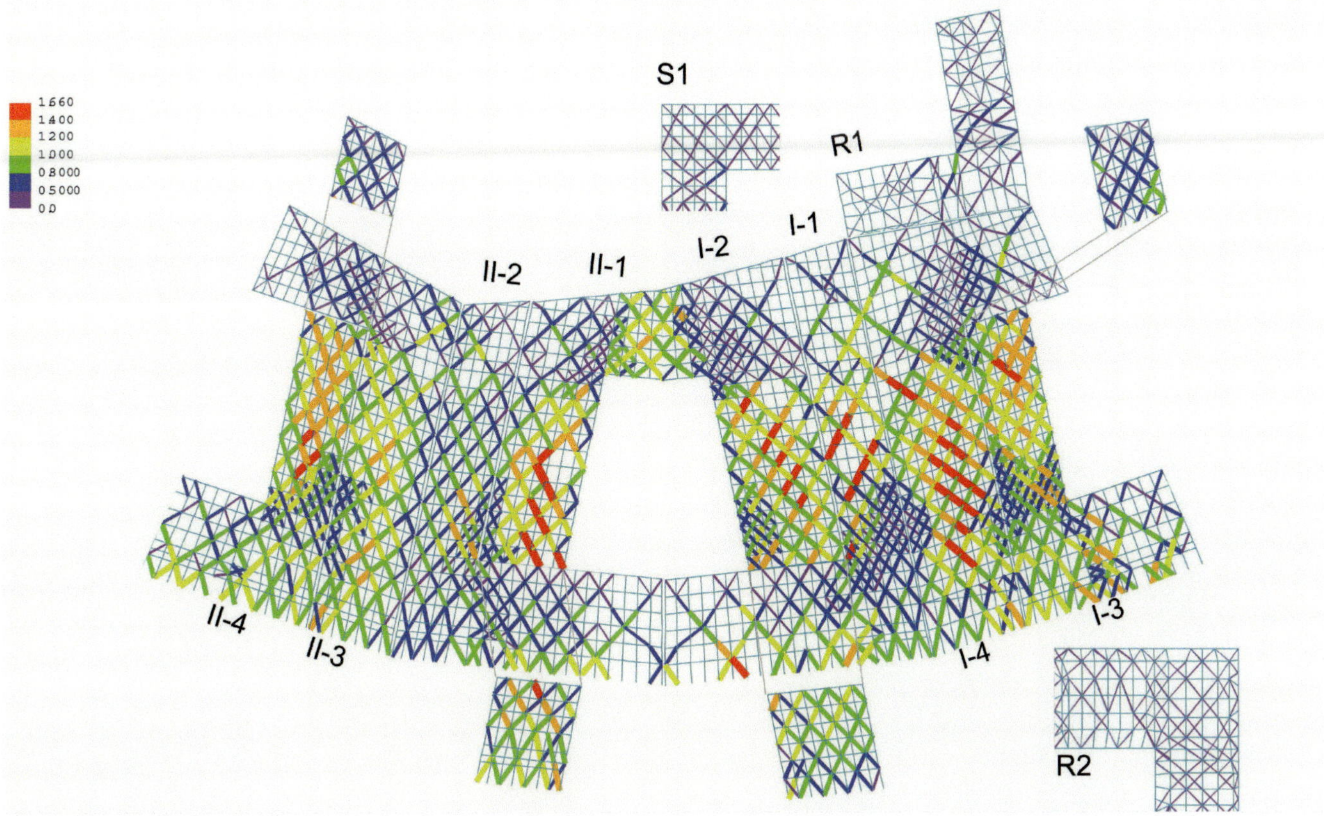

60 Curtain wall drawings for CCTV Headquarters facade showing the partial elevation with articulated recesses corresponding to structure within. The partial section shows the diagrid members clearly inboard and separate from the curtain wall that wraps the building and bypasses the slabs.

61 The "unrolled" elevations of the CCTV Headquarters color-coded to show structural demand capacity ratios, the comparison between forces acting on the element (demand) and the maximum load it can withstand before failure (capacity). The diagram of structural performance translated into the visual expression of the building facade.

62 The Faena Cultural Forum (2016) in Miami, was designed by Shohei Shigematsu of OMA NY to create a new venue for cultural events on Miami Beach. The exterior construction of the forum combines structural forces in compression and tension to give the building a unique patterned facade articulation.

63 Uniquely shaped window blanks used in the casting of the concrete shell.
64 The window blanks being installed between the rebar cages.

65 Internal formwork constructed to define the interior face of the concrete shell.
66 During the process of removing formwork and the window blanks, the effect of a thick perforated shell is apparent.

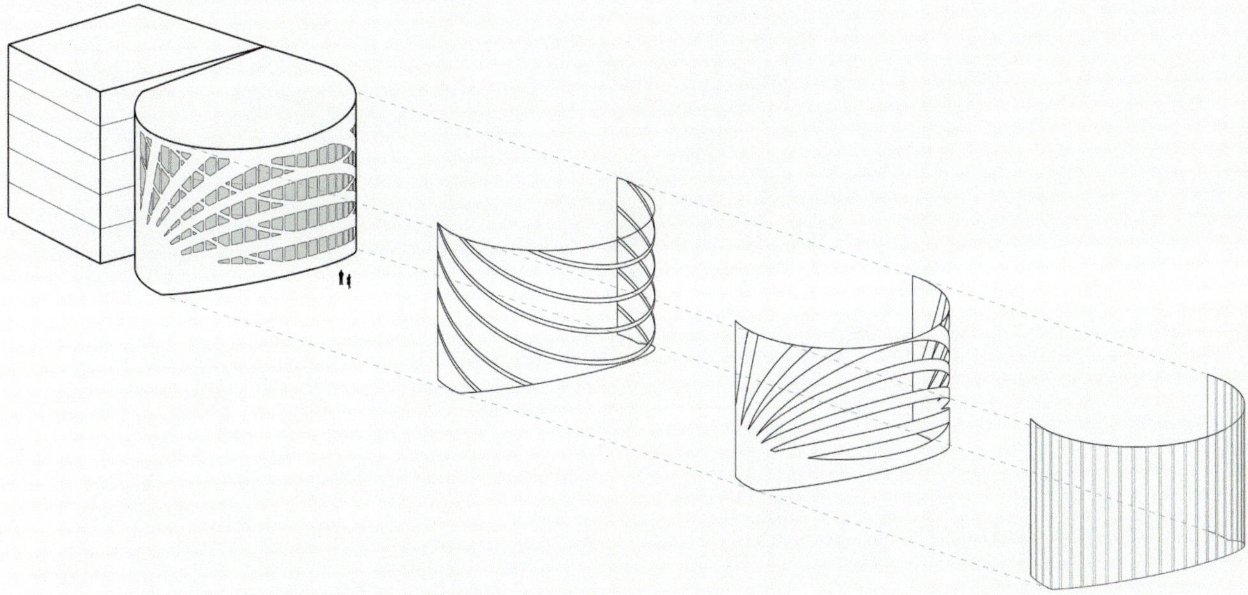

67 The facade design of the Faena Cultural Forum (2016) is derived from the overlaying of strap-like loops in tension (left), radial arches in compression (middle), and a vertical pattern intended to subdivide glass sizes for hurricane resistance (right). The composite pattern combines all three.

The Faena Cultural Forum was constructed with site-cast concrete as a load-bearing structural shell. 05 The building's 350 unique windows are set into 350 unique openings in the site-cast concrete wall. Specific forms allowed the window installer to create an accurate fit with the concrete once cast. 63 From the interior, the deep pattern of windows perforates the exterior wall, creating irregular areas of light and shadow and making its depth altogether more legible. 66 The overall effect of the window pattern is a uniquely diaphanous facade that is illuminated like a lantern at night. What appears as a pattern of palm leaves, or a seashell, is in fact a rigorous pattern of tension catenaries and compression arches. The composite expression of the forum's facade is representative of contemporary attitudes toward structural expression and novel form-making. The combination of constraints — structural cantilever, column-free interior, dramatic arrival sequence, site-cast concrete construction — all combined to make the irregular, crisscrossed catenary facade appear obvious and almost inevitable.

What is at stake with the selection of projects discussed in this chapter is the wide variability across modes of articulating force with form — specifically, the process of translation that transforms structural concepts and configurations into architectural gestures. Translation is rarely one-to-one, or strictly functional, but rather, like the principle of entasis, highly subject to conceptual negotiations between efficiency and expressivity.

Curvature and convexity, trussing and triangulation, regular and irregular patterns, the variation of material density, for example, all demonstrate the complexity of translation and complicate assumptions of a simplistic opposition between structure and decoration. Architects, engineers, and builders employ strategies of material disposition and organization. The architectures they construct demonstrate more than simply functional concerns or aesthetic sensibilities. Buildings that articulate force with form have the potential to shape and challenge popular preconceptions about the hard distinctions between the disciplines of architecture and engineering.

Though distinct in time and inclination, the buildings and structures included in this chapter bring structural design to the surface, and with it, the expressive capacity of architecture to relay what otherwise remains invisible — the variability of force. As such, these structures, colossal statues, soaring vaults, suspended surfaces, and densely gridded facades, train us how to see how the flows of forces and rouse a contemporary tectonic imagination.

1 The Climatron (1960) in Saint Louis, designed by MIT graduates Joseph Murphy and Eugene Makey, is a prototype for a "closed world" with a geodesic dome structure and an interior microclimate of a tropical rainforest.

Chapter 8 Climate and Conditioning

In Paolo Bacigalupi's 2015 cli-fi (climate fiction) novel, *The Water Knife*, the near-future world is plagued by drought, and water is the ultimate commodity.[1] Climate catastrophe has transformed the Southwestern United States into a lawless territory of gang warfare and eco-terrorism. The protagonist, Angel Velasquez, is a former gang member turned infrastructure saboteur working for the Southern Nevada Water Authority. Velasquez's company controls access to a water monopoly and has built a network of architectural megastructures, climate enclaves, where the wealthy one percent occupy lush interior landscapes and the other 99 percent fight for survival in the parched landscapes of a post-apocalyptic suburbia among the ruins of what used to be Phoenix.

In the opening scene, Velasquez and Braxton, a water authority lawyer, rendezvous in a lush, shaded interior landscape where water and plantings create a unique microclimate — an interiorized world where "sunlight filtered down from above, dappling bamboo and rain trees, illuminating tropical birds and casting pocket-mirror flashes on mossy koi ponds."[2] Bacigalupi refers to these fully enclosed spaces as "arcologies. "Within them, the temperature and humidity are carefully controlled, and the air is purified to filter out dust and smoke. He provides the specs: "Earthship sustainable arcology. Vertical farm, integrated housing, eighty-five percent water recycling."[3] In this speculative world, arcologies are megastructures — cities within cities, where the resource-rich live their lives separately from the collapsing cityscapes around them.

Bacigalupi borrows the name arcology from the visionary Italian American architect Paolo Soleri, who developed the concept of a synthesis of architecture and ecology and coined the term to describe his utopian model for a democratic society organized around principles of density, resource conservation, and communal living. Conceived in the 1960s, Soleri's vision for an urban alternative to consumerist sprawl was a high-density concentric megastructure embedded in the earth like a shipwrecked starship. A former student of Frank Lloyd Wright at Taliesin, Soleri imagined a futuristic network of intricate concentric structures housing thousands of people. Soleri's drawings, published in the 1973 book *Arcology, The City in the Image of Man*, map out 30 possible variations of arcology structures.[4] 2

In addition to the publication of his visionary drawings, Soleri also founded the experimental community of Arcosanti and embarked on the construction of an arcology. Arcosanti is located 60 miles north of Phoenix and consists of dozens of

[1] Climate fiction is a sub-genre of Science Fiction, including J.G. Ballard's *Drowned World*, and Kim Stanley Robinson's *The Ministry for the Future.*

[2] Paolo Bacigalupi, *The Water Knife,* First edition (New York: Alfred A. Knopf, 2015), p. 4.

[3] Paolo Bacigalupi, p. 57.

[4] Paolo Soleri, ed., *Arcology: The City in the Image of Man,* 2. print. paperback ed. (Cambridge, MA: MIT Press, 1977). Soleri's has been revised after his daughter, Daniela Soleri, accused her father of sexual abuse when she was a teenager. She states: "His work deserves recognition. But I believe its value should never negate his faults."

2 Paolo Soleri's fantastical drawing of a future city—an arcology—a combination of architecture and ecology.

low-rise structures sited at the top of a rocky hilltop. 3 Most of the buildings are partially embedded in the ground, drawing on vernacular building techniques of subterranean Native American structures like the kiva of the Pueblo peoples. Arcosanti's site plan consists of a series of clustered south-facing semicircular geometries. Discrete buildings house sleeping quarters, communal kitchens, dining rooms, and stepped outdoor common areas. Terraces are designed for gardens and greenhouses to support agricultural self-sufficiency.

Like Soleri's visionary drawings, Arcosanti relies on circular and radial geometries. Circular windows punctuate the residential buildings, and half-round arches span communal spaces. The bases of large, unfinished arches jut out from the foundations, anticipating future construction. 4 The intricate ensemble of arches, domes, and landform structures currently house about 50 people but was planned to grow into a small city designed for a population of 5,000.

Constructed of local stone and concrete, the thermal mass of the concrete walls absorbs heat during the day and reradiates it in the evening, when temperatures are cooler. Soleri's designs rely on several thermodynamic principles, such as the apse effect of shading the high-angle summer sun and inviting the low-angle winter sun. Greenhouses trap the warm air, while tunnels distribute it to other structures. Soleri also planned for convection towers that would employ the chimney effect to draw hot air out of the tops of tall structures.

Arcosanti initially broke ground in 1970. Since then, thousands of volunteers have contributed to its construction, but only a small fragment of the utopian megastructure has been completed. Today, the site's partially built arches, ribbing, and domed structures evoke the ruins of a previous era. Yet many of its principles of high-density building in harmony with its climate context, including passive design and natural ventilation, are highly relevant to architecture in the age of climate crisis.

PREMODERN TO MODERN

Ecologically conscious architecture has evolved with changing attitudes around technological progress and heightened awareness of how much buildings use materials, energy, and natural resources. Many premodern architectures employed passive practices of sun shading, thermal mass, and natural ventilation to make buildings comfortable in all seasons. Traditional Chinese architecture incorporated geomantic principles, such as Feng shui, that consist

3 The Arcosanti complex outside of Phoenix, Arizona, has been under construction for 50 years.

4 A speculative model of the future buildout of Arcosanti. The buildings in grey are the portions already completed.

of building practices that are in harmony with natural phenomena. Siting buildings on south-facing slopes, orienting them toward prevailing winds, and ensuring access to water, are all sound environmental design principles encoded within geomatic practices. The construction practices of the Pueblo peoples of the American Southwest were also carefully tuned to the arid climates and local ecosystems. Pueblo buildings consisted of thick adobe construction with small windows and large thermal mass, organized in clusters to reduce the amount of surface exposure. Around the world and across cultures, premodern builders arranged buildings, oriented openings, and employed overhangs and awnings to temper site-specific conditions and maximize the comfort of the inhabitants.

The emergence of modern architecture and city planning encouraged the use of new materials and building technologies, in many cases overtook passive practices, replacing them with new material systems and their associated technologies. With modern architecture, flat roofs and taut windows replaced pitched roofs and shading eaves, and mechanical systems were relied on instead of awnings and shutters. Modern mechanical systems were employed to provide temperature controls and thermal comfort, enabling a new architectural vocabulary independent of place or climate. An early advocate for mechanical conditioning, Le Corbusier was quick to embrace air conditioning and glazed facades, both novel technologies in his time. In 1930, he wrote polemically about a universal interior atmosphere, a *respiration exacte*, promising an ideal temperature of 18°C anywhere in the world.

"Every country builds houses for its own climate. At this moment of general diffusion, of international scientific techniques, I propose: only one house for all countries, the house of exact breathing. I draw the floors, and I set up the factory for the production of exact air. It is a small establishment, a few small spaces; I produce air at 18°C humidified according to seasonal needs. With a ventilator, I blow this air through carefully laid-out conduits. Means of expansion for the air have been invented to avoid drafts. Air comes out. This system of 18°C will be our arterial system. I have laid out a system of veins...The Russian house, the Parisian, at Suez or in Buenos Aires, the luxury liner crossing the Equator will be hermetically sealed. In winter it is warm inside, in summer cool, which means that at all times there is clean air inside at exactly 18°."[5]

Le Corbusier's 1929 design for the Cité de Refuge in Paris applied his theory of a mechanically conditioned building with a hermetically sealed

[5] Le Corbusier and Tim Benton, *Precisions on the Present State of Architecture and City Planning*, Reprint of the original American edition, MIT Press, 1991. (Zurich: Park Books, 2015). pp 64–66. Struck by Le Corbusier's emphatic 18 degrees everywhere claim, I published an essay entitled, "Thermal Space; 18 degrees inside/ 26 degrees outside," in *HK Lab 2: An Exploration of Hong Kong Interior Spaces*, (Hong Kong: Map Book Publishers, 2005). pp. 236–246.

5 The Cité de Refuge, in Paris, France, shortly after completion in 1933. Le Corbusier designed the building with a hermetically sealed glass facade and an "innovative" air conditioning system. The south-facing glass wall proved impossible to condition.

6 After the war, Le Corbusier renovated the Cité de Refuge and added an exterior sunscreen on the south-facing facade as a passive solar control strategy.

all-glass facade. 5 The building consists of a double-loaded bar in an east-west orientation, with its principal facade facing south. Le Corbusier originally proposed a double-glass facade that he called the *mur neutralisant*, a mechanically ventilated cavity that would serve as a thermal buffer between the indoor and outdoor temperatures. Ultimately, this element of his design proved too expensive, and the building was constructed with a single sealed glass facade.

As constructed, the Cité de Refuge was unbearable in the summer when the building's mechanical systems could not keep up with solar heat gain, and the sealed curtain wall made natural ventilation impossible. After the Second World War, Le Corbusier redesigned the south facade to incorporate a *brise-soleil* as exterior sunscreen. 6 The new concrete lattice mounted to the exterior was designed to limit the amount of solar radiation reaching the glass facade.[6] In his *The Architecture of the Well-Tempered Environment* (1969), architectural historian Reyner Banham recounts the shortcomings of Cité de Refuge as a result of Le Corbusier's "obstinate environmental misapprehensions." [7]

Lessons learned from Cité de Refuge and the undeniable necessity of external sunshades, or brise-soleil, became one of Le Corbusier's signature techniques. Architectural historian Daniel Barber has argued that brise-soleil is a kind of architectural media, an interface between the inhabitants and the larger urban and climate context. Barber's *Modern Architecture and Climate: Design before Air Conditioning* (2020) elaborates on the history of modern architecture and the evolution of environmental thinking before and after air conditioning. Describing the operable sun shading system on Le Corbusier's 1931 apartment building proposed for Barcelona, he writes, "The building facade is both a screen on which to watch environmental change and an industrial-material system from which to produce it."[8] He goes on to describe the his research agenda as, "a history of the facade as media and of the brise-soleil as a cultural technique — as a mediating device, selectively blocking the sun, and as a media device, rendering visible specific cultural relationships to climate patterns, as those relations and patterns change over time."[9]

Le Corbusier's design for Casa Curutchet in La Plata, Argentina, was built decades later, in 1954 and included a brise-soleil on its north, equator-facing facade. 7 The house contained both the doctor's medical practice and a private residence. Le Corbusier's scheme cleverly separates the two uses

6 Reyner Banham includes a recounting of the Cité de Refuge in his 1969 history of environmental design Reyner Banham, *The Architecture of the Well-Tempered Environment*, 2nd ed. (Chicago, IL: University of Chicago Press, 1984), pp. 155–158.

7 Reyner Banham, *The Architecture of the Well-Tempered Environment*, p. 156.

8 Daniel A. Barber, *Modern Architecture and Climate: Design before Air Conditioning* (Princeton, NJ: Princeton University Press, 2020), p. 9.

9 Daniel A. Barber, *Modern Architecture and Climate*, p. 10.

7 The Casa Curutchet (1953), in La Plata, Argentina, was designed by Le Corbusier with a *brise-soleil*, which soon became a signature design motif for a climate responsive architecture.

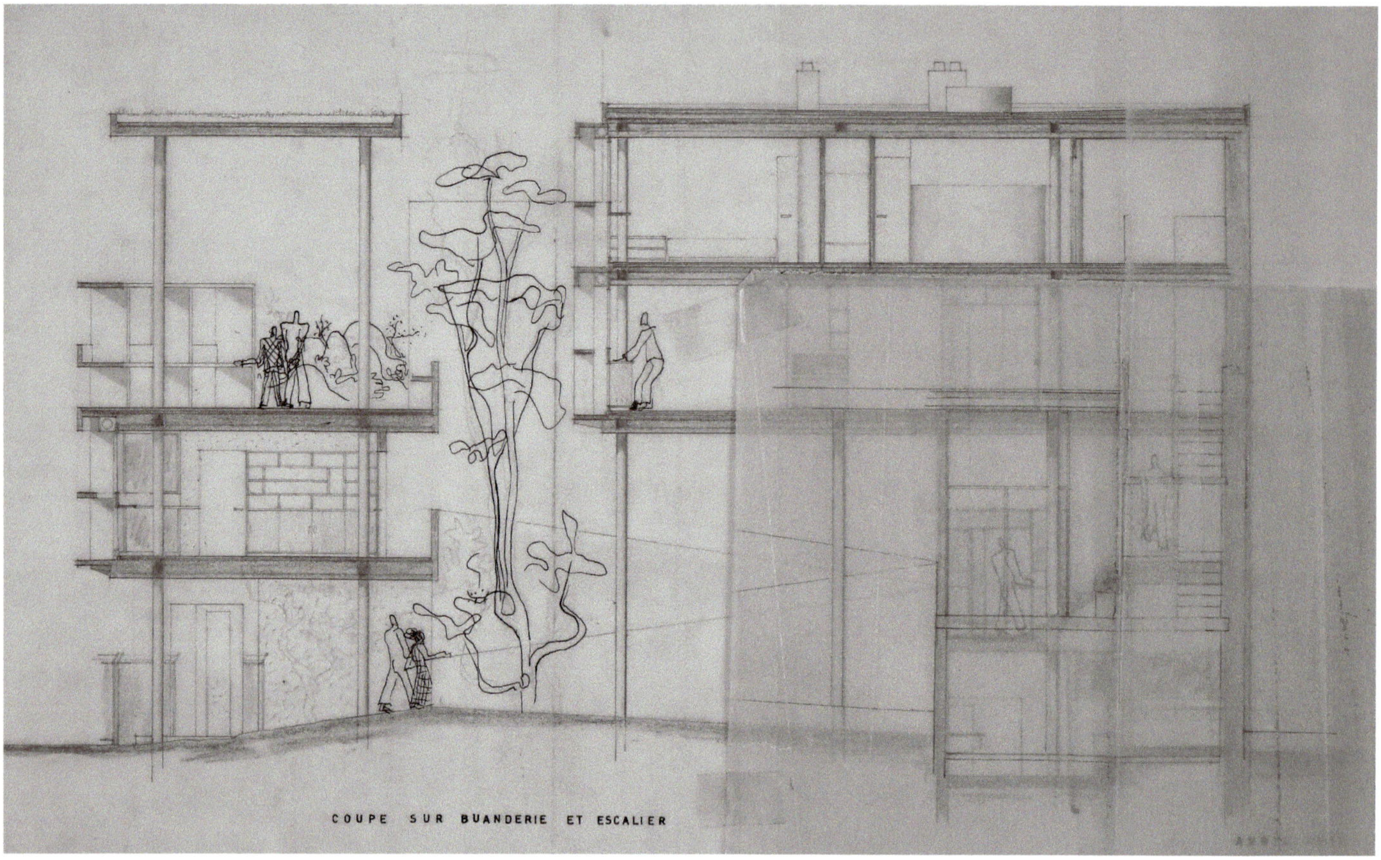

8 The north-south section of the Casa Curutchet shows the *brise-soleil* employed on the north-facing facades, screening the doctor's medical office as well portions of the main house.

within the narrow site and then stitches them back together with a switchback ramp. 8 The doctor's office fronts the public park along 53rd Avenue, while the residential apartment occupies the site's interior. Nestled between two traditional buildings with neoclassical details and punched windows, the delicate concrete framework of the screened facade broadcasts its modernity.

The brise-soleil at Casa Curutchet conveys not only the precision and logic of modern architecture, but also of science and modern medicine. The design of the lightweight concrete facade, including its height, depth, and the spacing of its vertical and horizontal members, are calibrated to their latitude in the southern hemisphere as well as their northwest orientation. In Le Corbusier's studio, his team included Bernhard Hoesli and Roger Aujame. In Buenos Aires he collaborated with Amancio Williams. Le Corbusier and his team developed a rigorous scientific method to align the building's facade with its orientation, altitude, and azimuth.

In 1949, when Le Corbusier was designing the Casa Curutchet, he had already worked extensively around the world. He had traveled to the USSR, North Africa, and Argentina and had proposed urban plans for Buenos Aires, Algiers, and Bogotá. In 1936, Le Corbusier had acted as an advisor on the Ministry of Education and Health building in Rio de Janeiro. 10 The design is distinguished by a dense brise-soleil grid on the north facade (facing the equatorial sun) and a taut glass skin on the south facade. In 1946, when asked to contribute to the design for the United Nations Headquarters, Le Corbusier advocated for a sun shading system for the west face of the Secretariat Building which was not adopted. Instead the Secretariat relies on tinted glass, that regularly overheats in the summer months.

Barber suggests that "the brise-soleil is the moment when modern architecture becomes global." Shading for comfort, a practical consideration as much as a deliberate design decision, intersects the central principles of modern architecture: human hygiene, health, comfort, and well-being with new materials and their articulation. In many ways, Le Corbusier's practice signaled the emergence of a global practice of architecture, where an architect in Paris was commissioned to design buildings in the Americas, North Africa, and South Asia. 11 The development of the brise-soleil as a globally applicable architectural device, calibrated to any site and orientation, and the tailoring of architecture to the latitude, elevation and exposure, serves to precisely locate the work of architecture on the curved surface of the earth.

10 Lucio Costa was the architect of the Ministry of Education and Health. Le Corbusier acted as design consultant.

11 The emergence of a global practice, and the rise of the "starchitect" is the subject for another book, but it is intersting to note the interrelatedness of the global practice with the emergence of an envrionmental awareness as articulated through architectural projects and the articulation of orientation-specific facades.

9 Victor and Aladar Olgyay, who pioneered the idea of bioclimatic design, photographed in 1956 with their physical model climate simulator, the Thermoheliodon.

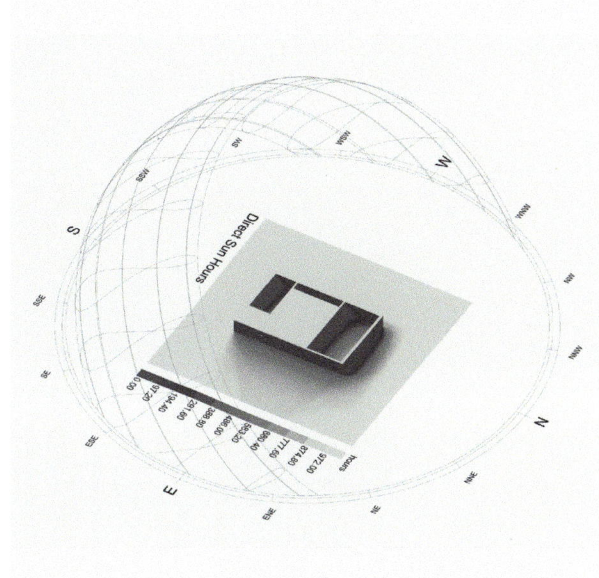

10 Solar simulation software that digitally reproduces some of the conditions of the Thermoheliodon is now widely available and part of many contemporary design processes.

ANALOG SIMULATIONS

Contemporary architects practicing today in our globalized, climate-changing world are taking up environmental design with new capacities for calibration and computation. Solar simulations and daylighting analysis software have made environmental analysis highly accessible. Software plugins for 3D modeling applications like Rhino are a standard analytical tool for contemporary design practice. Daniel Barber's cataloging of environmental design methods and techniques before the advent of mechanical conditioning, and, later, of sophisticated computational tools, seems to be asking: What was lost when we suddenly switched on a mechanically conditioned and standardized architectural interior atmosphere?

Two protagonists in this story are Hungarian American architects Victor and Aladar Olgyay, twin brothers and professors at Princeton University who worked to develop what we now know as environmental design. Their work relied on an analog machine, the Thermoheliodon, to test solar conditions, airflow, and humidity on scaled models. [9] The analog simulator consisted of a plexiglass dome over a circular table, four feet in diameter (1.2m), with a curved track for a mechanically adjustable 5000-watt light bulb. A mechanical fan with a directional screen simulated wind conditions. When operational, the Thermoheliodon would simulate climate conditions at a small scale and accelerated timeframe to test climate-responsive designs. The Olgyay brothers published their findings in their influential 1963 book, *Design with Climate: Bioclimatic Approach to Architectural Regionalism*.[12] Their conclusions include strategies for four different North American climate zones: New York, Minneapolis, Phoenix, and Miami, with practical recommendations including site planning strategies for building orientation, types of vegetation, window types, shading devices, and wall color and materials.

12 Victor Olgyay et al., *Design with Climate: Bioclimatic Approach to Architectural Regionalism*, New and expanded edition. (Princeton, NJ: Princeton University Press, 2015).

The influence of the Olgyay's research has waxed and waned with shifting attitudes toward environmental design, resource management, and the responsibility of architects, engineers, and the construction industry for ecological impacts of the built environment. The advent of readily available air conditioning and affordable energy coincided with a low point in public interest in passive strategies and regional approaches to managing shared resources.

Modern tendencies were further emboldened by an enthusiasm of the post-regional, enabled by technologies — light framed, air-conditioned comfort — regardless of orientation, climate or regional practices.

1 | 1 A Simple Factory Building (2012), in Singapore, designed by Erik L'Heureux, builds on the *brise-soleil* precedents of both Le Corbusier and local vernacular architectures.

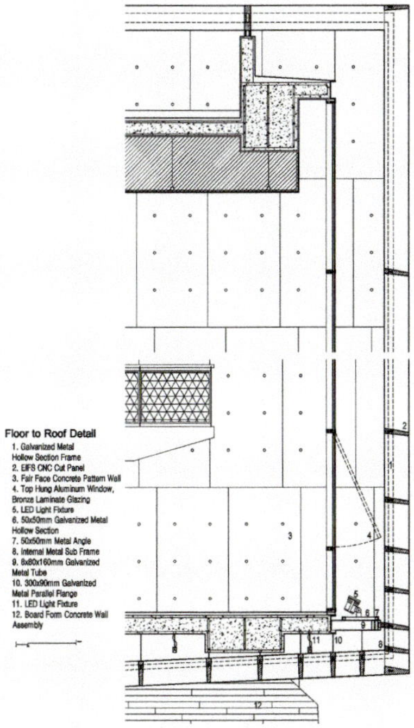

12 A partial wall section through A Simple Factory Building (2012), shows the external sunshades outboard of the building envelope.

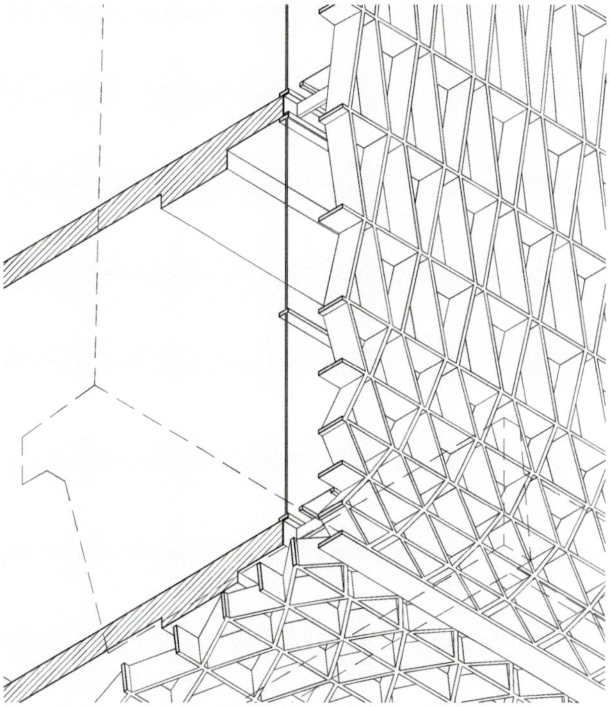

13 A section perspective of the external sunshades shows the pattern of shading elements compressing in proportion toward the bottom of the elevation, and then wrapping under the soffit to form a continuous loop.

Banham offers a lively critique of the "thermostat age, when modern architecture embraced the machine aesthetic without much thinking about the mechanical systems that conditioned those buildings."[13] New scholarship, including Barber's *Modern Architecture and Climate*, updates some of these important factors that seem buried in the recent history of late 20th-century architecture, and are newly relevant today. A renewed interest in passive design, includes *brise-soleil*, operable windows, breeze catchers, and convection ventilation.

Site selection, building orientation, material, and equipment specification are all key factors, many of which were touched on by the Olgyay brothers' bioclimatic approach to design research and practice that finds new relevance, and holds new promise as architects today think beyond the finished product. Increasingly, designers are as concerned with buildings as they are with resource extraction, relationships between humans and our natural worlds, material flows, labor practices, new modes of production, and material afterlife. Faced with ecological crises that render our natural and built environments virtually inextricable, design decisions determine whether this relationship is symbiotic and productive or mutually destructive. Climate literacy is today inseparable from design.

13 The protagonist in Paolo Bacigalupi's book, *The Water Knife*, muses about future times when archeologists might look back at the "present" and define the period by how people lived. "Maybe that's what archaeologists will call us. The Outdoors Period. For when people still lived outdoors." Paolo Bacigalupi, *The Water Knife*, p. 349.

14 L'Heureux notes the recent history of Singapore's rapid modernization and its adoption of air conditioning. He also quotes the first Prime Minister Lee Kuan Yew, who stated "air-conditioning was the best invention of the twentieth century," the reliance on conditioning and the intertwined relationship of Singapore's modernization with the ubiquitous technologies of air conditioning.

15 Erik G. L'Heureux and Grit Vltavsky, *Deep Veils*, First edition. (Novato, CA: ORO Editions, 2014), p. 8.

SOLAR VEILS

Architect Erik L'Heureux's practice could be understood as an inheritor of the bioclimatic imperative laid out by the Olgyay brothers. L'Heureux's work encompasses built projects, research, and scholarship, including his book, *Deep Veils* (2014), in which he situates his practice within the traditional uses of textiles and cladding as well as the discourses of Gottfried Semper Maxwell Fry and Jane Drew. Geographically, his work is located in Singapore's tropical climate, which informs the area's vernacular architectural traditions and the more recent history of mechanized conditioning and comfort.[14]

L'Heureux writes, "The idea of a deep veil is a counterpoint to the much ingrained transparent and spectacular qualities of glass that perpetuate much of the orthodox Modern architecture. I am interested in the spatial and physical qualities of veils that are primarily composed of opaque materials, layered, extruded, woven, stretched, and stacked at the architectural scale. The intent of the deep veil is then to destabilize normative ideas of enclosure and reconfigure our relationship with the tropical environment into one that is more porous and integrated."[15]

L'Heureux's projects explore the expressive capacities of solar control screens to provide shading

1
4

The new glass curtain wall of 633 Folsom, a renovation designed by Gensler, includes a solar shading "hood" on the vertical and head condition of the facade, giving the whole facade a distinctive solar-responsive texture.

15 From the interior of 633 Folsom, the solar shading "hoods" help to limit glare and solar heat gain while maintaining views.

for both environmental and architectural benefits. His design for A Simple Factory Building for OKP Holdings Ltd in a northern suburb of Singapore, for example, includes a single geometrical screen that wraps around the upper four stories of the compact building. 11 The screen is held off from the building's exterior surface by four feet (1.2m) and extends over the top, wrapping both the front and back facades as well as the roof garden and the underside of the soffit. 12 13 Singapore's equatorial location means that the noontime sun is directly overhead for 12 months of the year, which renders the north and south exposures largely equivalent and makes western exposure a significant design consideration. The continuous wrapping of the building is a means for unifying a stack of floors that vary in height and house various administrative functions. The spacing and geometry of the louvers changes in elevation, as does the subdivision of the horizontal bands, with the lowest bands being almost square across the bottom and highly attenuated toward the top. The variable angle of the members means that the upper bands are vertical, while the lowest bands are nearly 45 degrees, giving the crisscrossed members a sense of both weaving and attenuation at the upper portion of the facade.

The construction of the gradient veil consists of a galvanized hollow metal frame with CNC-cut EFIS segments mounted outboard of a window wall system. The cavity between the screening elements and the window wall allows maintenance access and operable windows to swing out in an awning configuration. L'Heureux describes the contradictory demands of the project as mitigating the tropical solar radiation while allowing for the openness of views sought by the office building tenants. This tension pervades most architectural strategies of solar control.

Compared to Singapore's equatorial location and tropical climate, San Francisco enjoys a Mediterranean-type climate characterized by mild, wet winters and dry summers. Gensler's 633 Folsom Street in downtown San Francisco offers a useful comparison of solar control strategies for a northern latitude. 14 The project consists of an addition and reuse of an existing building, as well as a recladding of the whole building. The original structure was originally built in the 1960s as a seven-story office building clad in a bronze-tinted curtain wall infilled between multi-story brick piers. The building owner, the Swig Company, opted to pursue a dramatic renovation rather than tear down the building. Gensler's redesign adds five new floors over the corner portion

16 The sun shading "hoods" on 633 Folsom are made of lightweight glass fiber reinforced polymer (GFRP) and fabricated over custom molds.

17 The custom fabrication of the GFRP shades allowed the designers to introduce a number "hood" types according to the orientation and solar exposure.

of the existing building, nearly doubling its height and dramatically altering its proportions. New micropile foundations allow for the expansion of the existing core, and a new lightweight steel structure supports the new floors that sit above the original building.

633 Folsom's facade plays a key role in the building's dramatic transformation, and "repositioning." The dated bronze facade was replaced by an all-glass, floor-to-ceiling curtain wall with expressive L-shaped sunshades. The 45-degree angle of San Francisco's south of Market Street grid means that all four faces of the building receive direct sun. The designers at Gensler developed the design through extensive daylight modeling simulations to limit glare and solar heat gain while affording views. 19 The L-shaped sunscreen combines an awning-like horizontal shade with a vertical blind to form an asymmetrical "hood" which fits onto the rectilinear curtain wall frame.

The sunscreen elements are made of lightweight glass fiber reinforced polymer (GFRP) and attach to the curtain wall with a custom Z-shaped bracket. The GFRP material can produce the supple geometry of the solar hood. 16 17 The precise shape of the hood varies depending on the orientation and the elevation of the building. There are 16 different shapes applied to the facade, with those near the base being shallower in depth, and taller as they near the top of the building. The interior effect of the shades is experienced as a highly transparent floor-to-ceiling glass facade with a tapered leg and awning to shield from solar glare. 15 The opaque shades make using clearer glass possible to increase visible light transmission while decreasing operational energy usage. Minimized glare means occupants can keep the windows open without lowering blinds during typical workdays.

The Kolon One & Only Tower in Seoul, South Korea, designed by Morphosis, similarly employs sculptural GFRP modules to provide sun shading with a three-dimensional pattern on the facade. 20 The slender west wing houses conference rooms and social spaces behind an all-glass sloped curtain wall. 21 The three-legged pillow forms are supported by aluminum brackets incorporated into the vertical mullions. Their convex geometries suggest that the light reflected from the sun shades would be scattered in all directions, while the large area of opaque shading elements reduces glare. The designers describe the facade as "both a performative and symbolic feature of the building." The modular elements have been designed with parametric software to "balance shading and views." The

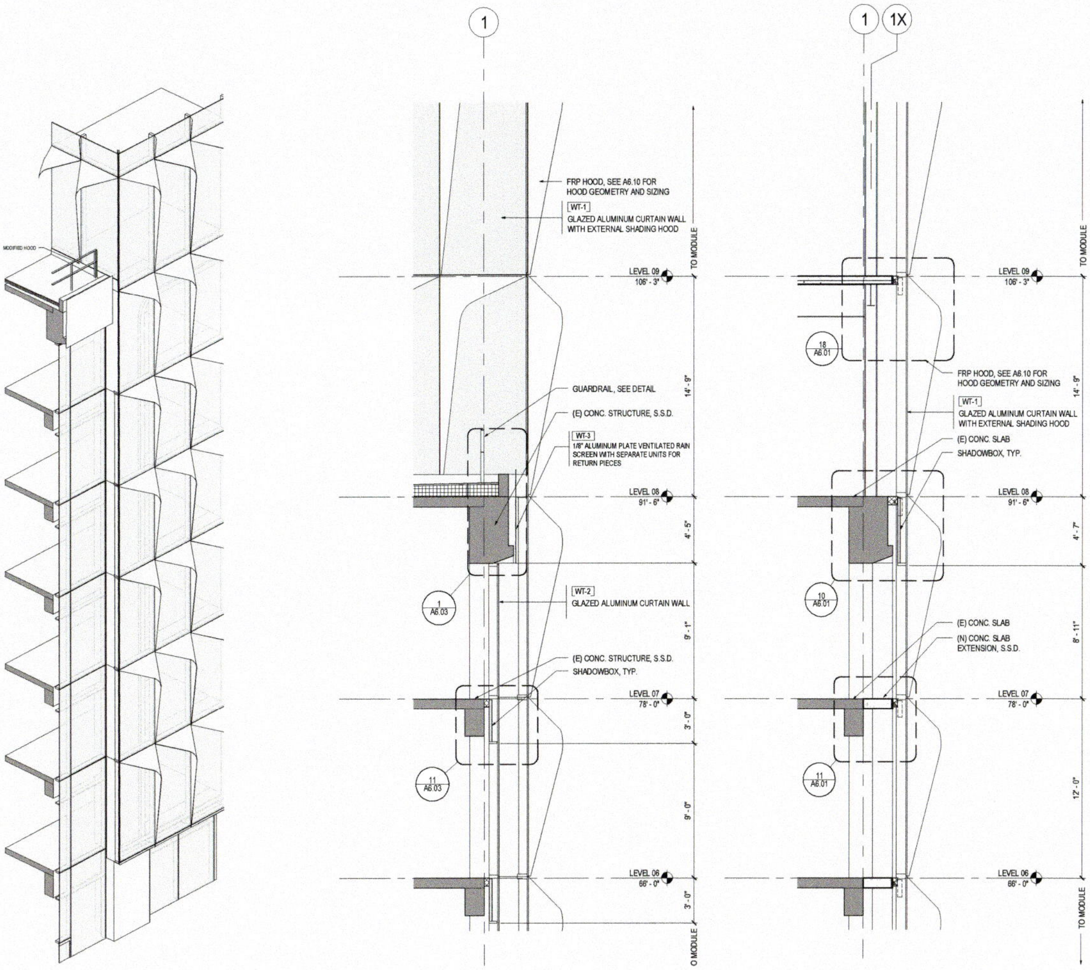

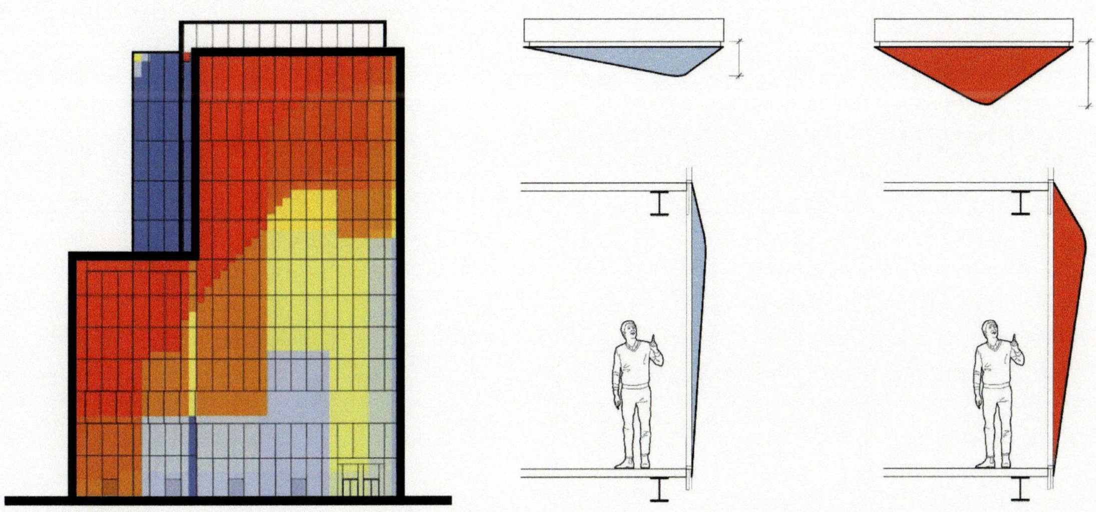

1.8 Axonometric drawing of the 633 Folsom curtain wall shows the solar shading "hoods" as they turn the corner, and section drawings show the extent of projection beyond the building face.

1.9 Solar exposure studies produced by Gensler showed the varying degrees of solar exposure and allowed the designers to apply different shade types to the different exposures.

20 The Kolon One & Only Tower (2018), in Seoul, South Korea, designed by Morphosis, clads the west facade with a sculptural pillow form derived from a knitted weave pattern.

21 The visual mockup of the Kolon One & Only Tower facade. The GFRP modules are supported by brackets that connect back to the building curtain wall.

GFRP material used was formulated to include one of Kolon's high-tech fabrics, thereby showcasing the company's own products in the design.[16]

16 From the Morphosis website, https://www.morphosis.com/architecture/261/

Making the building's relationship to the movement of the sun legible, the town hall in Freiburg im Breisgau, completed by Christoph Ingenhoven in 2017, utilizes a novel facade strategy for the ovoid-shaped building. 23 The six-story structure is arranged in a ring around a courtyard and sits on a corner site as a figure within a parklike setting. The offices around the building's perimeter face in all directions, with 360-degree exposures. 24 The architects employed a singular curtain wall facade with floor-to-ceiling glass. A horizontal shelf at the floor elevation of each story supports a dense layer of vertical sun shading and photovoltaic (PV) panels rotated toward an optimal solar orientation. 25 The continuous curvature of the building and the desire to face an optimal orientation means that the angle of each facade panel is unique. The north-facing segment of the facade is not equipped with PV panels, and in this area the verticals act as visual screens. The change of color, from the black PV panels to the lighter wood tone, indicates the zone that lacks direct sunlight.

The vertical wooden panels also give the building its unique architectural expression. The combination of sun shading and renewable energy generation makes Kolon One & Only a net energy surplus building, meaning it produces more energy than it consumes. While the development of photovoltaics has improved to the point where they generate energy within a range of orientations, the subtle rotations of the vertical panels do function to index an orientation of solar energy production. In this sense, the building acts like an analog sundial, gesturing in the direction of optimal solar power generation. 27

The difference between climate and weather is often articulated as, "climate is what you expect, and weather is what you get." Climate is understood as the accumulation of weather and temperature data collected over a long period of time to produce statistical averages and assumptions about dynamic phenomena. When we expect the temperature and humidity in Singapore to be hot and humid, that is climate. If a sudden thunderstorm creates flash floods, that is weather. Climate is average and habitual, while weather momentary and dynamic. Climate-responsive architecture relies on predictable solar and wind data. Passive strategies assume that the sun will be high in the summer and lower in winter. Sun shading strategies like L'Heureux's criss-crossing *brise-soleil* or Ingenhoven's pivoting solar panels are based on anticipated climate conditions.

22 The interior view from the Kolon One & Only Tower highlights the three-dimensional quality of the GFRP modules.

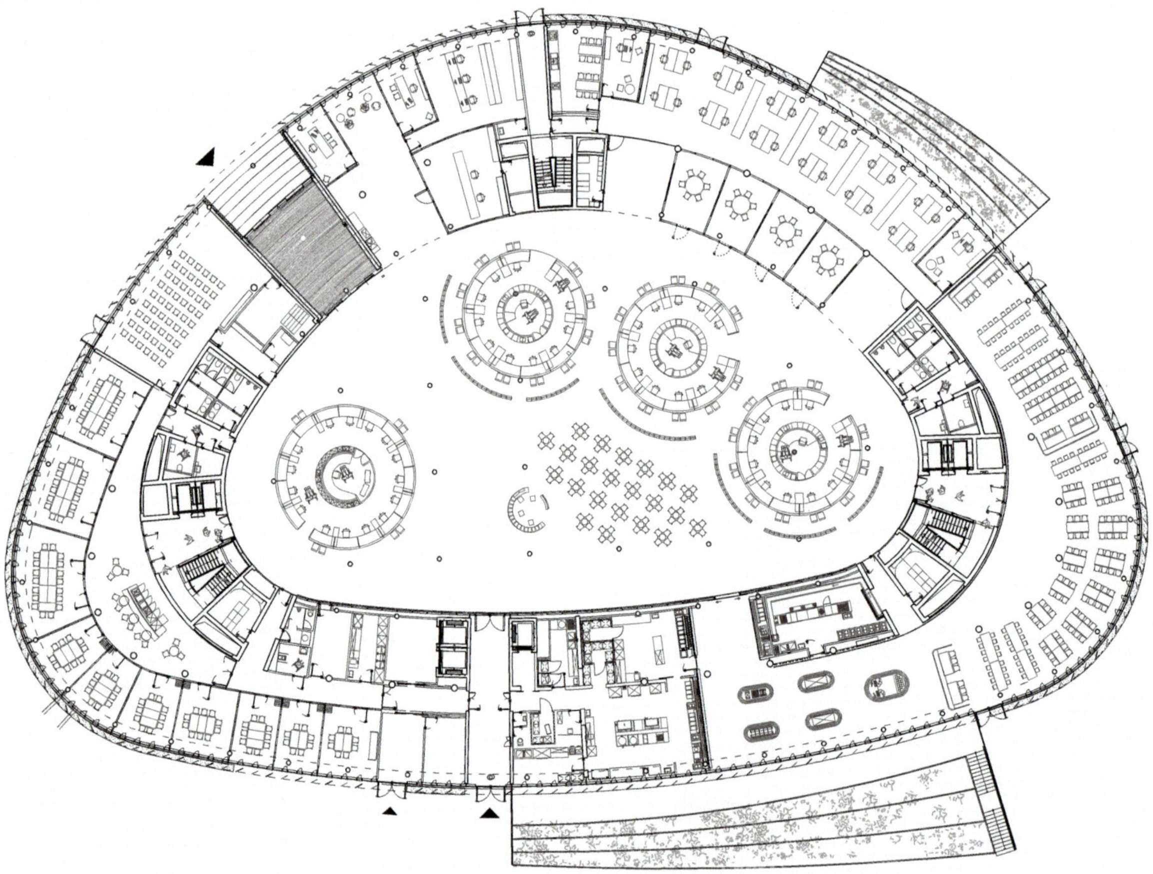

2|3 The Freiburg Town Hall (2017), designed by Christoph Ingenhoven, reimagines the civic building as a soft organically shaped building in the park. The facade strategy responds to solar orientation with a vertical panel oriented to simultaneously screen the direct sunlight and also produce power through photovoltaic panels.

2|4 The plan of Freiburg Town Hall (2017) shows the vertical panels mounted to all sides of the ovoid building. The north-facing panels are not equipped with photovoltaic panels.

25 The facade of Freiburg Town Hall (2017), consists of a curtain wall screened by a continuous system of vertical sunscreen panels, oriented toward the south.

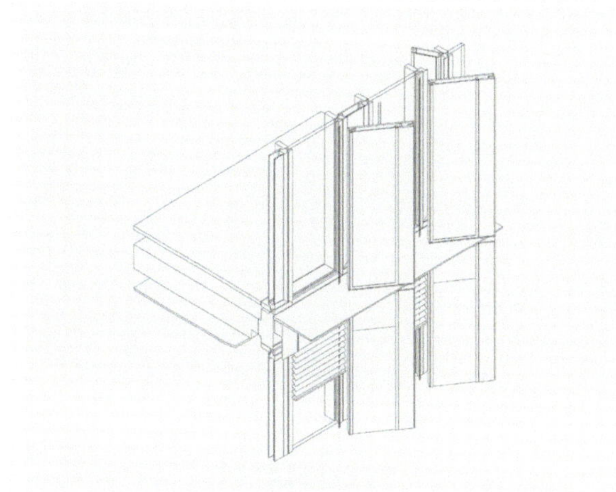

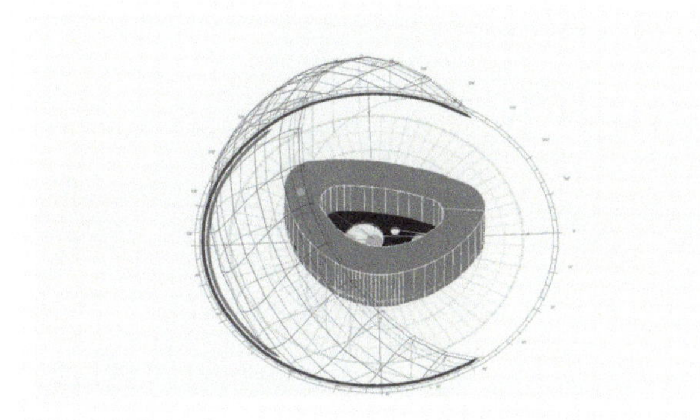

26 Axonometric of the curtain wall system shows the vertical sun shading system as well as operable panels and exterior Venetian blinds.
27 Solar exposure analysis for Freiburg Town Hall (2017).

MECHANICAL BLINDS

Architectural strategies for responding to environmental phenomena with a shorter cycle of change and variation might need to adapt to less predictable phenomena such as rainfall or cloud cover — i.e. weather. There are numerous examples of adaptive architectural elements that allow buildings to change over the course of the day. Retractable awnings may be deployed to cover outdoor seating for a sidewalk cafe, or window blinds lowered to limit glare in a south-facing apartment. Opening and closing windows for natural ventilation allows occupants to calibrate their indoor temperature and take advantage of prevailing winds.

The Institut du Monde Arabe (IMA) building by Jean Nouvel is an early example of an integrated kinetic facade designed to adapt to weather phenomena like sunlight and glare. 32 The building occupies an irregular site along the south bank of the Seine and the northern edge of the Jussieu campus of Sorbonne University. Its mass follows the curve of the northern site boundary shaped by the adjacent Quai Saint-Bernard, and its broad, southwest face aligns with the linear buildings of the Sorbonne campus grid. Conceived in the 1970s but realized in 1987, the building was one of the Grand Projects planned by French President François Mitterrand. Like many of the Grand Projects including the Centro Georges Pompidou and the Louvre Pyramid, Mitterrand encouraged the architects to design bold structures with a high-tech vocabulary to celebrate French culture and state institutions. Nouvel's design for the IMA also employed glass curtain walls and exposed structural elements to project an image of a progressive and future-facing France.

The detailing of the southwest facade consists of a taut glass envelope with a square plaid pattern of mullions, creating spandrel conditions at the floor slabs and column grid. 31 The square vision panels within the gridded lattice of the facade are occupied by a large mechanism resembling the diaphragm shutter of an early camera. 29 30 Small motors control hundreds of machine components, allowing the apertures to open or close in response to the sun's movement. Nouvel's high-tech design was celebrated at the time for being simultaneously a state-of-the-art kinetic facade while also evoking the geometric intricacies of mashrabiya patterns from traditional Islamic architecture. Architectural historian Annette Fierro reads Nouvel's glass-faced design, complete with the mechanical mashrabiya, as a "screen,"[17] an architectural surface

17 Annette Fierro, *The Glass State: The Technology of the Spectacle, Paris, 1981–1998* (Cambridge, MA: MIT Press, 2003), p. 128.

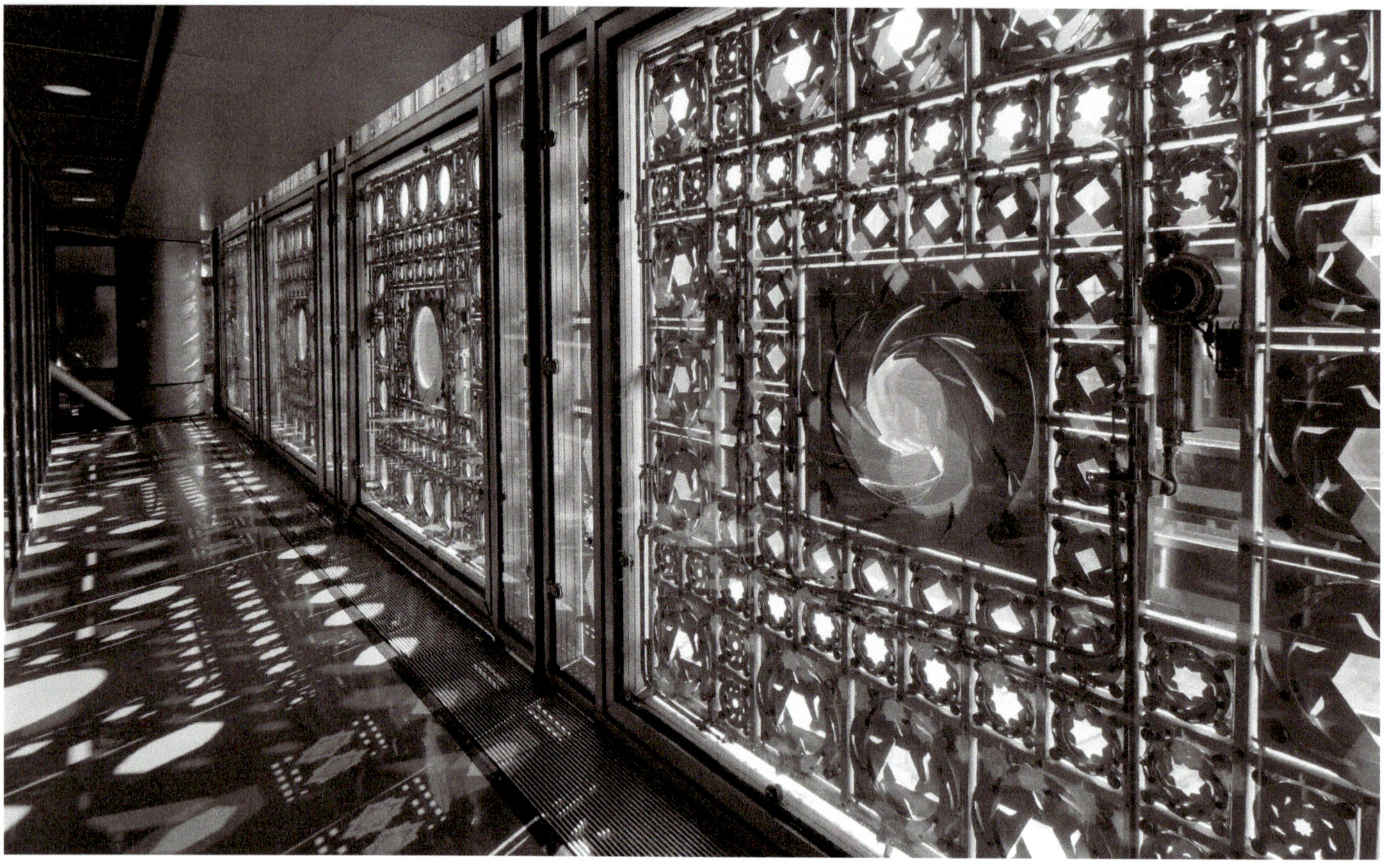

28 The Institut du Monde Arabe (1987) in Paris, designed by Jean Nouvel and Architecture Studio, included a pioneering kinetic facade made up of hundreds of mechanical diaphragm shutters that open and close in response to solar exposure, and create filtered patterns of light on the interior.

representation of Islamic culture in contemporary France. Fierro highlights the fraught politics of representation in the context of France's post-colonial legacy in North Africa and Nouvel's use of glass on the facade of the IMA as a "surface of transmission" and an "interface." She reads Nouvel's use of translucency as a reference, "not to Mies van der Rohe, but to Pierre Charreau; not to a reflective translucency of exteriority, but to the luminous, exotic interior translucency based on romantic postcolonial constructs."[18]

[18] Annette Fierro, *The Glass State*, p. 128.

The interior spaces of the library reading room behind the kinetic facade benefit from the intricate patterns of light and shadows produced by the geometry of the shutters. 28 Since its completion, the kinetic components of the southwest facade have been plagued by the maintenance challenges as a result of thousands of moving mechanical parts, and expensive refurbishment. Even so, the array of mechanical apertures produces distinctive patterns on the interior that mimic the qualities of light and shade found in vernacular Islamic structures, not by replicating specific motifs, but rather by simulating their environmental effects. Even in its inert state, the IMA facade effectively conveys the abstracted environmental qualities of an architectural tradition shaped by climate responsiveness while bringing that cultural expression into a contemporary architectural syntax by merging vernacular iconography and modern technology.

The Al Bahr Towers in Abu Dhabi offer a more recent example of a kinetic facade capable of responding in real-time to changing environmental conditions. 33 The twin towers, designed by multinational Aedas, are located along Sheikh Zayed Bin Sultan Street, a primary artery entering Abu Dhabi. Their slightly bowed cylindrical shafts give the towers an organic appearance, and the texture of their solar shading devices makes them distinctly iconic structures, appearing at times entirely faceted and opaque and at other times like a pair of woven vessels crafted of delicate fibers. 34

The architects describe the project as informed by regional patternmaking and climate-responsive vernacular traditions, underlying geometry of the towers' floor plates, and the mashrabiya-inspired lattice facades. The floor plates are based on a rounded triangle, derived from six superimposed circles. The towers' facades are subdivided in a pattern of equilateral triangles, with one edge oriented horizontally to align with the floor slabs. The building's envelope comprises an orthogonal curtain wall system with a triangulated sun shading system mounted to its exterior surface. Each module of the

29 A detail view of one kinetic diaphragm shutter.

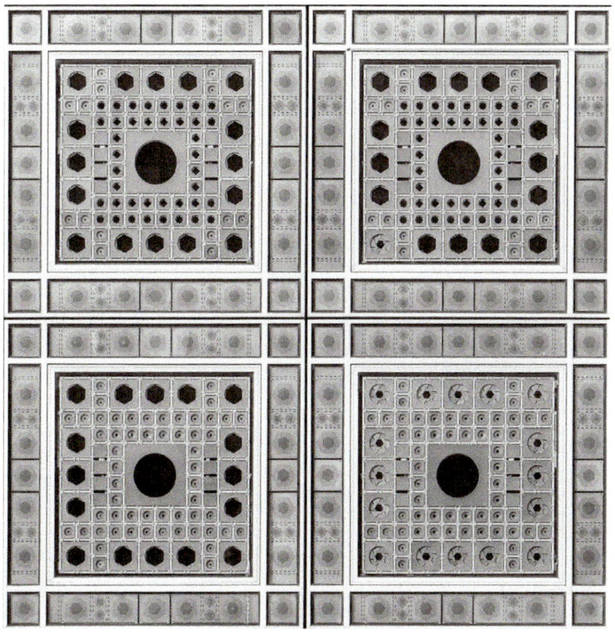

30 The exterior view of the IMA facade elements. The diaphragms have been likened to traditional Islamic *mashrabiya* screens that filter the sun and ensure privacy.

dynamic *brise-soleil* corresponds to the floor-to-floor height of the office tower, such that the corner support points for the triangular modules can be attached to the slab edge. Each triangle is then subdivided through its center into six equal facets of right-angle triangles. A linear actuator at the center of the triangle pushes the panels out of the plane, creating ridges and valleys and causing the opaque triangular panel to collapse into an equilateral three-pointed star. 35 36

This basic mechanical component, repeated across the facades, allows the screened towers to transform from opaque to transparent with the actuation of the center piston. There are 2,098 individually controlled units that open and close the faceted facade, responding to the movement of the sun. A central building monitoring system (BMS) operates the mechanical pistons that adjust to the solar conditions in real time. The typical operation of the facade includes opening the sunshades in the evening and shuttering the faces exposed to the sun over the course of the day. The rounded plan of the towers, like the rounded face of the town hall in Freiburg, results in a facade that is segmented along its curvature, such that a narrow portion of the building will always be directly facing the sun. At any given time, that zone of exposure is shaded by the deployed sun-shading system, while all other faces experience a less-than-direct angle of incidence. Shading systems with individually controlled units allow for fine-grain calibration in response to solar orientation and amount of direct solar exposure.

The implementation of the kinetic solar control system in the context of Abu Dhabi introduces several technical, operational, and constructability challenges. As with the Institut du Monde Arabe, the use of mechanical components requires ongoing maintenance to ensure the operation of its mechanical components. Unlike the IMA, where the moving parts are enclosed within a sealed interior cavity behind the glass, the Al Bahr sun shading system is exposed to the elements, subject to wind and dust as well as humidity and salinity, and therefore subject to wear over time. In proposing an exterior motorized sun shading system, the architects and engineers also had to develop a prototype for a custom facade application, as well as a testing protocol and maintenance schedule. The facade system underwent extensive testing with performance mockups prior to fabrication and installation. The extensive design and implementation process has been likened to that of the automotive industry, where components are designed, prototyped, and assembled from a global supply chain to deliver a working product. The critical

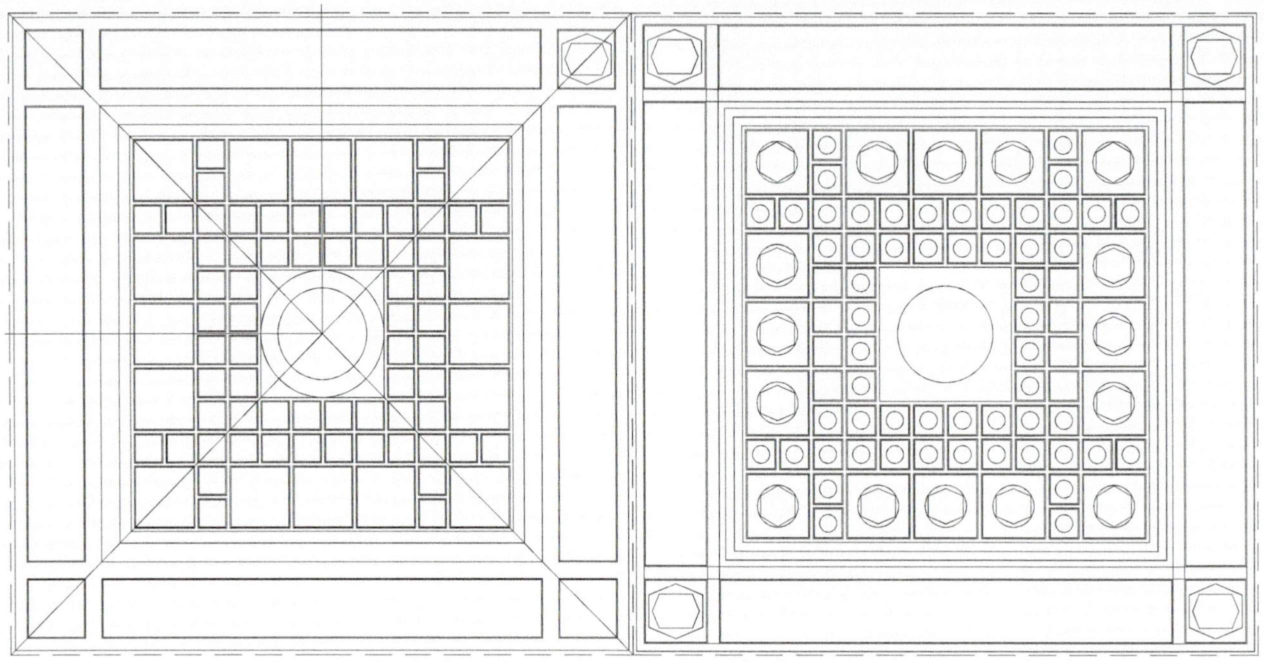

3 1 Partial elevation of the Institut du Monde Arabe (1987) facade.

difference is that research and development in the automotive industry produce millions of the same car, while in architecture, research and development inform one project, produce one building — or, in this case, two.

The completed Al Bahr Towers occupy a unique position in the development of climate-responsive architecture and conditioned interiors. Designed in 2009 and completed in 2012, the towers represent a moment where attitudes toward climate-responsive design aligned with the technical and financial capacity to execute a unique project in an extreme climate like Abu Dhabi. The clients for the Al Bahr Towers commissioned a "statement piece" of architecture equipped with state-of-the-art climate control systems. The "intelligent external shading" system applied to the prominent buildings signals a capacity to overcome the extreme climate conditions of Abu Dhabi through a combination of active and passive technologies.

These responsive design articulations display a range of approaches that combine environmental strategies with architectural expression. To different degrees, each project navigates the calibration to the specific movements of the sun and the degree to which they form a legible environmental representation. Returning to Daniel Barber's theory that the architectural facade can be understood as an environmental media device, the formal and geometrical articulation of sun shading strategies reflect the negotiated relationships between climate and culture.[19] The individually changeable, origami-like sun shades of the Al Bahr Towers respond to the sun's movement and to changes in atmospheric conditions like cloud cover, resulting in an architectural facade that registers the temporalities of weather patterns like a field of mechanical sunflowers, opening and closing with changing conditions in the sky above. Furthermore, the tessellation of the facades is highly visible to both occupants on the interior and viewers on the exterior, allowing for the environment and its architectural response to be immediately legible. The pixelated pattern of open and closed sun shading petals demonstrates environmental responsiveness, like an architectural semaphore broadcasting its current environmental conditions. It is also a clear signifier of a regional and national ambition to use architecture as a means for representing a modern, affluent gulf region that deploys design to address regional challenges in an age of global climate crisis.

The predictive capacities of contemporary building information models and environmental simulations are in a different order of magnitude in terms

19 Daniel A. Barber, *Modern Architecture and Climate.*

3.2 The south-west facade of the Institut du Monde Arabe (1987), designed by Jean Nouvel and Architecture Studio as a high-tech interpretation of a traditional Islamic *mashrabiya*.

3.3 The Al Bahr Towers (2012), in Abu Dahbi, UAE, designed by Aedas, offer a contemporary interpretation of the kinetic, solar-responsive sun shading device.

34 The near-cylindrical geometry of the Al Bahr Towers allows the buildings to track the sun almost a full 360 degrees.

of precision, than that of the earlier simulations of Olgyay brothers' 1950s Thermoheliodon. 9 10 Though equipped with similar environmental design tools, contemporary designers develop a wide range of architectural responses. The array of different facade expressions suggests that there is no singular or optimum approach to integrating climate conditions and design decisions. Contemporary climate-responsive design performs across a range of possible directions, seeking a tenuous balance between energy performance, occupant comfort, and view access, and with consideration to countless other factors, including the first cost of construction and life cycle analysis, payback periods, and the embodied energy of selected building materials. Taken together, these projects demonstrate how the contemporary architectural facade can be seen as a surface of environmental inscription, producing a built record of our shifting attitudes about climate and culture.

SKY EXPOSURE

New York City's 1916 Zoning Resolution is among the best-known examples of environmentally determined design parameters, and one that has produced thousands of unique design solutions. In response to the 1915 Equitable Building's excessive volume that blocked sunlight to the neighboring streets, New York City drafted a zoning resolution requiring new buildings to comply with a set of parameters determined by the specifics of each site. A building's mass was to fit within a geometrically defined envelope that combined vertical extruded faces and a sloped surface called the sky exposure plane. The 1916 Zoning Resolution was an early application of basic environmental design principles to ensure the penetration of sunlight to the public street.

The setback profile of New York City zoning is perceptible in the distinctive Art Deco profile of William van Alen's 1928 Chrysler Building and the shard-like, braced monolith of Jean Nouvel's 2019 midtown tower, 53W53. The architectural profiles of both are inscribed within the codified profile of New York's zoning.[20] The zoning resolution is not overly prescriptive, meaning it sets an outer boundary, and architects work within it. The "loose fit" between zoning geometry and architectural form has been the key to its successful implementation for over a century.

To ensure the penetration of sunlight to the street level, buildings must set back from the street at a prescribed angle regardless of orientation or solar exposure. Despite the solar basis of the sky exposure plane method, the geometrical principles

20 For more on the relationship between zoning and emerging skyscraper typologies see Carol Willis' *Form Follows Finance: Skyscrapers and Skylines in New York and Chicago*, 1st ed. (New York, NY: Princeton Architectural Press, 1995).

35 The closed position of a cluster of kinetic solar control shades of the Al Bahr Towers (2012).

36 The open position of a cluster of kinetic solar control shades of the Al Bahr Towers (2012).

are remarkably generic — the zoning setback is agnostic to solar orientation, with both south-facing and north-facing building volumes setting back with the same angle. It is as if the zoning commissioners were assuming a baseline condition of an overcast day, with diffuse solar radiation coming from the entire dome of the sky, or more likely, a more generic zoning envelope would be easier to implement than tailoring to the actual solar angle of altitude.

SOLAR SHAPING

Heliomorphism describes the practice of shaping architectural and urban form, in response to the sun.[21] The 1916 New York City Zoning Resolution could be understood as a generalized heliomorphism, as an early application of climate-responsive design. Contemporary computational tools allow greater precision in practices of solar shaping. Several recent projects by Studio Gang are explicitly shaped by the geometries of the sun: The Solstice on the Park in Chicago, the One Hundred Tower in Saint Louis, and the Solar Carve in New York all apply heliomorphic principles in their architectural massing.[22]

Solstice on the Park, completed in 2018, was the first of the series to employ Studio Gang's "solar carving" strategy to shape the building form. 40 The building facade slopes outward in two-story blocks to create a self-shading profile. The angle of the glass panels on the south facade is tilted at 72 degrees, matching the angle of altitude for the sun for Chicago's latitude on the summer solstice. An apartment building in Saint Louis, The One Hundred, employs the leaning glass technique in a more sculptural way, with four-story floor packages producing a repeating stack of inverted pyramid forms that resemble a pinecone. 41 The inclined glass facades in this application provide a measure of solar control as reflectivity is improved with the low angle of incidence; the sloped glass is more transparent while maintaining its energy performance. The heliomorphic design criteria produce the iconic form of the tower giving the building a unique silhouette in the Saint Louis skyline.

Situated between New York City's West Side Highway and the Highline, the Solar Carve building employs heliomorphic principles to shape the massing of the office building while allowing sunlight to reach the adjacent public park. 38 39 The dramatic corner expression gives the building its distinctive form. The Solar Carve building massing is arranged to hold the western edge and setback from the Highline with the roof of the podium located

21 The Office for Urbanization at the Harvard Graduate School of Design organized the Heliomorphism conference in September 15, 2016.

22 The term "solar envelope" was coined by architect and environmental design researcher Ralph Knowles. Jeanne Gang describes being inspired by Ralph Knowles' work on solar responsive architecture, Ralph L. Knowles, *Sun Rhythm Form*, (Cambridge, 1981).

37 Hugh Ferriss' "heliomorphic" renderings of the 1916 New York City Zoning Resolution, designed to ensure daylight access to public streets.

on the east side of the building, adjacent to the Highline. The designers point to New York zoning's intent to preserve daylight for public streets, while no provisions exist for public parks. The volumetric carving of the opposing southeast and northwest corners is intended to allow more daylight to reach the Highline while allowing park users to have expanded views of the Hudson River.

These three buildings designed by Studio Gang demonstrate heliomorphic operations at different scales that are combined with other motivations, including solar performance, enhancements to the public realm, and the provision of outdoor spaces, make iconic architectural form. Like the 1916 Zoning Resolution, so powerfully illustrated by Hugh Ferriss, 37 these buildings bring together environmental performance criteria and aesthetically motivated formal expression. In the era of increasingly precise simulation software, it would be a mistake to attribute the formal outcomes entirely to performance criteria. However, the negotiation between form and its performance is essential, enabling different approaches to climate-responsive architecture and the development of expressive form.

In extreme climate conditions, the tradeoffs between solar exposure and views produce more extreme design responses, with window-to-wall ratios requiring more opaque and insulated surfaces. The Shenzhen Energy Mansion project in Shenzhen, China, employs a sawtooth plan with alternating glazed and opaque surfaces that "turn away" from the sun and direct views toward the north. 42 The pair of office towers, designed by the Bjarke Ingles Group (BIG) for the Shenzhen Energy Company headquarters, showcase climate-responsive architectural strategies that highlight critical questions around energy use in architecture.

The facade and massing strategy for the pair of towers introduces a saw-toothed facade articulation described as a "pleated dress" that divides the building envelope into opaque surfaces that face the sun and glazed surfaces that face away from the sun. 46 The two buildings are sited on a large parcel with its primary site boundary facing west. The massing was predetermined by the district's zoning envelope requiring the towers to step down in height from north to south. BIG's design took the base massing, two rectangular towers joined by a shared podium, and introduced the highly textured facade as a climate-responsive strategy. Preliminary energy models demonstrated a 30 percent reduction in air conditioning costs.[23] This energy reduction involves no moving parts or complicated technology and is based purely on building geometry.

[23] Bjarke Ingels et al., eds., *Hot to Cold: An Odyssey of Architectural Adaptation* (Köln: Taschen, 2015), p. 109.

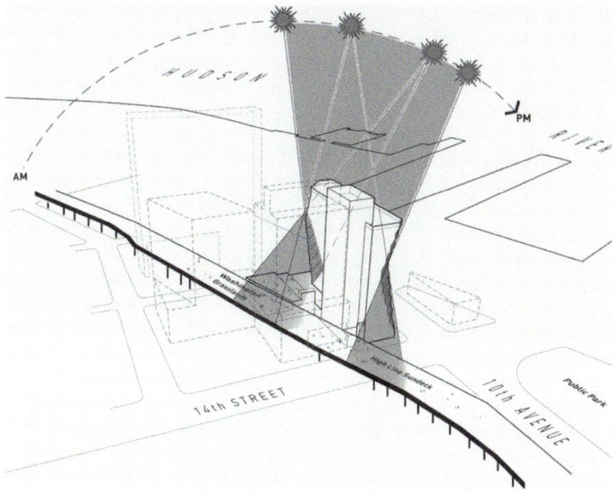

38 Studio Gang's "heliomorphic" solar diagrams of the Solar Carve building (2019), designed to ensure daylight access to the public park.

39 The faceted faces of Studio Gang's Solar Carve building (2019) give a formal expression to solar shaping, or heliomorphism.

It demonstrates passive environmental design strategies that achieve ecological goals, including reduced solar heat gain, improved quality of daylight, and improved occupant comfort, without consuming energy.

The Shenzhen Energy Mansion curtain wall consists of a 45-degree stepped plane with a nominal five-foot (1.5m) width that alternates between an opaque insulated panel and a fully glazed panel with a floor-to-ceiling vision glass and a flush glazed shadow box spandrel panel at the slabs. 44 The curtain wall is fully unitized, allowing for offsite prefabrication and highly precise installation. The modules are L-shaped with factory-sealed outer corners.

A special detail of the opaque panel is the micro saw-toothed texture of the face, which is achieved by joining three extruded panels together, rather than a flat aluminum panel or plate. The micro-scaled pattern recalls the large-scale sawtooth pattern of the overall building, while also creating the unique optical effects of striated light reflections across the panels. The articulation of the building facade into alternating vertical strips of glazed and opaque surfaces gives it a distinct vertical texture and attenuation. It also changes in appearance depending on the point of view. 24 Driving eastbound along Binhai Road, the buildings would appear largely opaque, whereas a motorist driving south along Jintian Road would perceive the two towers as almost entirely glazed.

24 Shenzhen Energy Mansion recalls the CBS headquarters building in Midtown Manhattan, designed by Eero Saarinen, which also employs a sawtooth plan. The Saarinen building, nicknamed "The Rock," appears solid from most oblique angles, with the glazing set back between extruded sawtooth piers. The difference with the Shenzhen Energy Mansion is the opacity bias is applied to a global orientation based on the sun exposures, whereas the CBS building glazing is parallel to the site and set between the vertical piers.

The optical effect of orientation-specific material strategies could be described as lenticular, meaning lens-like. Lenticular optical effects have been used to make an image appear animated as the perception of the image changes subject to the angle of view. The orientation-specific glazing employed by the Shenzhen Energy Mansion allows the building to flicker between glassy and metallic, transparent and opaque, and heat-absorptive and heat-reflective. The building's relative transparency or opacity enables it to be selective in the type of light it admits, like a low-resolution light filter.

The saw-toothed geometry also introduces an implicit grain to the building. The metal faces have a constant dimension like a ribbon. At the same time, the glass infill surfaces vary to create exceptional features within the logic of the facade system. These multi-story bulges are added at the corners and at the base to identify the entrance of the building, behind what appears to be a pulled-back curtain, or on the northwest corner to open a multi-story "bay window." 43 45 These architectural features are occupied by specialty program elements, like conference rooms

40 Solstice on the Park (2018), designed by Studio Gang, derives its outward sloping facade from the optimum 72-degree angle for Chicago's latitude to minimize summer heat gain.

41 The One Hundred (2020) multifamily residential tower in St. Louis also slopes outward to reduce energy loads and gives the building its iconic sculpted profile.

or corner offices, and give the two towers a rippling effect. The multi-story peel at the base of the towers introduces a gesture that extends 10 stories. It is like the skin of the building has been lifted, as if by an imaginary breeze. These local deformations, working within the global system of vertical stripes, give the building a distinctive appearance in the Shenzhen skyline. The iconic features of the two towers are neither additive (such as a crown or building top) nor subtractive (such as a sculpted building mass), but rather derived from a climate-responsive and orientation-specific building grain, and the geometric operations performed on the building are consistent with the implicit materiality of its grain. The heliomorphism of the Shenzhen Energy Mansion is not simply driven by solar optimization or energy efficiency, but rather as a combination of geometric operations set in motion by solar shading principles and their attributed material qualities.

In the introductory essay to his 2015 monograph, *Hot to Cold: An Odyssey of Architectural Adaptation*, Bjarke Ingels lays out a framework for the prolific output of his global office. Projects included in the book, designed by Ingels and his office are found at sites around the world and in a wide range of climate zones. Ingles describes the book's organization: "We have organized this book according to climate, like an odyssey across the globe from one of the hottest places on the planet, the Arabian Desert, to one of the coldest, the Finnish tundra."[25] He elaborates on this climate framework in an essay "Engineering without Engines,"[26] arguing that modern conditioning systems place buildings "on life support — supplemented by air conditioning, central heating, and mechanical ventilation. Machines replaced traditional considerations of the thickness of walls, solar orientation, proximity to windows, and the operability of windows."[27]

Ingels argues that architecture is always the result of two inescapable aspects: "the environmental and the social," and he proposes to read the body of work through those lenses. The Shenzhen Energy Mansion plays an important role in this line of argumentation. Ingles writes, "Originally powered by abundant energy for AC and ventilation, today the skyscraper must evolve into a new sustainable species. The Shenzhen Energy HQ is our first realized example of 'engineering without engines,' the idea that we can engineer the dependence on machinery out of our buildings and let the architecture perform in its place."[28]

In the discussion of heliomorphism and form-finding, or rather, "formgiving," as Ingels refers to it, the architectural design outcome is derived from a

25 Bjarke Ingels et al., eds., *Hot to Cold*, p. 9.

26 Ingles references the 1964 book by Bernard Rudofsky, *Architecture without Architects: A Short Introduction to Non-Pedigreed Architecture* (Albuquerque, NM: University of New Mexico Press, 1964).

27 Bjarke Ingels et al., eds., *Hot to Cold*, p. 9.

28 Bjarke Ingels, ed., *Formgiving* (Köln: TASCHEN, 2020), p. 156.

4|2 The Shenzhen Energy Mansion (2018), in Shenzhen, China, designed by Bjarke Ingels Group (BIG) employs a directional facade oriented to face away from the sun with a saw-toothed geometry.

4|3 The implied grain of the Shenzhen Energy Mansion facade allows for supple deformations in the grain to achieve peeled-apart entrances and bulging "bay windows."

basic formal operation. A single gesture typically organizes a large massing move. It may respond to the sun, to the context, to the program, or be entirely willful. The test of the hypothesis that the climate-responsive solar angle mapping essentially "designed itself" would be borne out in a test of the lived experience of the building and a post-occupancy evaluation of the project. Does the Shenzhen Energy Mansion actually consume 30 percent less operational energy than a similar building in a similar context? Do building occupants experience less glare, greater comfort, and consume less power?

The broad array of approaches to climate-responsive design strategies highlights the challenges in evaluating the environmental benefits achieved versus the claims made. The setback skyscrapers resulting from New York City's 1916 Zoning Resolution undoubtedly allow for broad environmental benefits of daylight penetration on an average day of the year, regardless of solar orientation, season, time of day, cloud cover, or other factors. In an age of averages and rules of thumb, before sophisticated simulation software and computational tools, the general approach to environmental considerations for design produced legitimate benefits and a recognizable New York skyscraper typology without over-determining or prescribing the shape of buildings. A hundred years later, with the wide availability of air conditioning, heat recovery ventilation, and desktop simulation software, questions of climate-responsive architecture persist. Can we design buildings that are better tailored to their contexts? Does the high-resolution data of a climate simulation allow us to better engineer structures where the estimated energy performance aligns with the post-occupancy performance? Presumably, "yes." Contemporary design tools help architects to simulate and shape predictable factors such as sunlight and heat gain. The question for many of these climate-responsive structures is: Does indexing the sun and responding to environmental phenomena like natural ventilation, and sun shading, introduce some value beyond the conservation of operational energy? Are the metrics of environmental design strictly quantitative, or could we articulate the qualitative effects of energy literacy and ecological consciousness? Or, is the gesturing to climate a particular form of greenwashing; of claiming environmental care without a meaningful impact. Taking the full spectrum of something as expansive and interconnected as climate in architecture means developing new tools to guide design decisions and a new language to describe and evaluate climate-responsive architecture.

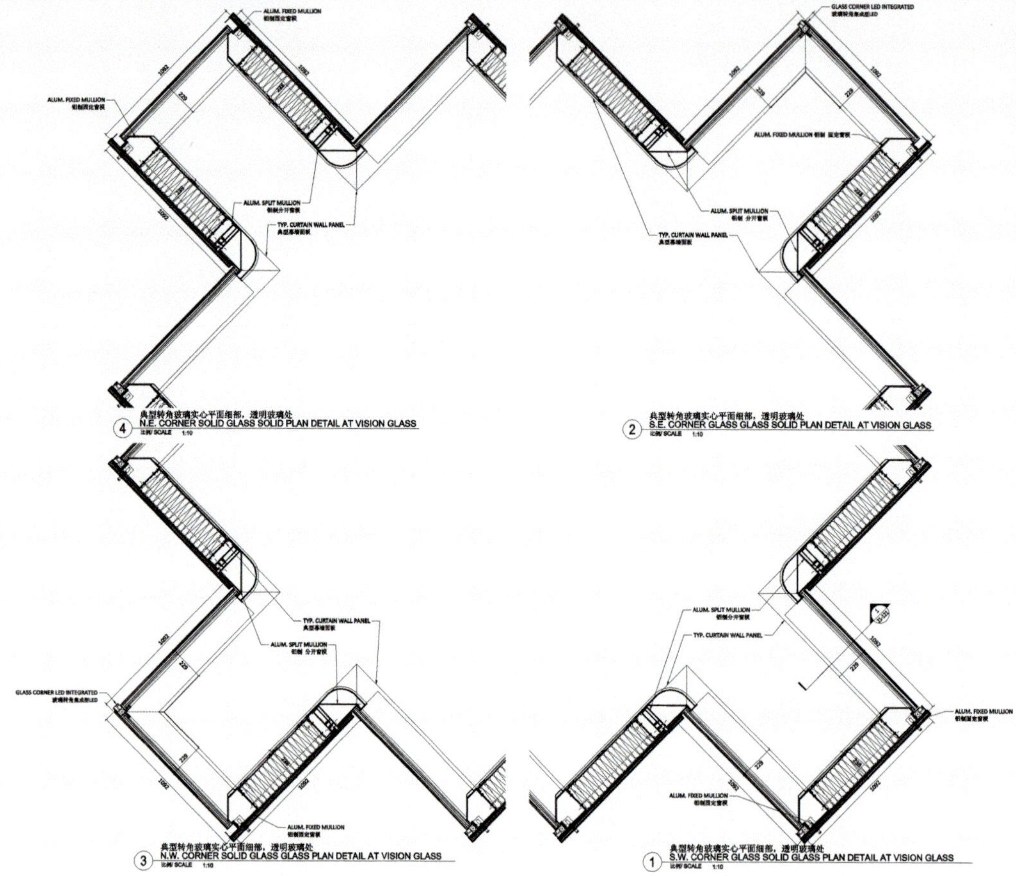

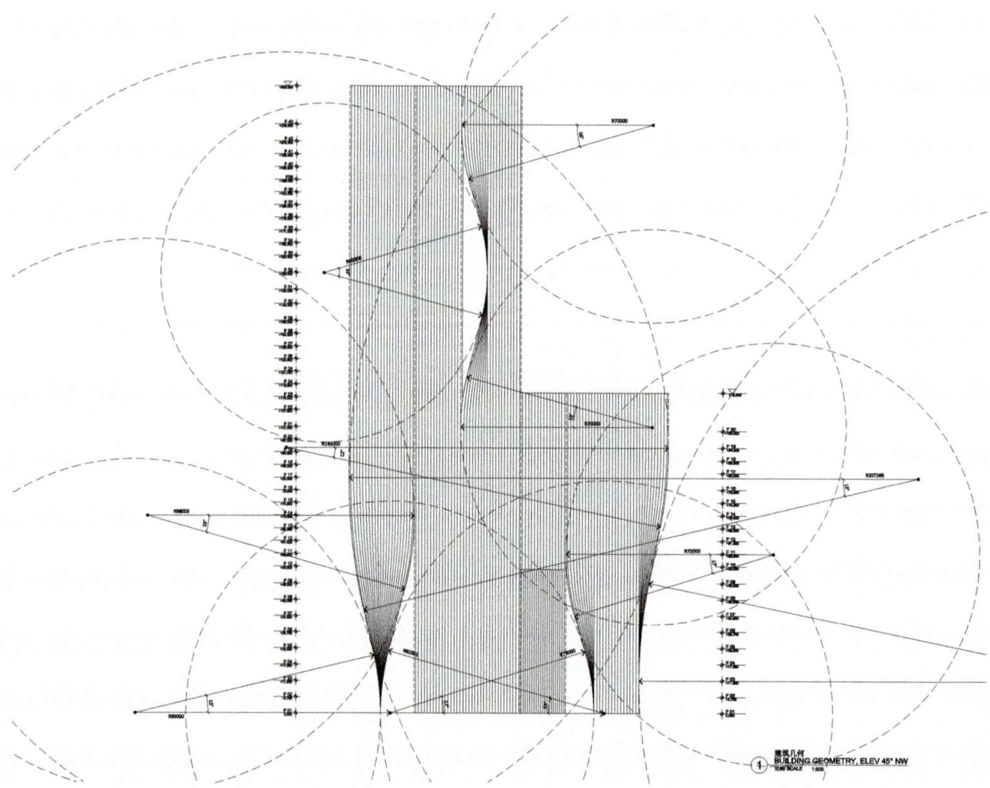

44 Partial plan details of the Shenzhen Energy Mansion (2018) curtain wall corners conditions show the bias for vision glass toward the north and east and away from the south and western exposures.

45 Shenzhen Energy Mansion (2018) elevation drawing defines the large-scale geometry of the facade peels and bulges.

46 Concept diagrams for Shenzhen Energy Mansion facade: Screen the sun and allow for directional views away from the sun.

INTERIOR CLIMATE

Searching for the origins of the concept of architecture as environment, the construction of the Crystal Palace to house the Great Exhibition of 1851 marks a milestone in architectural and engineering achievement. 47 Built in record time, the building demonstrated the possibility of architecture as an interior climate apart from its surrounding context, and it represents a critical turning point in the conceptualization of architecture and environmental design.

The glass hall, designed and constructed in nine months by the famed gardener and designer of greenhouses, Joseph Paxton, enclosed approximately 33 million cubic feet of air in a vast single chamber. The hall's enormous proportions and transparent glass enclosure mesmerized the public, who flocked to the exhibits for the duration of the exhibition. Sited in London's Hyde Park, the Crystal Palace was an extension of the surrounding environs to the building's interior. The presence of mature elm trees, temporarily enclosed by the iron framework, contributed to the surreal sense of being simultaneously inside and outside. 58 German critic Richard Lucae wrote at the time, "The barrier erected between us and the landscape is almost ethereal. If we imagine that air can be poured like a liquid, then it has, here, achieved a solid form, after the removal of the mold into which it was poured. We find ourselves within a cut-out segment of atmosphere."[29]

[29] John McKean, *Crystal Palace: Joseph Paxton and Charles Fox*, Architecture in detail (London: Phaidon, 1994), p. 32.

Paxton is said to have drawn the design of the giant glass shed as a single sketch 48 and then drafted the design in nine days, but as architect John McKean writes, "If Paxton's sketch is a few minutes doodle, it is one which he had already spent 20 years designing."[30] Indeed, Paxton's design for the Crystal Palace was based on the techniques and details he had developed in 1836 to construct the greenhouse known as the Great Stove in Chatsworth for the Duke of Devonshire. The central vaulted nave of the Chatsworth greenhouse was almost identical to the proposed nave of the Crystal Palace, and the ridge-and-furrow configuration of glass lites was also employed in his design for the Victoria Regina greenhouse for the tropical lily pad pool. In this sense, the overall design for the Crystal Palace was new, but the construction systems, cast iron, and glass details were not. This helps explain Paxton's confidence in submitting a design in just nine days for a structure of unprecedented scale, to be built in record time, for a fixed price.

[30] John McKean, *Crystal Palace*, p. 13.

The design of the Crystal Palace comprised thousands of repetitive, industrially produced parts

4|7 The Crystal Palace, designed by Joseph Paxton for the Great Exhibition of 1851 was one of the first demonstrations of a large-scale architectural interior that approached the scale of landscape, and produced its own interior microclimate – architecture as environment.

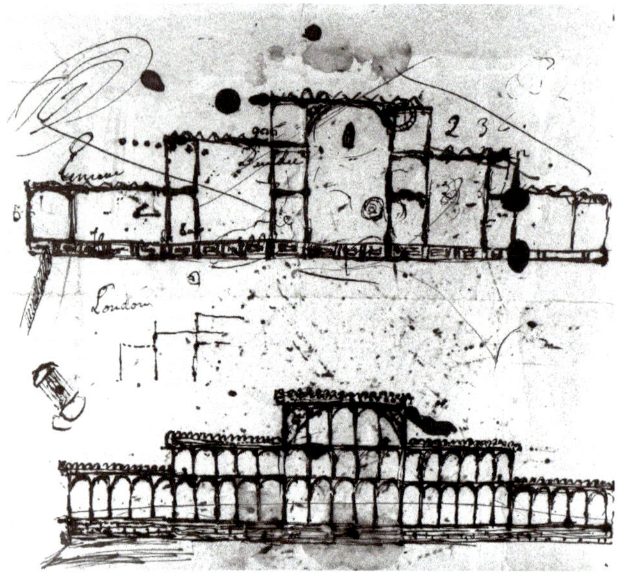

4|8 Joseph Paxton's original sketch of The Crystal Palace reveals his conception of the structure a modular construction of repeating bays of standard width.

that could be manufactured and delivered to the site in such a way as to maintain a continuous flow of materials for assembly, installation, and erection. The modular system devised by Paxton consisted of a 24 foot (7.3 m) square structural bay framed with cast iron columns that were eight inches (20cm) in diameter. The columns were tied together with horizontal cast iron trusses. The barrel-vaulted central nave was made up of wood and cast-iron half-circle arched sections. These were preassembled in the field and hoisted into position by teams of men using pulleys. As the metal frame was erected, teams of glaziers followed closely behind, installing 18,000 individual lites of glass, each 49 inches x 10 inches (120cm x 25cm). 4|9 Observers of the construction noted the incredible speed with which the enormous structure was erected and enclosed.

The Crystal Palace, designed to house the Great Exhibition of the Works of Industry of All Nations, was itself an exemplary work of industry, showcasing the potential of the processes of manufacturing, assembly, and installation applied to the task of housing the vast quantities of exhibits and the unprecedented crowds that came to witness the spectacular display. McKean writes, "The central focus here is not a building. Crystal Palace is not so much a form as a building process made manifest as a system — including conception, fabrication, shipment, assembly, dismantling, reuse."[31]

[31] John McKean, *Crystal Palace*, p. 21.

ARCHITECTURAL ATMOSPHERES

On opening day, May Day 1851, the Great Exhibition was inaugurated with an elaborate ceremony, and the Crystal Palace's 19 acres of interior space, totaling 800,000 square feet (7430sm), were occupied by 30,000 people. Visitors were struck by the other-worldliness of the new and vast architectural interior. Saint Paul's Cathedral (1710) would have been a frame of reference for many visitors. The Crystal Palace, at 1848 feet (563 m) long, and 456 feet (139 m) wide, was nearly three times Saint Paul's length, and almost twice as wide.

The precursor to the Crystal Palace was not the cathedral or the basilica, although the terms nave and transept were used to describe it. The precedent for the Crystal Palace was an entirely new building type, the enclosed plant nursery — the greenhouse. At the time, the study of botany and the natural sciences was an area of growing interest for a particular class of gentleman scholars.

British colonial expansion to the farthest reaches of the globe resulted in a collection of

49 Workmen assembling the "ridge and furrow" glazing of the barrel-vaulted central space.

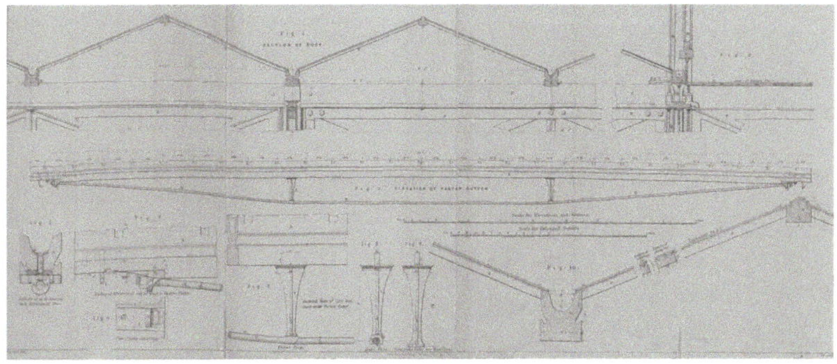

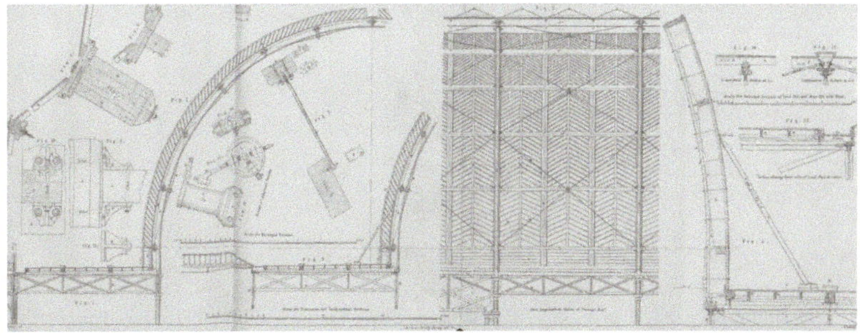

50 Section details of the "ridge and furrow" skylight glazing system showing the U-shaped profile of the wooden beam that acts as a gutter in the "furrow" section. Also note the grooves on the side of the beam that form gutters to manage condensation.

51 Section and elevation details of the barrel-vaulted section of The Crystal Palace.

seeds and specimens that were housed in the new architectural microclimates of greenhouses. A visitor to Paxton's Great Stove in Chatsworth described the interior as "a tropical scene with a glass sky."[32] Other accounts described the new nature of the greenhouse as "the Glass Garden of Eden at Chatsworth where plants flourish without rattlesnakes, and sugar-cane with no yellow fever."[33] The glass houses of the Victorian era were able to maintain tropical plant species by carefully calibrating the new artificial interior environments. Victorian glasshouse builders experimented with temperature and ventilation, the angle of the glazing, and furnaces to provide heat via subterranean tunnels.[34]

Paxton's design for the Crystal Palace was paradigm-shifting on many levels: the materials and prefabricated components launched a new era of standardized industrial parts, environmental design, critical path procurement, standardized assembly, and division of labor, which applied the theories of Adam Smith and anticipated the assembly line of Henry Ford. However, the significance of the Crystal is its awe-inspiring demonstration of architecture as a conditioned environment, at the scale and scope of the natural environment. It is a new nature, enclosed and compartmentalized, subject to its interior microclimates, and demanding its conditioning mechanisms for temperature monitoring, sun shading, and ventilation. According to scholar Susan Buck-Morss, "The Crystal Palace blended old nature and new nature — palms as well as pumps and pistons — in a fantasy world that entered the imagination of an entire generation of Europeans."[35]

Paxton succeeded because he understood his task was not really the creation of a work of architecture but rather, a temporary artificial environment designed to shed water, manage solar heat gain, channel condensation, and provide passive ventilation and fresh air. Paxton anticipated the effects of moisture, temperature, plants, and humans within the artificial environment of the Crystal Palace. His details incorporated methods for managing rainfall, condensation, heat buildup, and ventilation. Paxton's ridge and furrow glazing arrangement channeled rainwater to the cast iron columns, which did double duty as downspouts.

The primary spanning members between columns were composite beams comprising of routed-out wooden beams trussed on its underside with a one-inch (25mm) tension rod. 50 The beam's profile was carved on the top to act as a gutter for rainwater, while its sides were grooved to carry

32 John McKean, *Crystal Palace*, p. 14.

33 John McKean, *Crystal Palace*, p. 29.

34 Michael Wigginton provides a detailed account of the evolution of conservatories and glass houses in his book, *Glass in Architecture* (London: Phaidon, 1996), pp. 30–45.

35 Susan Buck-Morss, *The Dialectics of Seeing: Walter Benjamin and the Arcades Project*, 8. print. (Cambridge, MA: MIT Press, 1999), p. 85.

52 Aerial photo of the Crystal Palace during its second installation in Sydenham Hill, where it was occupied until it was destroyed in a fire in 1936.

53 The lush vegetation and greenhouse atmosphere of The Crystal Palace.

condensation, which Paxton knew would inevitably form on the underside of the great glass shed. 50

Throughout the Great Exhibition, which spanned six months from May to October, Paxton monitored the temperature with thermometers distributed throughout the extensive interior. He also compared the indoor temperature to the outdoor temperature. During hot days in the summer months, the temperatures inside the Crystal Palace reached 100 degrees Fahrenheit (38 degrees Celsius), and many reported the interior conditions to be uncomfortable. Paxton contemplated the use of calico fabric to cover the south-facing glass walls to reduce the solar heat gain and diffuse the light.

In addition to the solar heat gain, the presence of thousands of bodies and operating steam machinery on display produced internal heat loads. To mitigate those effects, Paxton incorporated horizontal bands of operable louvers, called ventilators, in the spandrel areas of the multiple tiers. 54 55 "The louvers within the low dado, and the depth of the roof girders are six-and-a-half-inch (16.5cm) wide, gently curved, galvanized iron blades; attached by levers, rods, and cranks in groups for mechanical adjustment."[36] These exterior shading and passive ventilation strategies were adapted from several greenhouse designs. Other techniques of wetting the fabric and introducing mechanical fans were contemplated, but it is unclear if they were employed.[37] Architectural historian Henrik Schoenefeldt notes that Paxton's plans for regulating the heat gain within the Crystal Palace were ultimately not fulfilled. Still, he considers the building "a first step toward the kind of glass building with a mechanically-controlled indoor climate which formed the core of Paxton's socio-environmental vision."[38]

The Crystal Palace occupies a unique place within the trajectories of modern architecture. It is the first large-scale implementation of environment design in service of creating an interiorized world with its own microclimates. Adapting techniques from greenhouses that effectively partition the interior from its surroundings, the building could theoretically be designed to mimic the Mediterranean climate in Southern Italy, or that of tropical India. On display within its expansive interior were artifacts from around the globe, collected and presented for the entertainment of the visiting crowds, allowing them to see the world without leaving the Crystal Palace. The structure was the product of an industrializing imperial power,

[36] John McKean, *Crystal Palace*, p. 25.

[37] Accounts of the actual thermal comfort within the Crystal Palace vary. Henrik Schoenefeldt's essay, "The Crystal Palace Environmentally Considered," identifies the building as a milestone in the introduction of environmental considerations and credits Paxton with employing a number of passive environmental control strategies. His review of the recorded temperatures within the building leads him to conclude that the means of controlling the indoor temperature were unsuccessful. Henrik Schoenefeldt (2008). "The Crystal Palace, environmentally considered." *Architectural Research Quarterly*, 12, pp. 283–294.

[38] Henrik Schoenefeldt, "The Crystal Palace, Environmentally Considered" p. 292.

5|4 Partial elevation of exterior wall of The Crystal Palace showing the band of horizontal ventilator louvers over the arches.

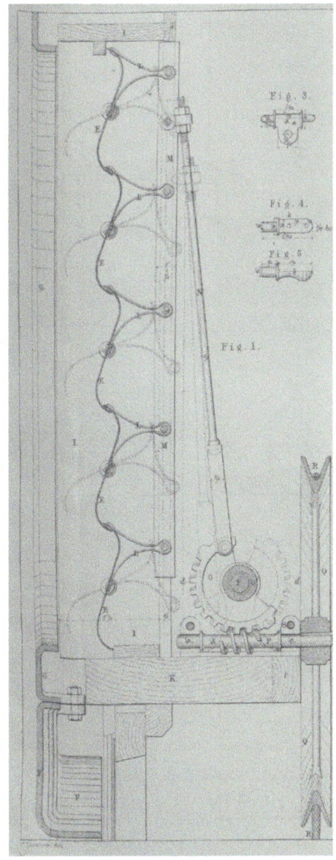

5|5 Detail section of the operable ventilator louvers designed to be opened or closed to regulate the interior environmental conditions.

expanding its global reach through a worldwide system of colonialism and open markets. At the same time, the interior world, contained by its thin glass envelope and comprised of regulated lighting, humidity, and temperature, acts as a parallel, independent interior environment — a "cut-out atmosphere," where it is "forever springtime."[39]

Part greenhouse, part basilica, part railroad shed, the Crystal Palace debuted a new architectural type that redefined architecture as the provision of an artificial environment and an atmosphere apart. It was a product of a global system even as it contained worlds within it, where visitors could view artifacts and experience climates from distant lands while also catching an oblique glimpse of an architecturally conditioned future.[40]

After the English parliament voted to dismantle the Crystal Palace, Paxton proposed alternate uses as a winter garden and a place for the public to recreate within the atmosphere of clean air and artificial climate. According to Schoenefeldt, "Paxton's proposal represented an attempt to utilize glasshouses as a means of reinstating a healthy human habitat within the environment of the emerging industrial cities which... were confronted with poor hygienic conditions, overcrowding, lack of daylight, air and water pollution."[41] Paxton, a gardener and inventor, conceived of a structure that would redefine both architecture and environment, raising issues of climate, health and wellness that continue to redefine architecture today.[42]

CLOSED WORLDS

Architectural historian Lydia Kallipoliti has been investigating the history of design concepts and case studies that she calls "the architecture of closed worlds."[43] Exploring the resonance between the present moment of climate crisis and recent past, when the fragility of the planet and its ecosystems pushed architectural agendas toward explorations with enclosed and self-sufficient structures, her work investigates the testing of the aqualung — a wearable apparatus invented by Jacques-Yves Cousteau that SCUBA divers later adopted — and meteorological experiments like the FNRS balloon that allowed scientist Auguste Piccard to reach the stratosphere in 1931 with the use of a "nacelle," a pressurized, aluminum precursor to the space capsule. Specific architectural experiments include the geodesic domes of Buckminster Fuller (1952), the Climatron in Saint Louis 1 by MIT graduates Joseph Murphy and Eugene Makey (1960), and Biosphere 2 (1991), a hermetically sealed research compound in Oracle, Arizona. 5|9

[39] John McKean, *Crystal Palace*, p. 29.

[40] Architectural historian John Hix has noted the technology transfer from greenhouse design to architecture, or the structures for housing plants to the structures for housing humans, in his book, John Hix, *The Glasshouse* (London; New York, NY: Phaidon, 2005).

[41] Henrik Schoenefeldt describes Paxton's winter garden proposals as a biosphere. "Paxton conceived the interior as a type of self-contained biosphere, in which plants and animals, including human beings, mutually participate in the sustenance of an internal carbon dioxide and oxygen cycle. Animals functioned as producers of carbon dioxide, which plants need for photosynthesis, while plants in turn maintained the oxygen levels of the atmosphere." Henrick Schoenefeldt, "The Crystal Palace, Environmentally Considered," p. 285.

[42] Reyner Banham, taking Joseph Paxton as an example, notes that "an intelligent commercial glasshouse operator today, judiciously metering temperature, moisture and carbon dioxide levels in the atmosphere around his out of season chrysanthemums, has more environmental knowledge at his fingertips than most architects ever learn." Reyner Banham, *The Architecture of the Well-Tempered Environment*, p. 145.

56 Exterior view of the Crystal Palace at Sydenham Hill showing the "ridge and furrow" glazing and the guttering system at the base of the barrel vault.

57 Rendering of the interior of the Crystal Palace at the original Hyde Park location, showing the mature elm trees that were temporarily enclosed within the glass enclosure. Gottfried Semper called it "the glass covered vacuum."

Kallipoliti does not feature The Crystal Palace in her closed world catalog, but it could be understood as a historical precursor to these 20th-century experiments. Its encapsulated interior landscape of distinct microclimates tracked by a series of temperature readings to monitor the comfort of its inhabitants. The Crystal Palace was the first large-scale closed-world experiment, and the thousands of exhibition visitors were its unwitting test subjects.

The most visible and perhaps most extreme closed world prototype is Biosphere 2, a scaled-down, simulated ecosystem that was sealed off from the outside world for 24 months while its occupants, called biospherians, lived and conducted a range of experiments from air and water quality to natural biomes, physical fitness, diets and physiology. 59 The name Biosphere 2 indicates the beginning of a series, which begins with Biosphere 1, planet Earth. Biosphere 2, then, seeks to produce a new and improved microcosm of our increasingly threatened biosphere — Earth 2.0, in today's terms.

The compound, built in the Arizona desert, was designed to completely compartmentalize interior and exterior. The space frame structure encloses 3.14 acres (1.27 hectares) and contains several distinct biomes, including a rainforest, a desert, a savannah grassland, a mangrove wetland, an ocean including a coral reef, and two anthropogenic biomes: an agricultural area and an area for human habitation. 60 Biosphere 2 was intended to test the potential for human inhabitation of a self-sustaining space colony, where plants, animals, and humans would form an interconnected ecosystem, where water and waste would be processed and cycled back into the system.

Biosphere 2 was a hermetically sealed alternative atmosphere, relying on architecture to maintain its absolute separation. The glazed long-span space frame roof structure, designed by Peter J. Pearce, an associate of Buckminster Fuller, achieved a near-airtight enclosure. A unique "multi-hinge," node-less space frame was developed to minimize thermal movements and leakage. The glass assembly consists of two layers of glass and plastic laminate, doubly sealed in two colors to make the redundant enclosure legible.[44] Additionally, a steel liner in the foundation was installed to ensure a tight seal below. The siting of Biosphere 2 in the Sonoran Desert simulates conditions on the surface of Mars. As such, its architecture, like the hull of a space shuttle, was intended to encapsulate the interior atmosphere as a prototype for a space colony.[45]

43 Lydia Kallipoliti, *The Architecture of Closed Worlds: Or, What Is the Power of Shit?*, (Zürich: Müller, Lars, 2018).

44 Janette Kim and Erick Carver, "Biosphere 2's Contested Ecologies," *Volume*, CILab Bootlegged Edition Urban China, 29-33. Retrieved from https://issuu.com/gsapponline/docs/01_crisis_interior_final_copy/79

45 The use of Biosphere 2 as a prototype for a future space colony was an explicit part of the founding mission and "historic imperative." Biosphere 2 founders John Allen and Mark Nelson describe the relationship between biosphere and the space program. "Biospherics opens up, together with astronautics, the ecotechnical possibilities, even the historic imperative, to expand Earth-life into the solar system and beyond that to the stars and then time's good opportunity to the galaxies." Quoted in Natalie Koch, *Arid Empire: The Entangled Fates of Arizona and Arabia* (London; Brooklyn, NY: Verso, 2022) p. 132.

8|5 Painting of the interior of the Crystal Palace by Owen Jones, who designed the color scheme and interior details. In this view the elm trees frame the view down the barrel-vaulted "nave" of the exhibition hall.

59 The Biosphere 2 project (1991), in Oracle, Arizona, was a prototype to test survivability in a hostile environment like Mars, or an environmentally damaged planet Earth.

The Biosphere 2 project failure was due to technical and operational failures, like when oxygen levels dropped below projected levels, and the disagreements between the occupants, the research institution, and the primary sponsor. As an architectural project, Biosphere 2 is an extreme case of architecture as a contained and encapsulated world with its own internal systems and microclimates. As Kallipoliti points out, the motivations behind many closed-world experiments were varied, ranging from scientific and medical research to 1960s counterculture and back-to-the-land movements. Some projects were government-sponsored research for military purposes. What they all share is a desire to test the limits of architecture to produce an insular and autonomous alternative to the outside world.

The search for a new architectural solution to the existing biosphere anticipates the inhospitability of the earthly environment, or the fear of its inevitable deterioration and decline. Indeed, Kallipoliti notes the contemporary resonance of closed-world thinking and current architectural concerns with the climate crisis, high-performance building envelopes, and the post-COVID-19 pandemic — along with the new relevance of air changes per hour, indoor air quality, filtration, and the relationship between architecture and occupant wellbeing.

AIR AS MATERIAL

A common thread through all these projects is air. Air is the medium of an architecture of closed worlds and high-performance building envelopes. Jacques-Yves Cousteau's aqualung is a breathing apparatus. Biosphere 2's technical equipment included an expandable lung designed to regulate the air pressure in the chamber to account for the expansion and contraction of air due to diurnal temperature fluctuation. Air is what would make a Mars space colony habitable. Airtightness is the essential quality of Passive House design and a key factor in energy conservation in heating and cooling contemporary buildings. Air filtration and quality are among the many critical concerns raised by airborne disease transmission, which the COVID-19 pandemic highlighted. In considering an architecture of climate and conditioning, air is a material, as much as concrete, steel, and glass.

Although glass is the material most associated with modern architecture, air is the substance that is central to an environmental, or climate architecture. Glass windows admit daylight, but they also create a greenhouse effect that traps heat. As discussed in Chapter 5, Paul Scheerbart imagined a material-specific utopia where cities are made of layers

60 The interior of Biosphere 2 contained distinct biomes for rainforest, desert, savannah grassland, and mangrove wetlands.

of colorful glass, and, importantly, a cavity of air. Scheerbart's writings influenced many German architects, including Bruno Taut and Mies van der Rohe. Taut produced a series of drawings that mirrored Scheerbart's ideas, illustrating crystalline buildings evocative of mountain formations, which he labeled Alpine Architecture. 61 Taut realized a prototype of his Alpine Architecture in Cologne at the 1914 Deutscher Werkbund exhibition. 63 His temporary pavilion, called the Glashaus, was constructed of glass and concrete, and featured a pointed dome of diamond-shaped glass panels. The prismatic dome evokes the faceted geometries of crystals and represents the search for a new materially driven architectural syntax.

The German glass industry funded the Glashaus in Cologne to promote the use of glass in buildings. In 1914, the use of glass in greenhouses was already well understood, but its applications in architecture were only beginning. Taut and Scheerbart were two advocates who believed in the material's potential to transform architecture and society at large. Taut associated the Glashaus pavilion with Gothic architecture, and the effects of colored light would have been akin to a new type of stained-glass window. The optical and specular effects of glass were associated with spiritual awakening.

Visitors to the Glashaus remarked on the interior effects of the colored light that streamed into the space through the glass dome above. The interior was lined with glass blocks, and an inner dome was glazed with colored glass. The effect was an immersive experience of space through colored light and lighting effects. Visitors to the pavilion would have also felt a change in temperature and humidity. As with any greenhouse, the glass structure inevitably admitted solar radiation, which would have elevated the indoor air temperature. The Glashaus demonstrated some of glass's most desirable optical and specular properties but also illustrated some of its inevitable challenges.

The building's dome had two layers of glass; the outer layer was glazed with reflective glass, and the inner layer was glazed with colored lites, forming a double-glass facade with a buffering air cavity. The concrete frame of the dome essentially served as a heavy mullion system for a diagrid glass envelope. 62 The sizeable exterior glass panels were approximately 12 feet (3.6m) tall and 7 feet (2.1m) wide. They were mounted on the outside to the concrete frame. The inner layer of glass employed smaller panels and was assembled like a stained-glass window in three vertical strips and set between the concrete mullions. The depth of the glazing cavity is visible in a historical

61 Bruno Taut's illustrations of Alpine Architektur (1917), evoke geometries of crystals and geological formations.

photo probably taken during installation, which shows a glass diamond with one of the interior strips removed and the figure of a man in a suit standing in the opening. 63 The X-shaped figure visible behind the man is the structure on the other side of the dome. Even though the pavilion was meant to showcase the use of glass as a building material, it also highlights the concept of an air cavity as an assembly and as a potential environmental design strategy.

Scheerbart was an author capable of imagining fantastical new worlds, complete with new architecture, rituals, and behaviors for a new society. He also was quite practical in thinking through the details that would allow his visionary writings to make the leap from imagination to implementation. Writing on the question of thermal buffering, "Since air is one of the worst conductors of heat, all glass architecture needs this double wall. The two glass skins can be a metre, or even further, apart... convectors and radiant heaters should not be put between the two skins because too much of their output will be lost to the outside air. Obviously, glass buildings are only suitable for building around the temperate zone, not in polar or equatorial areas."[46]

Building with glass also meant building with air. Air as a non-conductive, or rather, as an insulating material, was one of the central ideas of Scheerbart's 1914 utopian imagination. Both Scheerbart and Taut were admirers of the new conservatory at the Botanical Gardens at Dahlem in 1907, which housed a tropical greenhouse enclosed in an iron and glass shell with an interior span of 98 feet (30m). Scheerbart was critical of its single-layer glass construction. "The worst thing, though, is that the walls are single and not double; in consequence, the expenditure on winter heating is simply enormous."[47]

The use of translucent materials to create architectural microclimates for agricultural purposes can be traced back to the Romans, who, in the first century AD, according to Pliny the Elder, employed a cleft stone, *lapis specularis*, as a provisional glazed enclosure to grow cucumbers.[48] English garden architect John Claudius Loudon experimented with double walls and air cavities and used various angles of glass and orientations to maximize daylight penetration in the early 1800s. His pioneering published work on sun angles, glass assemblies, ventilation, and condensation, informed subsequent glass house designs. The origins of double glazing in architectural applications are found in 19th-century Northern European cities, where double windows,

46 Paul Scheerbart and Bruno Taut, *Glass Architecture and Alpine Architecture* (London: 1972), p. 42.

47 Paul Scheerbart and Bruno Taut, *Glass Architecture and Alpine Architecture*, p. 41.

48 John Hix, *The Glasshouse*, p. 10.

49 Alejandro Zaera-Polo, *The Ecologies of the Building Envelope: A Material History and Theory of Architectural Surfaces*, print edition. (New York, NY: Actar Publishers, 2019), p. 311.

50 Architectural historian Daniel A. Barber includes an account of the development of the IGU as part of the research and development of solar architecture. Daniel A. Barber, *A House in the Sun: Modern Architecture and Solar Energy in the Cold War* (New York, NY: Oxford University Press, 2016), p. 24.

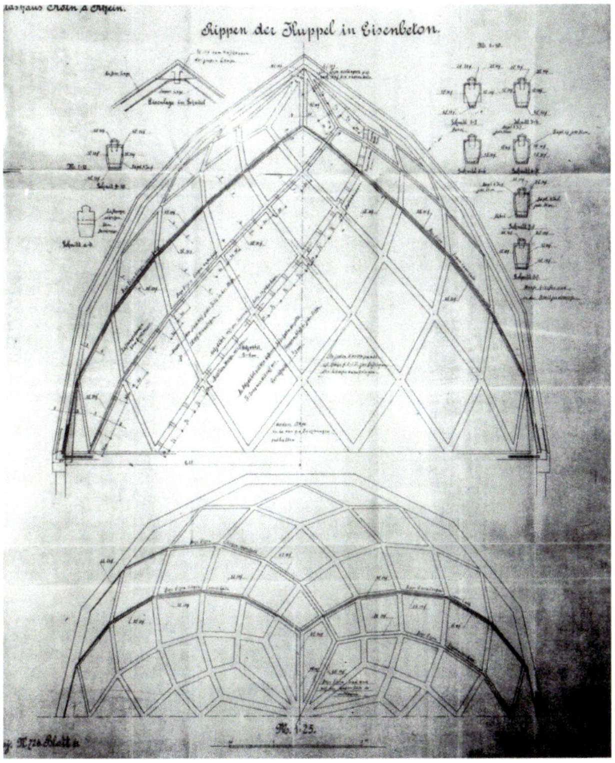

6|2 Bruno Taut's drawings for the Glashaus pavilion constructed as part of the 1914 Deutscher Werkbund exhibition in Cologne, Germany.

6|3 The constructed Glashaus pavilion (1914), with a double skin glass cladding mounted to the concrete frame. Note the scale of the man in the window in the center, and the subdivision of the inner lites into three strips.

known as *kastenfenster*, used a chamber of air to buffer the effects of heat loss.[49] These "box window" solutions were a precursor to the residential storm windows added to conventional windows to improve energy performance in cold climates.

The air cavity as an insulator is vital to an effective facade system. The cavity can vary in depth and still be effective. From as much as several feet (wide enough to be inhabited by a person) to half an inch (as is typical for an insulated glass unit, or IGU), the compartment of trapped air significantly improves the insulating qualities well beyond a single pane of glass. The IGU was developed and marketed by the Libbey-Owens-Ford (LOF) glass company as Thermopane in 1937.[50] Two layers of glass of equal size were separated by a half-inch spacer at the perimeter and sealed in the factory to create an integral assembly. Further developments beyond Thermopane have improved the material of the spacer and the seal at the perimeter to further reduce the inevitable thermal bridging at the perimeter. The IGU is now a standard product used in most buildings, and energy codes typically demand an energy performance for glazing that all but requires the double- or triple-pane assembly. Histories of modern architecture have been so focused on glass as transformative material and transparency as visual effect that they overlooked the elemental substance essential to its environmental performance: the compartmentalized air.

51 Lydia Kallipoliti, *The Architecture of Closed Worlds*, p. 21.

52 Lydia Kallipoliti, *The Architecture of Closed Worlds*, p. 282.

53 Describing the use of glass at the Toledo Glass Pavilion, Japanese architect Toshihiro Oki writes, "Glass walls are used both as physical separators and visual connectors, bringing far to near, outside to inside, and molten glass to exhibited art, and they allowed for further laying of the complexities of light." Michael Bell and Jeannie Kim, eds., *Engineered Transparency: The Technical, Visual, and Spatial Effects of Glass*, 1st ed. (New York, NY: Princeton Architectural Press, 2009), p. 119.

COMPARTMENTALIZED AIR

The hermetic dream of autonomous, interiorized worlds might serve as a cautionary tale. Early examples of mechanically conditioned and hermetically sealed buildings, like Pietro Belluschi's 1948 Equitable Building in Portland, Oregon, were susceptible to a host of problems, including poor indoor air quality, moisture, mold, and what became known as "sick building syndrome."[51] Kallipoliti included a coda to her book, *The Architecture of Closed Worlds*, written after the 2016 presidential election, where she likens the closed world thinking of architectural modernism to the bubble mentality of closed social media communities that self-affirm shared political opinions by creating "echo chambers" and avoiding or excluding different perspectives.[52]

One of the paradoxes of glass architecture is that it simultaneously separates and connects.[53] For example, the large-format single pane lites of glass installed at Mies van der Rohe's Farnsworth House produce the illusion that nothing separates the inside

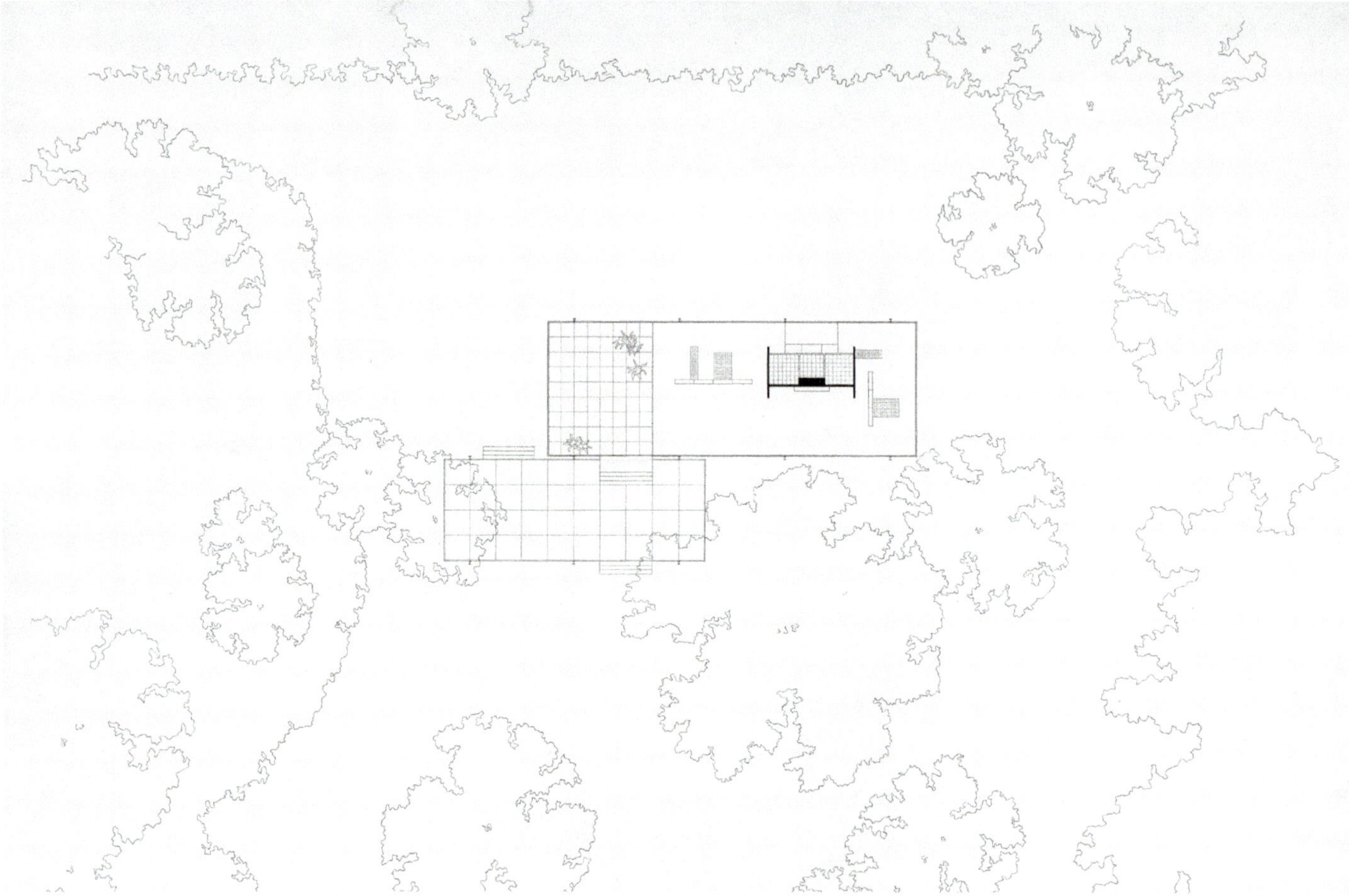

64 Plan of the Farnsworth House (1951), designed by Ludwig Mies van der Rohe, shows the house embedded in a clearing surrounded by mature trees.

domestic environment from the outdoor landscape. 65 Van der Rohe famously sought an architecture of "*beinahe nichts*," or almost nothing.[64] The Farnsworth House is understood as a reduction of every element to its essence, to the point where the architecture seems to disappear. Standing inside the house, looking south, a thin line of trees screens the waters of the Fox River that glide by. The view is framed by the gridded travertine pavers of the floor and the smooth white surface of the ceiling. 75 A white-painted steel mullion divides the panorama and seems to bind the floor to the ceiling, in tension, as if the ceiling was in danger of floating away. The slightly elevated perspective, produced by a platform held by eight steel columns, makes the middle ground disappear, leaving only the foreground floor and ceiling, and the background of the landscape beyond. The house's interior is visually collapsed with, yet separate from, its natural setting. 65

The house as a conditioned capsule thrust into a pristine landscape, akin to a view from a space station in an alien world, is a common trope of modern architecture. The thin glass membrane, and the chamber of air within is, as Lucae wrote of the Crystal Palace, "a cut-out segment of atmosphere." Yet, as Lydia Kallipoliti and others point out, closed-world thinking is a detrimental misconception of being in the world, doubly damaging to both the interior world and the broader environment.[65] It is more productive to think about the embedding of worlds within worlds, as well as the transactions and negotiations between its various layers.

NESTED TRANSPARENCIES

The Glass Pavilion at the Toledo Museum of Art shares many qualities with the Farnsworth House, as both are one-story, glass-enclosed structures nested in their landscapes, 66 yet their nuanced differences are instructive. Constructed 55 years apart, they share a common schema: a pavilion-like structure surrounded by the natural environment, a horizontally proportioned space captured between a floor and a ceiling, enclosed in a thin membrane of glass. The architect of the Glass Pavilion, Kazuyo Sejima, however, has stated that transparency was not the intent:

"All the glass is transparent but there are so many curved layers that the building has an opaque feeling to it. You cannot grasp if the reflection is from one layer of glass or another, or whether you are seeing the other side of the museum. The building produces a completely different feeling

[64] Philip Johnson, *Mies van der Rohe* (New York, NY: The Museum of Modern Art, 1947), p. 140.

[65] Kiel Moe's book *Insulating Modernism*, offers a sharp critique of the attitudes shaping contemporary approaches to architecture, and their insistence on a strategy of isolation. "Misguided propagations and permutations of isolation are evident in standing heat transfer practices for building envelopes, as well as in contemporary agendas for 'sustainable,' 'self-sufficient,' 'energy-independent,' 'autonomous,' and most impossibly, 'net zero energy' buildings." Kiel Moe, *Insulating Modernism: Isolated and Non-Isolated Thermodynamics in Architecture* (Basel, Switzerland;

65 View of the Farnsworth House (1951), designed by Ludwig Mies van der Rohe, hovering above the ground and enclosed in a thin membrane of quarter inch glass, to produce the impression of a dissolved boundary between inside and outside.

66 View of the Glass Pavilion at the Toledo Museum of Art (2006), designed by SANAA, standing on the ground and enclosed in multiple layers of half inch thick laminated glass, to produce the impression of a deep layered opacity.

6 7 Filleted corner of the Glass Pavilion, with deep a cavity, layers of curved glass, and subtle distortions.

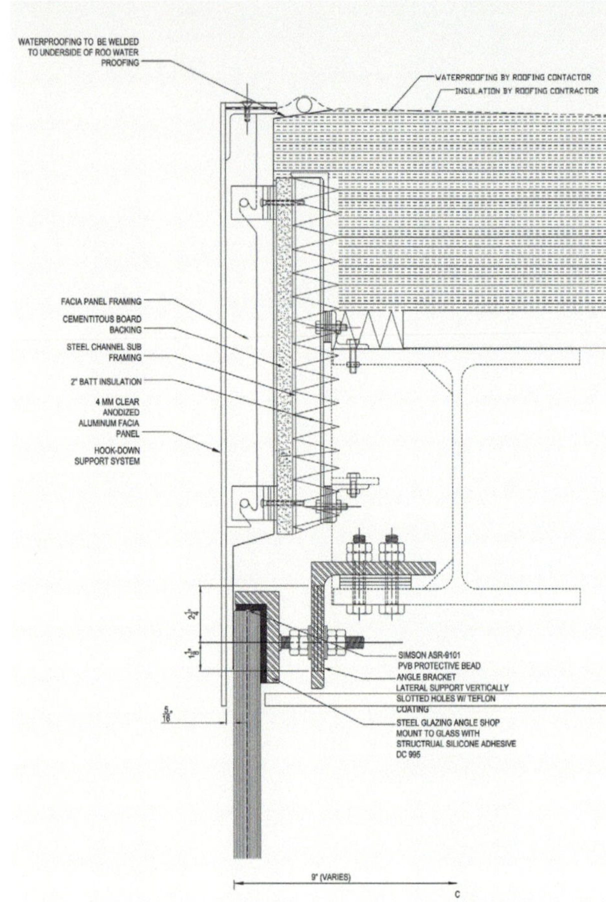

6 8 Section detail at the head of the exterior wall showing the laminated glass makeup and near-flush aluminum plate fascia panel.

of transparency. You can see through it, but it is opaque."66

For Mies, a single layer of quarter-inch (5mm) glass separates inside from outside, and a view through the whole house would traverse two layers of single pane glass. The Glass Pavilion's exterior walls comprise two lites of half-inch-thick (12mm) laminated glass. Interior partitions are similarly laminated but are made up of three-eighth-inch (10mm) glass. A view through the museum would traverse 12 glass walls, or 24 lites of glass. The accumulation of transparent layers produces a particular kind of opacity.

The rounded corners of the Glass Pavilion also create unique distortions and optical effects that make the building far from nothing. 6 7 Sejima's insistence on a material logic that multiplies layers and propagates effects makes the building a very specific something — a distinct material presence within the landscape of the museum campus. Sejima and the design team developed details, mocked up full-scale prototypes, and worked through procurement strategies and construction sequences to achieve the desired architectural effects.

The plan of the Glass Pavilion resembles a conceptual bubble diagram, a provisional arrangement of programs abstracted as filleted figures until their final arrangement could be decided. In taking the bubble diagram as the final design, arresting the initial moment of potential, Sejima produced a building that hovers between the concept and the constructed. The shorthand representation of provisional programs becomes the actual architecture. 6 9

The floor-to-ceiling curved glass introduces multiple challenges in procurement and installation. The glass is necessarily taller than the dimension between the finished floor and the finished ceiling. It sits in an aluminum "shoe" channel in the slab and projects into a "header" channel at the ceiling. 6 8 The steel frame structure and glazing channels were installed before the glass walls, making their installation a precise series of maneuvers. To install a full-height glass lite, the glaziers rotated the glass into the header and pushed it into an oversized pocket before lowering it into the shoe to rest on a pair of hard plastic or rubber setting blocks. After the glass was installed, the ceiling soffits were constructed flush with the bottom of the channel flanges. The glass appears to be stretched taut between the floor and the ceiling, with no visible frame or mullions.

The large air spaces between layers of glass are part of the building's unique optical and environmental strategy. The spacing of glass partitions

Boston, MA: Birkhäuser, 2014), p. 18. "Buildings are not isolated, autonomous, or self-sustaining systems. They are radically contingent, gradient dependent, dissipative systems," p. 21.

66 Kazuyo Sejima, in "Liquid Playgrounds." Fragments of a Conversation. *El Croquis* 121/122, p. 19.

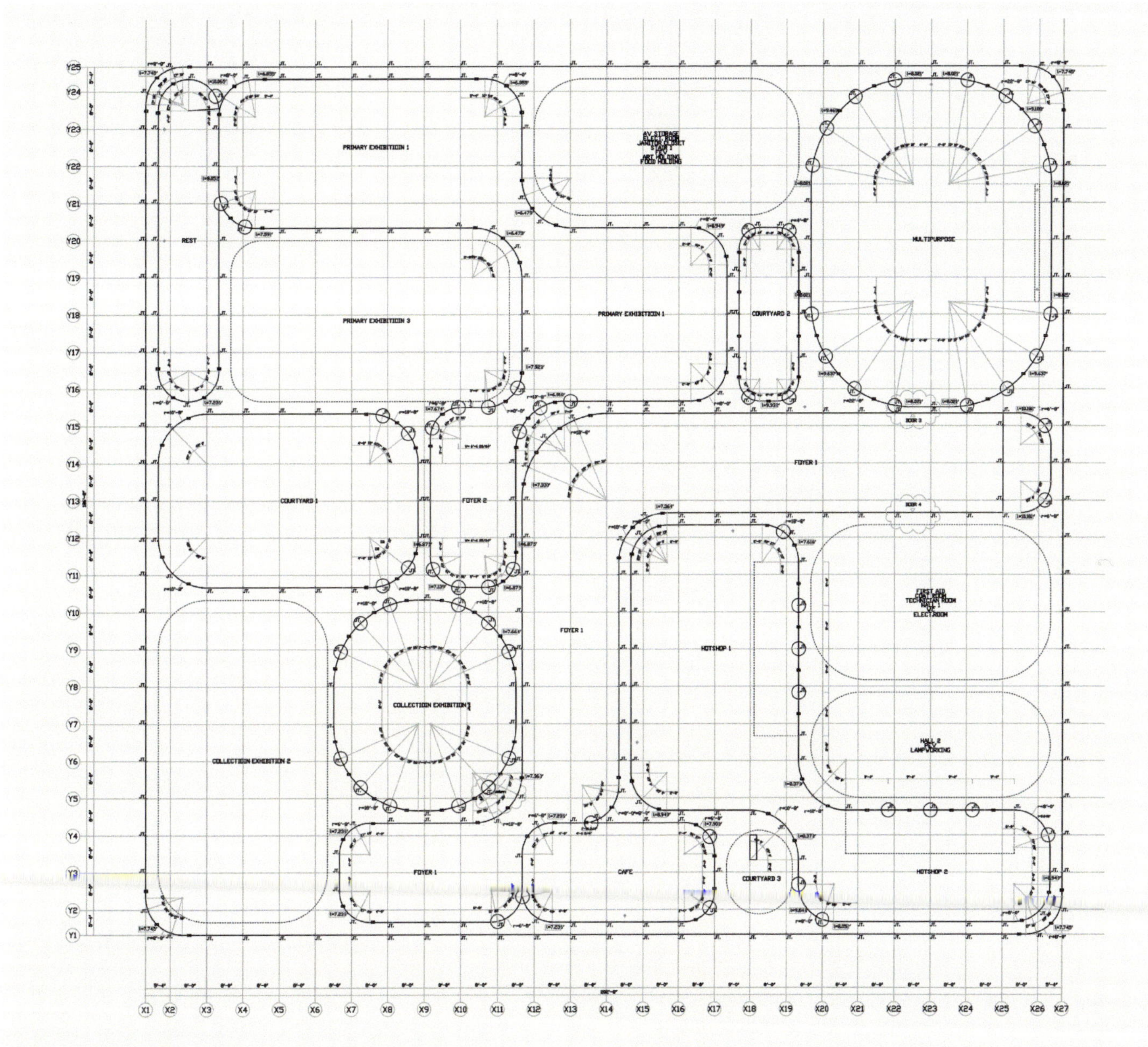

69 Glass geometry plan of the Toledo Glass Pavilion showing glass sizes, and the geometry of the various filleted corners.

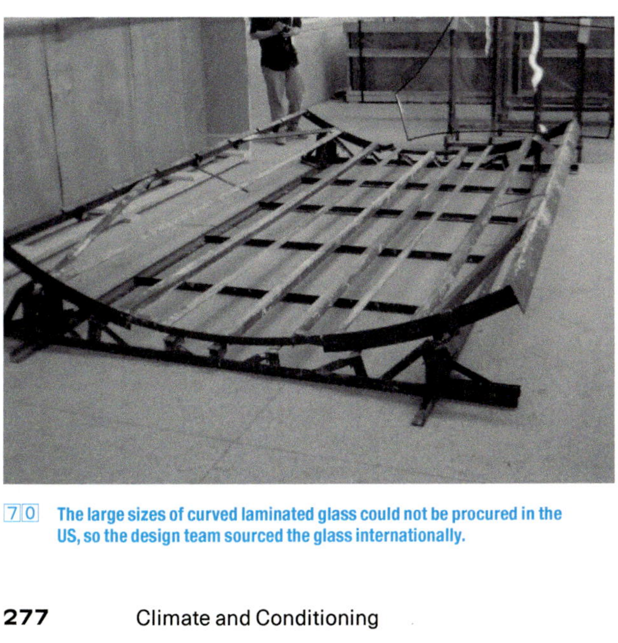

70 The large sizes of curved laminated glass could not be procured in the US, so the design team sourced the glass internationally.

71 The glass for the Glass Pavilion was produced in Austria, curved and laminated in China and shipped and installed to Toledo, Ohio for installation.

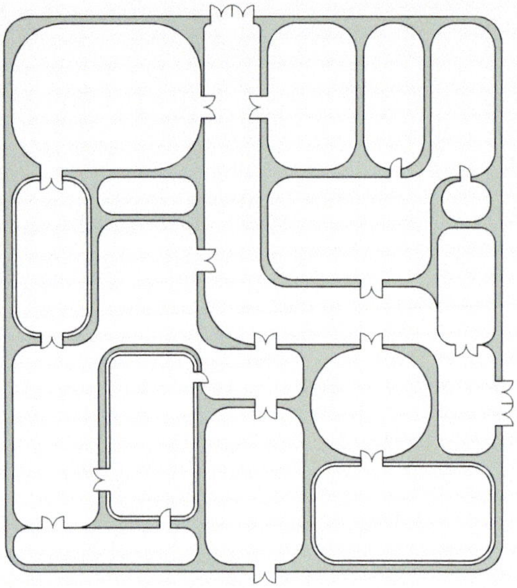

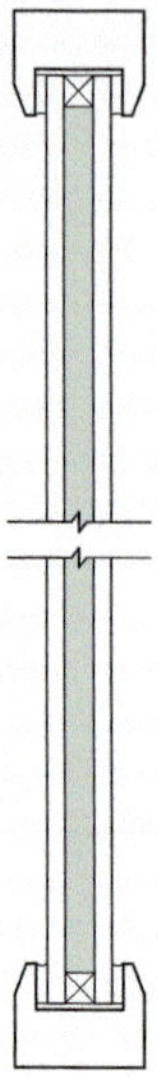

72 Plan diagram of the Glass Pavilion highlighting the space of the air cavity.
73 Section diagram of an Insulated Glass Unit (IGU) highlighting the space of the air cavity.

appears to multiply the building envelope, producing the layered visual opacity described by Sejima. The nesting of interior spaces also produces a series of cavities conceived as thermal buffer zones. Rather than specify insulated glass for the exterior envelope, Sejima proposed layers of laminated glass with an oversized airspace of varying dimensions. The building could be understood as a large, inhabitable IGU, where some places expand the typical half-inch (13mm) air gap to three to six feet (91–183cm). 57 72 73

57 The large air spaces in the Glass Pavilion are designed to be accessible for maintenance purposes, so their minimal dimension is three feet (91 cm).

The air gap in a typical IGU is hermetically sealed to achieve thermal buffering of the non-conductive air. The air gap in the Glass Pavilion's occupiable glass unit is conditioned to create a thermal buffer between the inside and outside. Flush radial diffusers in the cavity space supply conditioned air to maintain the temperature in the buffer space. The glass-blowing furnace used for demonstration purposes at the museum, allows the building's mechanical systems to reuse "waste heat" from the furnace to condition the buffer cavity.

While they may appear similar, the Farnsworth House and the Glass Pavilion articulate vastly different environmental dispositions and constructional logics. While Mies seeks to minimize the barrier between inside and outside to a quarter-inch (5mm) of transparent material, Sejima multiplies and maximizes the layers that make up that barrier, producing a nested and complex relationship between the interior and exterior.

The *kastenfenster*, the IGU, and the occupiable air cavity share an understanding of air as a material with physical properties. Air is never still. It is constantly heating and cooling, rising and falling. As it encounters occupants, it warms and rises. Eventually, it cools and settles. If you stand by a tall glass wall in the winter, you can feel a downdraft of cool air "falling" down its face. Air flows constantly in architecture, at the scale of the room, or at the scale of a half-inch (13mm) air cavity in an IGU. Air is a dynamic substance and a primary medium of climate design.

MUR NEUTRALISANT

In 1928, Le Corbusier won an international competition for the Centrosoyuz Building in Moscow, a large administrative building to house 3,500 employees of the Central Union of Consumer Cooperatives. 78 Le Corbusier saw Centrosoyuz as an opportunity to apply his theories at a large scale, including his experimental environmental systems for double-skin

7.4 The experience of the Toledo Glass Pavilion is of layers of glass and large inaccessible gaps between layers.

7.5 The experience of the Farnsworth House is of the planes of the floor and ceiling and the almost invisible glass membrane between inside and outside.

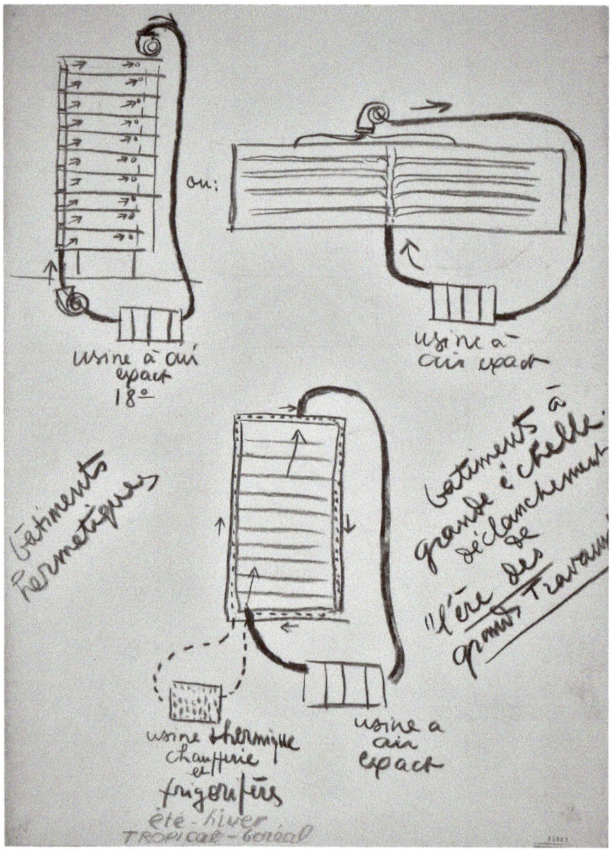

7.6 Le Corbusier's sketches of the Centrosoyuz Building mechanical systems for air distribution and the proposal for what he called a *mur neutralisant*, a conditioned cavity between the double facade.

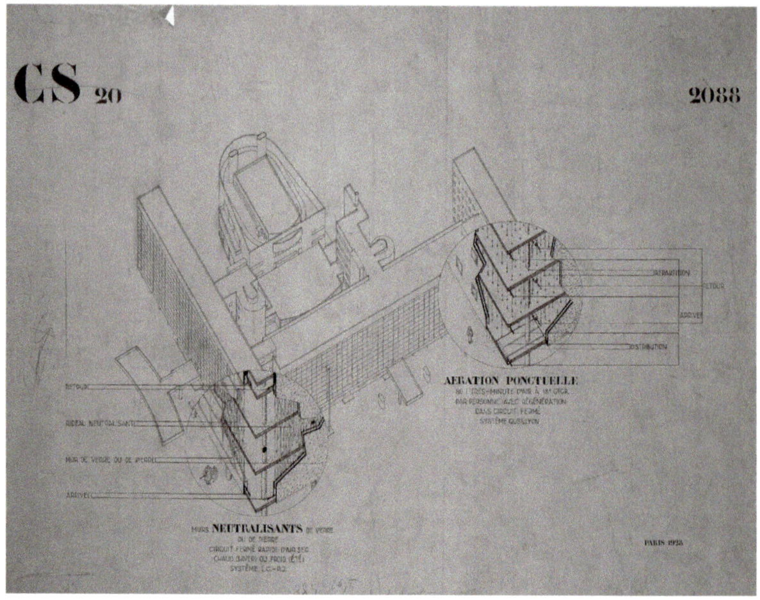

7.7 Axonometric drawing of the Centrosoyuz Building highlighting the layers of the *mur neutralisant* proposed for the building envelope.

facades. The building massing consists of three narrow, bar-shaped buildings arranged like an I-beam, and a large horseshoe-shaped auditorium volume with articulated circulation elements and bridge-like connectors. 7.7 Large sections of the massing are raised on *piloti* columns with Le Corbusier's signature ramps connecting floors.

Le Corbusier proposed a fully glazed and hermetically sealed double facade for the bar buildings, which he called the *mur neutralisant*, or neutralizing wall. Convinced that modern technologies could deliver a uniform interior climate, a *respiration exacte*, by architectural means, Le Corbusier developed his conditioned cavity wall as a thermal buffer. 7.6 The air in the cavity was to be heated or cooled, depending on the season, and ventilated at a high rate such that the occupants would enjoy a stable interior temperature of 18 degrees Celsius. Le Corbusier consulted with conditioning experts from the American Blower Company, who advised that the proposed system would be expensive and inefficient, "requiring four times as much steam and twice as much mechanical power as would be necessary with methods currently employed in our country under comparable atmospheric conditions."[58]

[58] Jean-Louis Cohen, *Le Corbusier and the Mystique of the USSR: Theories and Projects for Moscow, 1928–1936* (Princeton, NJ: Princeton University Press, 1992), pp. 91–92.

[59] Although Le Corbusier's concept of the hermetically sealed, conditioned cavity of the mur neutralisant was never realized, there were several installations of double-glazed facades and naturally ventilated air cavities realized by Soviet architects, such as the 1932 Narkomfin Building in Moscow by Moisei Ginzburg.

Undeterred, Le Corbusier insisted on his scheme and delivered architectural drawings showing fully glazed facades for all primary street elevations, regardless of orientation. Each of the primary administrative bars comprises a primary glazed facade and a secondary solid facade with punched windows. The solid facades were constructed of red sandstone blocks that are not evident in the black-and-white photographs of the time. Material shortages in the USSR delayed construction, and the project took eight years to build. Le Corbusier's mur neutralisant concept was rejected by the client as expensive and risky. The double-glass facade was eventually replaced with a more practical window wall system with operable windows for natural ventilation and a radiant heating system in the cavity.[59] Frustrated by the compromises to his original vision for the project, Le Corbusier never returned to the USSR to see the project completed.

It took another 50 years for the double-glass facade to be tested again, this time for an office building in Niagara Falls, New York. The Hooker Building is a compact nine-story cube designed by Mark Mendell of CannonDesign. 7.9 Built in 1980, with the effects of the 1970s oil crisis still in recent memory, the investment in a double facade was presumably made to reduce heating and cooling

7 8 The Centrosoyuz Building as realized, did not include the double facade or the *mur neutralisant*, but did reali:e a large, glazed curtain wall on the broad faces and stone cladding on the narrow faces. ,

79 One of the first double curtain wall facades, the Hooker Building, completed in 1980, wraps all four sides of the nine-story building in a four-foot (1.2m) wide cavity.

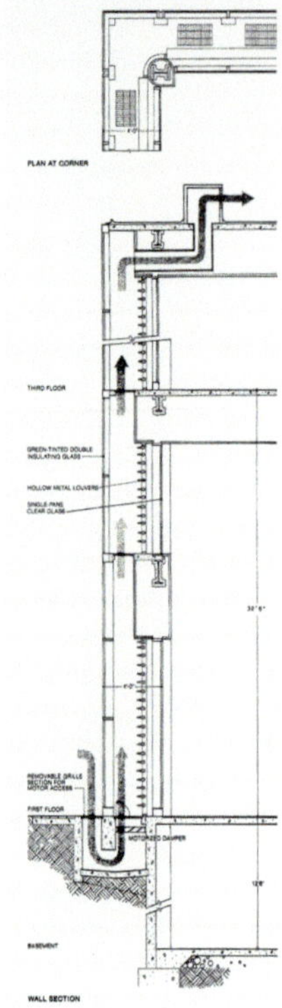

80 The cavity of the Hooker Building allows for buoyancy ventilation in the summer and can be closed to provide the thermal buffer in the winter. Sun shading louvers in the cavity form an additional layer.

81 Section diagram of the buoyancy ventilation in summer mode.

costs. The Hooker Building relies on a four-foot (1.2m) ventilated cavity with operable dampers at the street-level intake and the roof level exhaust. In the heating seasons, the dampers are kept closed, and the air cavity forms an extra thermal buffer outside the building. In the cooling seasons, the dampers are opened, and the cavity uses convection, or buoyancy ventilation, allowing heated air to rise naturally and exhaust at the roof level. 81 In addition to a thermal buffer, the cavity originally contained a series of operable horizontal sunshades that would open and close to provide additional shading as needed, or open to allow for more views.

The Hooker Building was pioneering in many ways, introducing passive strategies such as buoyancy ventilation in concert with adjustable solar shades integrated into the facade cavity. 80 The building also employed solar cells on the roof to track the sun and automatically adjust the louvers, introducing a level of environmental tracking and responsiveness at a time when most buildings were still conceived as hermetically sealed, inert envelopes. After it opened in 1980, the building was celebrated as an "energy tour de force" by *Progressive Architecture*, which reported that the mechanical systems contract had been cut in half and the gas-fired boiler had never been used for heat.[60] Due to poor maintenance, the operable louvers no longer function, and the lack of tenants makes the future of this important building uncertain.

The 1990s saw several significant double-facade buildings realized, particularly in Germany, due to strict energy codes and high energy costs. Spanish architect Alejandro Zaera-Polo describes building envelopes in evolutionary terms, with certain species of facades propagating under the right environmental (and political) contexts. "Reborn as an assemblage attuned to site-specific conditions, the double-facade species would finally see a wide proliferation in 1990s Germany and has remained a popular technology in Northern European countries, where the convergence of affluence, cold weather, dark skies, strict environmental policies, and progressive labor laws had created an optimal environment for these species to develop."[61]

Among the double-glass facades to emerge in 90s-early 2000s Germany are the RWE Tower in Essen by Ingenhoven Overdiek und Partner (1996), 87 the GSW Headquarters in Berlin by sauerbruch hutton (1999), 82 and the Post Tower in Bonn by Helmut Jahn (2002). 85 Each of these projects employs a version of the naturally ventilating air cavity wall system with variations in height and

60 Scott Murray, *Contemporary Curtain Wall Architecture*, 1st ed. (New York, NY: Princeton Architectural Press, 2009), p. 54.

61 Alejandro Zaera-Polo and Jeffrey S. Anderson, *The Ecologies of the Building Envelope*, p. 323.

8 2 GSW Headquarters (1999) in Berlin, designed by sauerbruch hutton, employs a full height double curtain wall facade on the broad face of the tower, as well as colorful sun shading panels in the cavity.

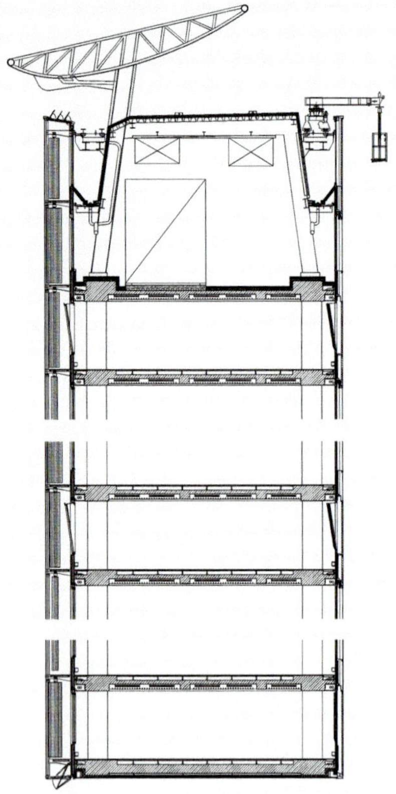

8.3 Building section of the GSW Headquarters (1999) showing the double facade with operable panels at the top and bottom on the west facade, and the operable windows on the east facade.

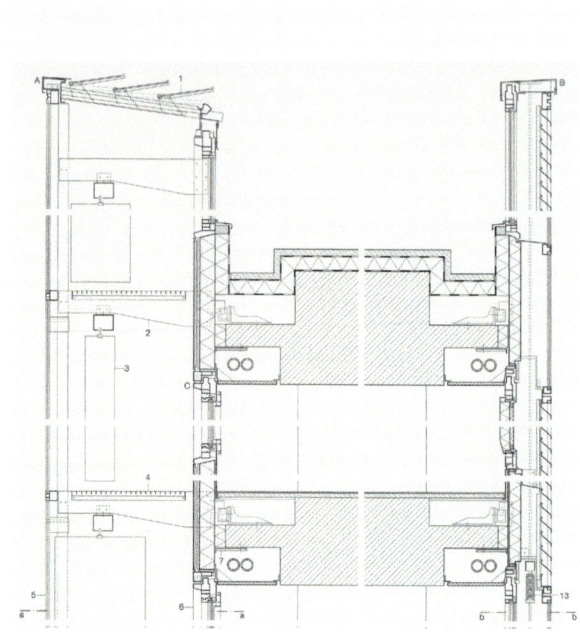

8.4 Section details of the GSW Headquarters (1999) showing the operable hatch at the top of the cavity, the sun shading panels in the cavity, and the metal grating for maintenance access to the cavity.

depth of the cavity, operability of the interior glazing, solar orientation, and sun shading systems.

The double facade relies on the thermodynamic behavior known as the chimney effect, with heated air rising within the cavity's "flue." This passive strategy works irrespective of scale — whether a single floor, or 20. The RWE Headquarters employs a single-story cavity, so air is drawn in from the bottom through a custom extruded aluminum vent and exhausted at the top of the single-story cavity. 8.8 Occupants can access the cavity by opening an interior glass partition, taking advantage of the ventilation of the fresh air moving through it in the summer. The relatively shallow cavity is maintained from the inside.

The GSW Headquarters' double facade extends the full height of the 20-story tower. 8.3 8.4 A hydraulic actuator opens a panel at the base of the tower, and a similar vent opens at the top, drawing air in and pulling it through the full height of the building. An open grille catwalk allows for maintenance access while allowing the air to pass through. Brightly-colored sunshades occupy the cavity, allowing occupants to regulate the sunlight.

The Post Tower in Bonn also employs a continuous cavity the full height of the tower on the north side, but the shingled profile of the south-facing outer facade allows for vents to intake air at every level, while operable windows on the inner facade allow the occupants to introduce fresh outside air into their office spaces. 8.6

A cost-benefit analysis of the double-facade system needs to weigh the high initial cost against the operational energy savings, which could take years to realize. Additionally, the loss of leasable square footage to the maintenance cavity zone makes the double facade inefficient from a real estate point of view, though the high visibility of the double facade does make a statement about sustainable design and broadcasts a certain progressive environmental stance. The financial viability for double facades is extremely rare, limiting the number of double-facade projects in North America.

VENTILATION AND VIEWS

An exception to the rule, the 2009 three-story addition to the Cambridge Public Library by William Rawn employs a modestly-scaled double facade. 8.9 9.0 Rawn conceived of the new library with a long reading room along the southern edge as an extension of the park. To achieve this goal, he set the ground level coplanar with the lawn beyond and ensured that the floor-to-ceiling glazing allowed the ground plane to

8 5 Helmut Jahn's Post Tower (2002) in Bonn, Germany, employs a glass double facade on both the north and south facing portions of the tower.

8 6 The south facing cavity wall of the Post Tower (2002) employs a shingled outer layer and a Venetian blind shading system in the cavity.

8 7 RWE Headquarters (1999), in Essen, Germany, is a small floorplate office building with a double facade.

8 8 The double facade of the RWE Headquarters employs a "fish mouth" mullion at the floor slab to regulate the airflow into the single-story cavity.

89 One of the few North American double facade projects, the Cambridge Public Library (2009), designed by William Rawn, employs a double facade on the south facade facing the park. The horizontal light shelves and Venetian blinds in the cavity work to limit the glare on the south facade, while encouraging views out to the park beyond.

90 The naturally ventilated air cavity of the Cambridge Public Library is open to the exterior in the summer mode allowing the cavity to draw warm air up through and out of the top.

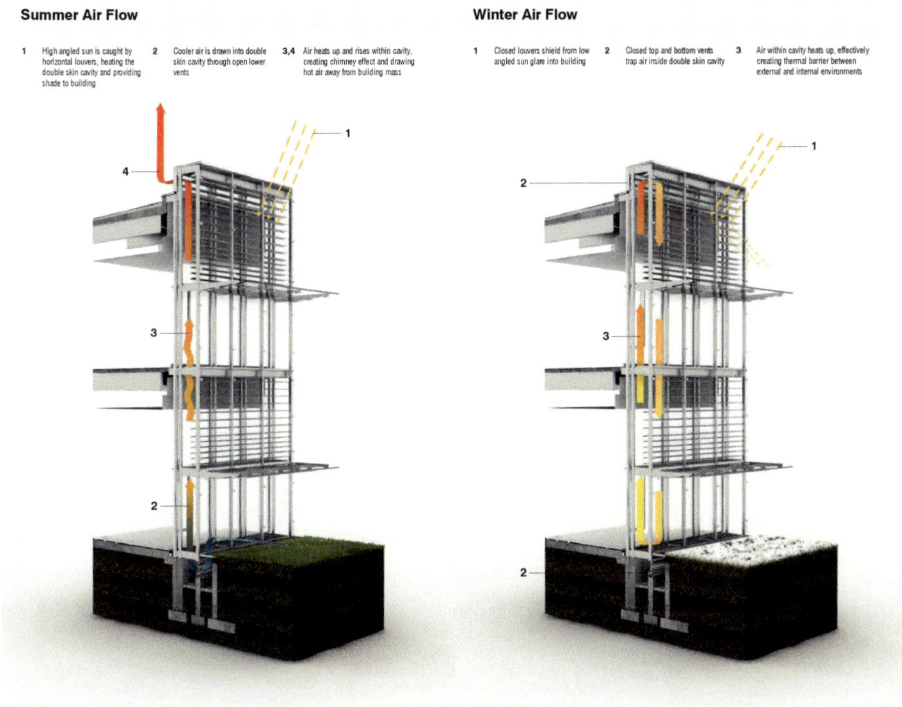

91 **Sections through the Cambridge Public Library double facade in both summer mode and winter mode. The chimney effect draws out warm air in the summer months, while the trapped air acts as a buffer in the winter months.**

appear to continue from outside to inside. Concerns about glare and heat gain led him to develop the double-facade strategy while maintaining visual continuity.

The ventilated air cavity relies on dampers at the top and bottom to allow the cavity to act as a buffer or chimney, depending on seasonal needs. Representations of these systems typically involve illustrating the different modes of operation, both in the winter and the summer modes, with what Barber refers to as "seasonal sections."[62] 91 The two-story cavity is wide enough to be accessible for maintenance, and operable windows on the inner layer allow for fresh outside air to be introduced to the library. The cavity houses a mechanically operated Venetian blind system for shading. Unlike many European examples, exterior sunshades are uncommon in North America, though they are most effective when placed outside the building. Maintenance concerns are likely the reason for the lack of adoption, and putting the sun shading in a protected, accessible cavity is one solution.

[62] Daniel A. Barber comments during a lecture titled, "Out of time," at the Harvard Graduate School of Design, on September 27, 2024.

FILM FACADES

A new generation of double facades is utilizing materials, such as ethylene tetrafluoroethylene (ETFE) and polyethylene tetrafluoroethylene (PTFE) films, allowing for thin pillow-like cavities that trap a pocket of air between the layers of film, and expand the expressive range of the double facade beyond a flat, gridded glass surface. ETFE envelope projects include the Eden Project in Cornwall by Grimshaw and Partners (2001) 93, the Allianz Arena in Munich by Herzog & de Meuron (2005) 97, and the Kaplan Institute at the Illinois Institute of Technology (IIT) by John Ronan (2018) 96. Each project employs a pneumatically pressurized, large-format ETFE film within a metal frame to achieve a distinctive, lightweight air cavity building envelope.

Related to the early Victorian greenhouses of the early and mid-1800s, the Eden Project is a next-generation conservatory conceived and constructed around the millennium as an environmental education destination. The complex includes eight intersecting geodesic domes that house two distinct biomes: a tropical rainforest and a Mediterranean climate. 95 The complex was designed by Grimshaw Architects, whose high-tech, long-span structures include rail stations, airports, and conference centers. The Eden Project was one of the first large-scale architectural uses of ETFE, which is extremely lightweight — weighing one percent of the weight of glass — and is not limited

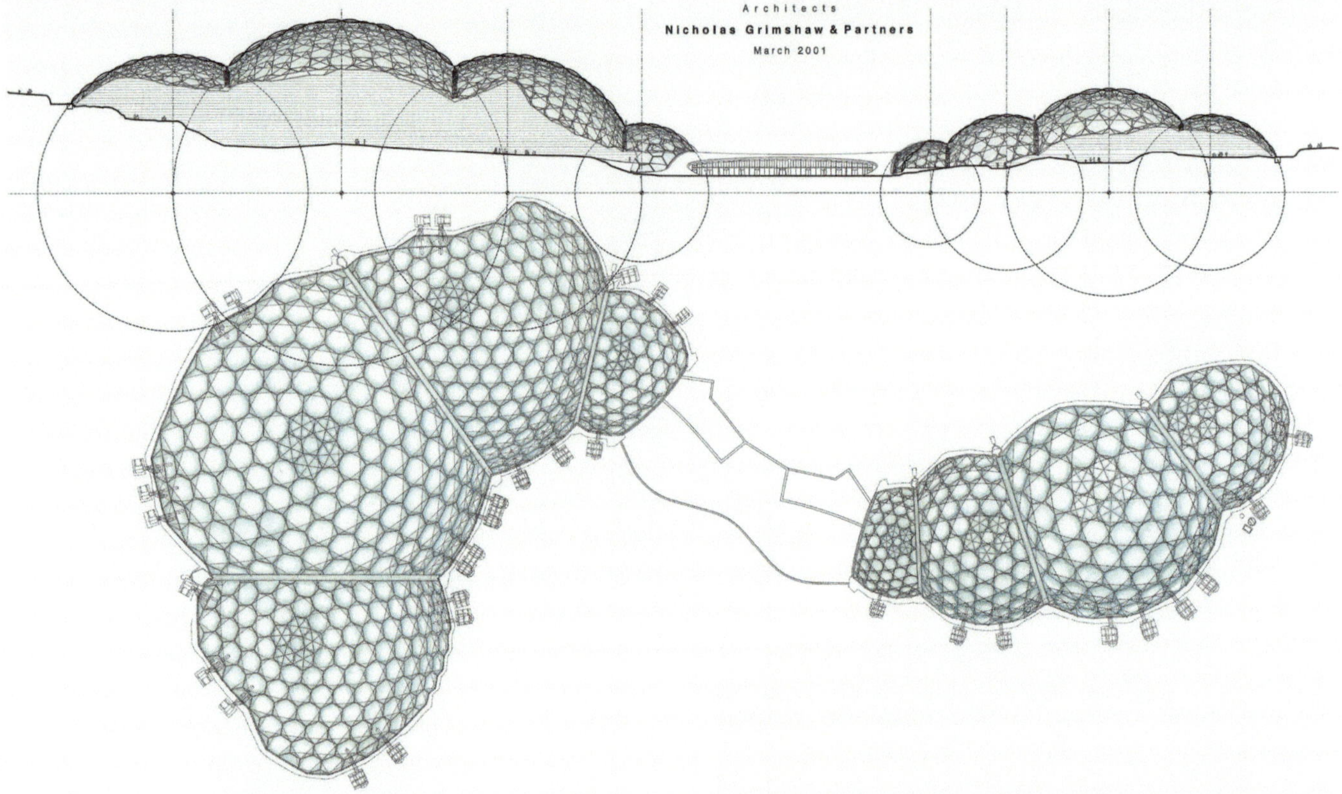

9.2 Section and roof plan of The Eden Project (2001), designed by Nicholas Grimshaw as a series of intersecting "soap bubbles" that conjoin to form two primary volumes housing two distinct biomes. The geodesic dome geometry was chosen to allow it to intersect with the uneven site conditions of the former clay quarry.
9.3 The spherical geometry of The Eden Project (2001) refers to the geodesic dome studies of Buckminster Fuller (1960s), but are here scaled up to form significantly longer spans with larger hexagonal cells.

9|4 The geodesic cells are infilled with pressurized ETFE pillows that are able to span the large openings in the geodesic frames. Some of the largest frames are 36 feet (11m) wide.

9|5 The interior of the Eden project is a lush rainforest biome.

by size, as it ships in a roll and is field-fitted to the frame. The hexagonal and pentagonal tiles of the Eden Project vary in size, with the largest cells spanning 36 feet (11m). 9|4 The triple-layer ETFE cells are individually pressurized with air hoses to achieve its expanded pillow profile. The lightweight geodesic structure is capable of extremely long spans, creating a voluminous interior suitable for tall rainforest trees. As with historic greenhouses, the building creates unique microclimates within the shell tailored to specific plant species and ecosystems. Learning from the conservatory projects of the past, like Paxton's Crystal Palace and Saint Louis' Climatron geodesic dome, the Eden Project domes are capable of ventilating through operable cells at their apexes, as well as through louvered openings around the perimeter.

John Ronan, designer of the Kaplan Institute at the Illinois Institute of Technology (IIT) had the difficult task of integrating a new academic building into a historic campus, master planned and designed almost entirely by Mies van der Rohe. Mies designed 20 buildings as part of the historic South Chicago campus. More recent buildings on the campus include the McCormick Tribune Campus Center by OMA from 2003. Ronan's Kaplan Institute building occupies a site that was formerly a surface parking lot, creating a new building footprint analogous to other low-rise rectilinear buildings at IIT that are planned on a 24-foot (7.3m) structural grid. The massing adheres to the global orthogonal campus buildings, but instead of the bronze plate glass typical of Mies buildings, he uses an ETFE pillow for the building envelope. Instead of another taut, dark glass box, the Kaplan Institute's facade is milky white and bulges out in three horizontal bands around the second-story volume. 9|6

The use of the ETFE film facade does something that none of the other IIT buildings can do: it curves. Ronan's curves, however, are not willful or gestural, but material deformation resulting from the internal pressure of the air pushing against the rigid frame of the gridded facade. Ronan, who teaches at IIT in Mies' Crown Hall building, describes his choice of materials, "Mies had described Crown Hall as 'skin and bones' architecture, and I wanted to carry on that tradition, but using materials that were not available to Mies. The ETFE was a lightweight material, a tiny fraction of the weight of glass, and less costly. It was relatively new to architecture and would give the building a weightless, cloud-like appearance."[63]

Ronan also finds an innovative approach to the corner condition that dialogs with the adjacent

63 John Ronan et al., *Out of the Ordinary: The Work of John Ronan Architects* (New York, NY: Actar Publishers, 2022), p. 179.

96 The Kaplan Institute (2018), designed by John Ronan as an addition to the Mies-designed campus, is clad in large ETFE pillows offering a contrast to the dark bronze rectilinear massing of the majority of the IIT campus.

97 The Allianz Arena (2005), designed by Herzog & de Meuron, uses ETFE pillows to give the stadium a singular unifying texture.

buildings. Mies' architecture is a system made up of repetitive orthogonal components arranged according to a universal grid. Mies' corner is where the universal grid of the structure and the aligned grid of the building envelope combine to form a unique expression of an articulated modular corner. [98] For Rohan, the choice of ETFE as a building envelope meant that the module was not limited to standard sizes that would be transportable to the construction site on a truck. As a sheet good, the ETFE material's length is almost limitless, and its flexibility means that it could wrap around the corner. Rather than interrupt the horizontal banding with a mullion on the corner, Ronan designed a clear acrylic cylinder corner mullion so the ETFE pillow could wrap the corner smoothly without creating a dark shadow of an opaque material. The soft filleted corners are distinctly un-Miesian, but they are true to the inherent logics of the new film material. [99]

The ETFE pillows trap a significant amount of air in between the layers, and the large format enabled by the sheet material maximizes the ratio of envelope to frame, thus limiting the amount of thermal bridging at the frames. To reduce the heat gain in the summer, the four layers of ETFE film are printed with opaque dot screen patterns. [100] The patterns offset and misaligned but are installed so that changes to the internal pressure between the different layers can change the relationship between patterns, making them transparent or opaque. Ronan's response to the history of mechanical louvers and ventilators with fans and actuators is variable opacity with no moving components — using only air, under pressure.

MASSIVENESS

Architect and author Kiel Moe describes buildings as "vast vortices of matter and energy,"[64] combining short-term energy cycles like heating and cooling over the course of a day, with extremely long cycles like energy embodied in building materials. Arguing against conventional practices of conceiving of buildings as isolated and discrete objects, Moe uses the laws of thermodynamics to illustrate how buildings are contingent systems, constantly exchanging energy with their surroundings. "Buildings are not isolated, autonomous, or self-sustaining systems. They are radically contingent, gradient-dependent, dissipative systems."[65] For Moe, buildings are inevitably thermodynamic entities, consuming energy in their construction and operation. Rather

64 Kiel Moe, *Insulating Modernism*, p. 19.

65 Kiel Moe, *Insulating Modernism*, p. 21.

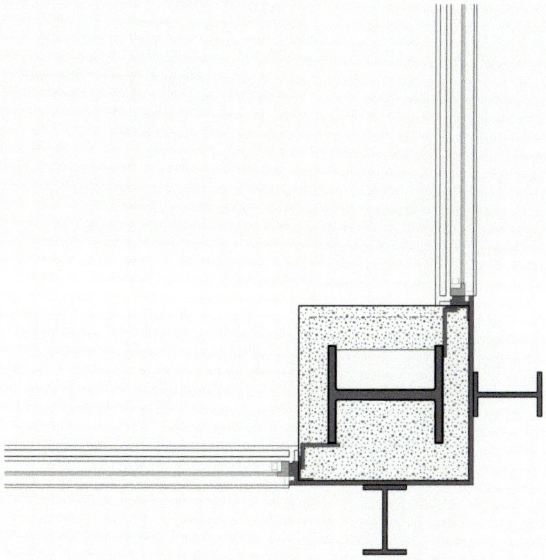

98 Corner plan detail of Ludwig Mies van der Rohe's Crown Hall (1956) at IIT. For Mies, the corner is where the building systems are expressed, and the thickness of materials are legible. Mies marks the column grid centerlines with the additive I-beams.

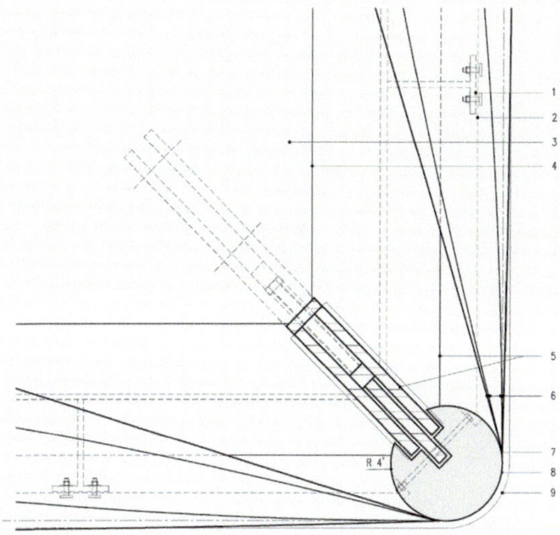

99 Corner plan detail of John Ronan's Kaplan Institute (2018) at IIT. For Ronan, the corner also celebrates the material properties, which for ETFE is its ability to wrap around the corner. Ronan uses a clear acrylic rod to maintain the sense of clarity and lightness at the corner.

than attempt to isolate buildings from their contexts, Moe urges architects to reconceptualize architecture as part of a vast flow of energy, with different cycles and velocities of energy dissipation. Conceptualized thus, a building with thick exterior walls, operable windows, and thoughtful siting relative to solar orientation can use its thermal mass to its advantage, capture the warmth of the sun on southern exposures, and allow users to regulate its temperature with natural ventilation. Moe calls this approach the "Massivhaus" paradigm.[66]

As an example of a building with no mechanical systems employing thermal mass as a climate strategy, the 2226 Building by Baumschlager Eberle Architects offers an alternative approach to climate design. 102 The six-story building, located in Lustenau, Austria, is used by the architects themselves as their office space. The compact building massing is square in plan, and the exterior of the building consists of a grid of vertically proportioned punched windows. The building relies on its load-bearing perimeter walls for both structure and enclosure. Two layers of 15-inch-thick (38cm) masonry units create the massive exterior wall. 101 The inner layer is load-bearing, and the outer layer provides insulation. Finally, an outer layer of plaster over the masonry gives the building a uniform exterior finish.

The building has no heating or cooling system, relying on the thermal mass of the building to gain and retain heat during the day and dissipate it in the evening. Building occupants and equipment also produce heat, which contributes to the thermal conditions of the interior. According to the designers, "The only heat sources it contains are there for other purposes, from the users themselves — each person gives off an average of 80 watts of heat due to the use of lights, computers and photo-copiers, even coffee machines. Living in harmony with the elements."[67]

The name of the building, 2226, refers to the temperature range of the building's interior: between 22 and 26 degrees Celsius. The operable windows also allow occupants to regulate the temperature with natural ventilation, if necessary. The vertically proportioned punched windows allow daylight to penetrate deep into the shallow floor plate, allowing the occupants to minimize their energy consumption.

The architects describe the building as "a functioning part of its environment rather than just 'architecture.'"[68] This project represents a significant deviation from architecture that has

66 Kiel Moe, *Insulating Modernism*, p. 252.

67 https://www.baumschlager-eberle.com/en/work/projekte-details/2226-lustenau/

68 https://www.baumschlager-eberle.com/en/work/projekte-details/2226-lustenau/

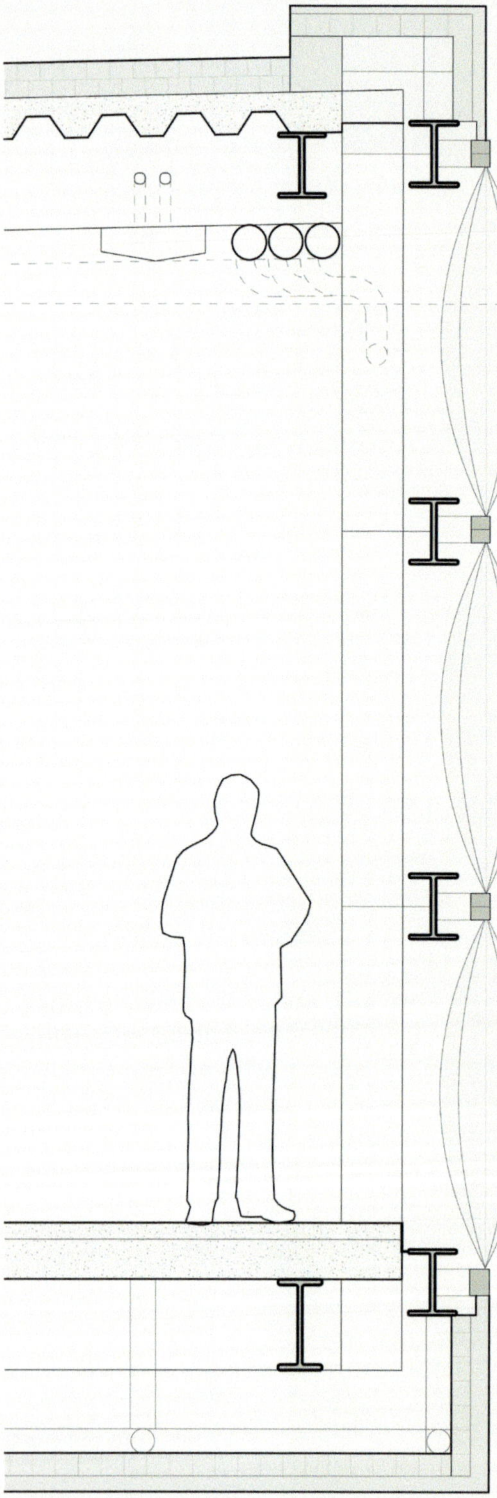

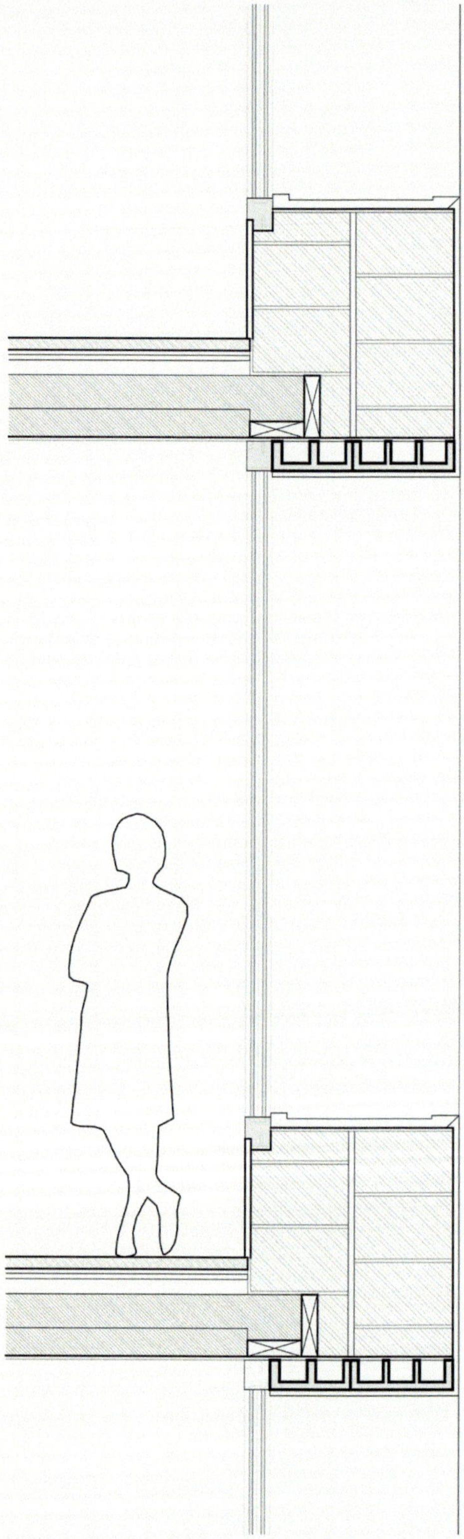

100 Wall section of the Kaplan Institute at IIT showing the steel frame construction and the lightweight ETFE pillow facade that uses air as an insulating material.

101 Wall section of Building 2226, designed by Baumschlager Eberle, showing the massive thickness of the exterior wall that uses the thermal mass of material as an insulating strategy.

102 Building 2226 (2013), in Lustenau, Austria, designed by Baumschlager Eberle, embodies the principles of passive design, with no mechanical systems it relies on the thermal mass of the exterior wall construction.

sought to isolate buildings from their surroundings through material systems, insulation, and active and passive conditioning systems. A low-tech, high-thermal mass approach acts as a productive reset in a building culture that has moved toward an increasingly specific, layered systems approach to building envelopes and the mechanical conditioning of interior atmospheres. The straightforward construction of the 2226 building and its high-quality materials is an argument for durability as a sustainable building strategy. The architects project the building to last 200 years, well beyond the lifespan, and 5- to10-year warranties of energy-consuming, high-maintenance, mechanical systems.

Conceiving of architecture in terms of climate and conditioning, as both the interiorized Crystal Palace microclimate or the thermodynamics of the 2226 building, requires approaches that understand environments as dynamic and ever-changing, and a long-term perspective on durability and operations. This approach runs counter to the prevailing practices and contemporary cultures of high-tech systems and designed obsolescence, yet it is a crucial and necessary paradigm shift, if we are to reimagine architecture in the age of climate crisis.

1 The Elbphilharmonie (2017) in Hamburg, designed by Herzog & de Meuron, is a heroic case study for the adaptive reuse of an existing warehouse in an industrial port area of Hamburg that transformed both the warehouse and the district with a catalytic development of a cultural anchor program.

Chapter 9 Afterlives

Architectural historian Charles Jencks declared the death of modern architecture precisely, if incorrectly:[1] "Modern architecture died in St Louis, Missouri on July 15, 1972, at 3:32 p.m. (or thereabouts) when the infamous Pruitt-Igoe scheme, or rather several of its slab blocks, were given the final *coup de grâce* by dynamite. Previously, it had been vandalized, mutilated, and defaced by its black inhabitants, and although millions of dollars were pumped back, trying to keep it alive (fixing the broken elevators, repairing smashed windows, repainting), it was finally put out of its misery. Boom, boom, boom."[2]

Jencks' declaration was accompanied by black-and-white photos of the implosion of the Pruitt-Igoe housing blocks only 16 years after completion. [2] The now infamous photo, captured from a neighboring building, shows Block C-15 engulfed in clouds of dust, lurching toward its collapsing middle section. The ends of the building appear intact, windows and all, tipping inward like a capsizing boat. Beyond the central building, we see rows of identical buildings awaiting the same fate. The Gateway Arch is visible on the skyline in the upper left-hand side of the photo.

The highly publicized demolition of Pruitt-Igoe, and Jencks' opportunistic rhetoric solidified in the public's mind the triple myth of failure: the failure of subsidized housing in the US, the failure of the welfare state, and the failure of modern architecture to serve a social purpose. For Jencks, the photo of the apartment building, frozen in freefall, was a convenient milestone from which to declare modern architecture dead and launch a new movement, that of postmodernism. For the residents of Pruitt-Igoe, who had suffered a housing project plagued by mismanagement and lack of maintenance, the demolition of was one of a series of displacements, from the 1950s urban renewal to the 1970s demolitions. For local politicians, demolition was a powerful means for signaling change, and the 33-building complex in Northeast Saint Louis was a convenient prop.

The images of the collapsing buildings are difficult to forget. The photos were published in *Life* magazine and circulated widely. The event was broadcast live on television across the country, turning it into a national spectacle and a reckoning for publicly funded housing. Video footage of the collapsing buildings was also included in Godfrey Reggio's experimental 1982 film, *Koyaanisqatsi*. Panning in slow motion and overlaid with music composed by Philip Glass, the symphony of buildings slowly being dragged down by gravity in a cloud of dust is awesome and tragic.

Built in 1951 and designed by Japanese American architect Minoru Yamasaki, the repetitive

1 The demolition in the photograph took place in May, not July. For Jencks, the correctness of the date was less important than the specificity of a time and place. Jencks sought to narrate the demise of modern architecture, taking on the language of a detective novel. In doing so he also revealed his own intense prejudices and lack of concern for actual events.

2 Charles Jencks, *The Language of Post-Modern Architecture* (New York, NY: Rizzoli, 1977), p. 9.

2 The televised implosion of Pruitt-Igoe housing blocks was used to signify the end of public housing as well as the end of a vision of modern architecture that was intended to address social issues. Their demolition highlights the way obsolescence functions both materially and culturally.

slabs of apartment buildings were initially heralded as a modern alternative to the various tenement buildings that they replaced as part of "urban renewal" and "slum clearance." Tenants who occupied the new buildings marveled at the elevators, wide corridors, and ample space between buildings, but systemic mismanagement combined with "white flight" marred the utopian beginnings of Pruitt-Igoe and led to abandonment, vandalism, and vacancy. Their demolition in 1972 transformed them into symbols of a system gone wrong.

As the Pruitt-Igoe towers were brought crashing down through controlled demolition, another pair of towers, also designed by Minoru Yamasaki, were just topping off. In New York, the World Trade Center's Twin Towers opened in April of 1973, claiming the title of the world's tallest building. 3 An engineering marvel, the towers rose 1,368 feet (417 m) from the street in a single, uninterrupted line. No setbacks. No base, no middle, no top, as the traditionally regarded organization of the tall building. You could stand at the base and take in their full height, a quarter mile of parallel lines of their anodized aluminum panel facade. The Twin Towers stood at the tip of Manhattan for 28 years before they, too, came crashing down on September 11, 2001.

In the weeks following the attacks on the World Trade Center, the identity of the hijackers who commandeered the planes that crashed into the buildings emerged. The leader of the group, Mohamed Atta, was found to have studied architecture in Cairo and engineering in Hamburg University of Technology. Did his familiarity with structures somehow enable the unprecedented terrorist act? And did the event as a whole, in some way, expose a terrible symmetry between construction and destruction?

The destruction of Pruitt-Igoe and the Twin Towers were two of the most visible building demolitions in history, a coincidence not lost on writer Justin Beal. He writes, "Neither Pruitt-Igoe nor the World Trade Center will be remembered as masterpieces, nor should they, but it is hard to imagine another pair of buildings which, in their lifespan — from conception to construction to spectacular, violent destruction — have exerted a greater influence on the course of American architecture. It is difficult to imagine any story that has shaped the culture and politics of architecture in the last 80 years more than those told by the handful of photographs of Pruitt-Igoe tumbling to the ground and the thousands of images of the towers of the World Trade Center collapsing under their own weight."[3]

[3] Justin Beal, *Sandfuture* (Cambridge, MA: The MIT Press, 2021), p. 252.

3 At the same moment that the Pruitt-Igoe blocks were coming down, the World Trade Center Twin Towers were going up. Both projects were designed by Japanese American architect Minoru Yamasaki.

4 The American Folk Art Museum (2001), designed by Tod Williams and Billie Tsien, was enthusiastically received when it opened in 2001, only to be demolished 13 years later to make room for the MoMA expansion.

A MATERIAL MOMENT IN TIME

Despite how exceptional these two events were, the practice of demolition is surprisingly common. Demolition might be better understood as the norm than the exception.[4] The historic preservation movement in the US was launched as a reaction to the 1963 demolition of Penn Station despite its status as one of the great architectural works of New York City. More recently, the demolition of the American Folk Art Museum, 4 designed by Tod Williams and Billie Tsien, built in 2001 and demolished in 2014, highlights the rapacious appetite for newness and the precarity of architecture in the hottest real estate market in the world. The 52-story Union Carbide Building 5 in Manhattan by SOM's Gordon Bunshaft and Natalie de Blois has acquired the dubious title of the tallest building ever demolished. It has been replaced by Norman Foster's 60-story JP Morgan Building at 270 Park Avenue.

In considering the afterlives of architecture — after needs change, systems fail, and materials wear — it is important to understand architecture as embodying a material moment in time. It is constructed of materials with degrees of durability, finite lifespans, and warranty periods. The degree to which architecture is valued by its users, owners, and society generally correlates with its longevity. In the 1960s, under the spell of post-war consumer culture, architects imagined instant cities that could transform to suit their inhabitants' wishes. This architecture of maximum flexibility and disposability is still present in contemporary culture despite a heightened sense of environmental consciousness.

Though counterintuitive, a possible antithesis to disposable architecture is the ritual construction and reconstruction of the Ise Jingu Shrine in Mia prefecture, Japan. 6 The rebuilding of the shrine every 20 years has been ongoing for centuries, with some estimating that the shrine is now 2,000 years old. The process of continuous construction is a means for preserving not only the building but also the community around it and the skill of the craftsmen involved in its construction.

What makes a building so significant that a community commits to its reconstruction every 20 years for 2,000 years? Conversely, how could a building in Midtown Manhattan be slated for demolition after only 13 years? Architecture is simultaneously a material and cultural artifact. As a material artifact, it has a finite lifespan subject to wear and tear, weathering, and decay.[5] As a cultural artifact, its lifespan is subject to cultural criteria for valuation. A building may become culturally or socially obsolete

[4] For the cultural history of demolition see Jeff Byles, *Rubble: Unearthing the History of Demolition*, 1st pbk. ed. (New York, NY: Three Rivers Press, 2006).

[5] For an essay on the material of architecture and weathering, see Mohsen Mostafavi and David Leatherbarrow, *On Weathering: The Life of Buildings in Time* (Cambridge, MA: MIT Press, 1993)

5 The Union Carbide Building (1960), designed by Gordon Bunshaft and Natalie de Blois, is the tallest building ever demolished.

6 The Ise Jingu Shrine is painstakingly rebuilt every 20 years, a ritual that has been repeated for up to 2,000 years.

before it is materially obsolete.[6] In fact, obsolescence is integrally tied to notions of modernity and progress. Demolition is a corollary to construction, just as obsolescence is the flip side of modernity.

DESIGNED OBSOLESCENCE

Architectural historian Daniel Abramson argues that obsolescence is integral to understanding architecture under capitalism, as architecture is recast not only as a means of wealth creation via its value as a real estate commodity, but also as a depreciating asset from a tax point of view. Modernity relies on the concept of obsolescence as an essential component of the process. Abramson claims, "The discourses of obsolescence and progress are partnered and mutually implicit."[7] And this, in turn, creates a perpetual demand for newness in a world of constant change.

Entropy and obsolescence are both forces that demand renewal and refurbishment. The hotel interior needs to be "refreshed" every five to ten years to maintain a sense of up-to-datedness. A building's roofing membrane may need to be replaced every 20 years. The trusses under a railroad bridge need to be inspected every five years and are expected to have a service life of 70 years.

[6] I am indebted to Susan Buck-Morss for this interpretation of Walter Benjamin's theory of fashion.

[7] Daniel Abramson, "Obsolescence: Notes Towards a History," in *Praxis* 5, Architecture After Capitalism, 2003, p. 110.

Half of the buildings in the world were built after 1950, and many are now 70 years old.[8] The building stock of many North American buildings constructed in the mid-20th century is aging and has rapidly become "obsolete." Given the waste that often goes along with demolition, the expense of procuring new materials, and as a society capable of thinking at least decades into the future, we could afford to think not only of design for construction but also for deconstruction — for disassembly, reuse, renewal, and adaptation. The selection of projects and research that follow illustrate different approaches to design for an architectural afterlife, as historic preservation, adaptive reuse, or radical transformation of existing structures.

CONGLOMERATE STRATEGIES

The original Santa Caterina Market was built in Barcelona's dense Ciutat Vella neighborhood in 1845. Its neoclassical facade is a single-story masonry structure with a repetitive rhythm of unadorned arched openings and a raised central bay with a pedimented profile. The market was completely rebuilt in 2005, with a design by Enric Miralles and Benedetta Tagliabue that preserved the historic facade, excavated two levels of underground parking, and

[7] The rebuilt Santa Caterina Market (2005), designed by Miralles Tagliabue EMBT, in Barcelona, Spain, represents a mode of historical preservation and adaptive reuse that selectively retains elements, demolishes some, and adds others, to form a new "conglomerate" architecture.

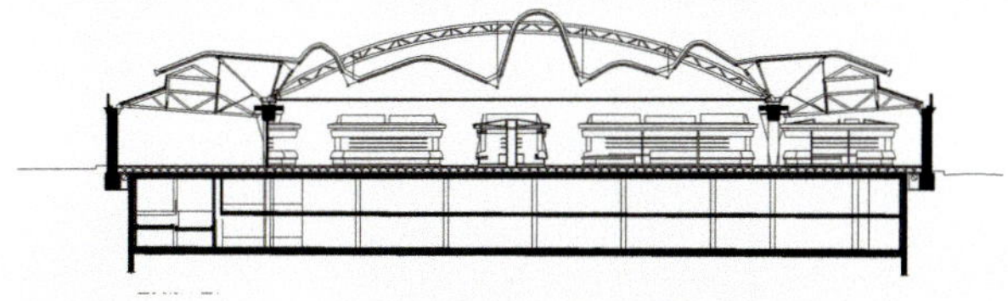

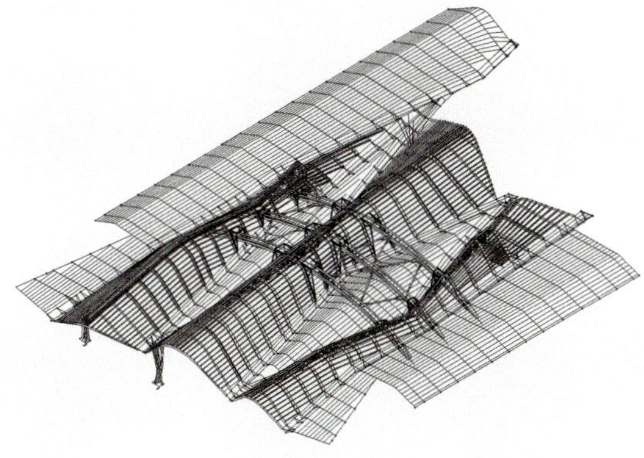

[8] A transverse section through the Santa Caterina Market (2005) shows the two levels of below-grade parking and the long span central bay of the new roof structure.

[9] An axonometric drawing of the roof structure of the Santa Caterina Market (2005) illustrates the three arched trusses that support the undulating surface of the new roof.

added a new colored canopy. The reconstructed roof consists of a dramatic undulating surface draped over the existing hall and hovering above the historic masonry walls. The gap between the old and new maintains a critical separation, and the contrasting forms, colors, and materials remind us of the tension between the two. [10]

Miralles and Tagliabue, describe their approach to working within historic contexts. "The first mistake is to talk about old and new. Whatever has managed to survive into the present is current, useful, and contemporary. And it permits us to move back in time to continue forward. To be permanent is contrary to existence. Things are forever changing."[9]

In contrast to the grounded language of the historic market, Miralles and Tagliabue's canopy soars overhead with a supple curvature that seems to billow in the wind. [7] The structure's timber beams are visible in the underbelly of the canopy, and the long-span structure is articulated as a series of bent tubular trusses supported by columns made of bundled tubes.

Seen from above, the current market roof is a splash of brightly colored glazed tiles, arranged to form an image of fresh produce; red tomatoes, oranges, and leafy greens are discernable in the pixelated image. [11] Local craftspeople set 325,000 individual tiles on the roof surface. The colorful roof becomes a kind of fifth facade and resembles a pixelated print on a textile surface.

The roof's geometry includes a density of pleats and folds that do not align with the structural bays. The gathered fabric reading from above suggests seven or eight ridges and valleys, bending and creasing along its length. The structure bays beneath are organized into one central, long-span bay and two flanking bays. [8] The entire market is built above a new, below-grade parking structure, so the column grid of the subterranean order is coordinated with the roof above. The result is a significant clash between a long-span arched truss and the undulating roof form, with the trusses puncturing the roof surface, appearing and disappearing on the interior.

The Santa Caterina Market illustrates the complexities of working with historic structures and producing continuity with the existing context by maintaining its use and preserving its discrete components like the whitewashed masonry wall while adding new facilities like the below-grade parking structure and substantial architectural elements like a long-span roof. The authors describe their efforts to avoid freezing a structure in a specific moment in time, but rather to "become a part of it."

[8] "Between 1945 and 1980 the volume of the global built environment more than doubled relative to all that had been constructed in the course of human history. The vast quantity an uneven quality of this heritage — much of which falls into the category of ordinary everyday modernism (OEM) — has tested the limits of traditional preservation practice." Thomas C. Jester and David N. Fixler "Modern Heritage: Progress, Priorities, and Prognosis," in *APT Bulletin*, Vol. 42, No. 2/3 Special Issue on Modern Heritage (2011) pp. 3–4.

[9] EMBT statement quoted in EU Mies Award website. See https://miesarch.com/work/682, accessed 4.23.2022.

10 The new roof of the Santa Caterina Market hovers over the historic walls of the original market, supported by bundles of bent tube columns that are distinct from the historic structure.
11 An aerial view of the Santa Caterina Market highlights the colorful tiles that evoke the bright colors of the fresh produce sold within.

1 2 The renovation and recladding of the Cité du Grand-Parc (2016), designed by Lacaton & Vassal, retains the original structure while adding a new balcony zone, improving the thermal performance, refurbishing the 1960s institutional look of the project, and offering an alternative to demolition.

For them, the project "superimposes new architecture on old, mixes them, and comes up with a conglomerate, a hybrid that accentuates usefulness and is contemporary."[10]

ADDITIVE STRATEGIES

In some cases, a building slated for demolition can be saved by a new facade, as with the Cité du Grand Parc apartment buildings in Bordeaux renovated by French architects Lacaton & Vassal. 1 2 The original structures, built in the 1960s to accommodate a growing population on the outskirts of the city, are slab-type housing blocks that were constructed economically and repetitively. A half-century later, these housing blocks show their age largely due to poor insulation and deferred maintenance. Often, the default is to tear buildings like these down and rebuild them with lower-scale structures. Anne Lacaton and Jean-Philippe Vassal put forth a radical proposal to avoid demolition, suggesting the strategic addition of an architectural liner that would replace the aging facade, but also serve as a balcony and a winter garden.

Lacaton and Vassal offer the following logic for their additive approach as basic arithmetic: "Demolition and reconstruction is always (1-1) + 1 = 1. Adding is 1+1 = 2. The figures are clear. Cost-effectiveness is the key."[11]

In 2001, the architects published a PLUS manifesto, arguing against demolition and for addition. "Never demolish, never remove or replace, always add, transform, and reuse." Their proposal for Grand Parc consisted of a twelve-and-a-half feet (3.8m) wide self-standing structure constructed in front of the existing slab building. The original facade was removed and replaced with a new window wall system with sliding glass doors. The outer skin is now made up of a combination of glass and corrugated polycarbonate sheets, and the new winter garden zone expands living space significantly while improving the thermal performance of the building envelope.

The renovation of the existing buildings runs counter to the trend of demolishing and rebuilding. It was also completed for a fraction of the cost of demolition and reconstruction. Notably, the new construction work was performed while the building remained occupied, avoiding the significant disruption and displacement of residents that a "tear down and rebuild" approach would entail.

The technical challenges of their additive approach included managing the logistics of

10 Project description from Benedetta Tagliabue FMBT website, http://www.mirallestagliabue.com/project/santa-caterina-market-renovation/ accessed 4.23.2022.

11 *Lacaton & Vassal*, Editions HYX, 2009, p. 158.

13 The Dominicanen Bookstore (2007), designed by Merkx+Girod as an adaptive reuse of a historic Gothic church in Maastricht, Netherlands, reactivates the unused structure through the insertion of a new bookstore program.

14 The construction of the Dominicanen Bookstore (2007) involved the addition of a new steel frame structure designed to contrast with the historic masonry of the church, and presumably, it is also reversible at a future date.

construction around an occupied apartment building, as well as bridging the existing structure and new construction. The 16-story structures were surveyed and documented for tolerances and irregularities. The new construction had to be independently supported to not impose additional loads on the existing buildings. Because existing balconies were supported by pairs of columns, one at the edge of the slab and one at the face of the original buildings, the new footings were designed to avoid the existing foundations and the new columns converge to form a V-shape at the ground floor.

Lacaton & Vassal's additive approach has been recognized as an effective and replicable alternative to demolition — and a new approach to sustainable design practice that acknowledges the significance of the investment and embodied energy that existing architecture always already represents. According to the architects, they are attempting to address the problem of "how to make sustainable what already exists." This adapt-and-evolve approach is in clear contrast to the waste and spectacle of the demolition of the Pruitt-Igoe housing slabs from the same era.

INSERTION STRATEGIES

The Dominicanen Bookstore in Maastricht, Netherlands, reactivates a historic Gothic church with a delicate "ship in a bottle" approach to inserting a new program into a historic shell. 13 Originally built in 1294, the building consists of a high central nave and two side aisles. Over time, the original church fell into disuse and vacancy. The proposal to adaptively reuse the historic landmark as a bookstore offered to breathe new life into a significant structure that was materially and structurally sound but socially and culturally obsolete.

The architects for the renovation project, Evelyne Merkx and Patrice Girod, designed a lightweight steel frame that slides in between the columns of the original nave, evoking the structure of steel-framed book stacks that blur the line between furniture and architecture. The completed bookstore renovation consists of a three-story steel platform that appears to float within the space. The platform cantilevers from the new steel posts set between the building's original columns. The addition is shifted off the center of the nave to allow the central axis to remain visible. 15

New foundations were built within the church to support the new steel column bases. Prefabricated

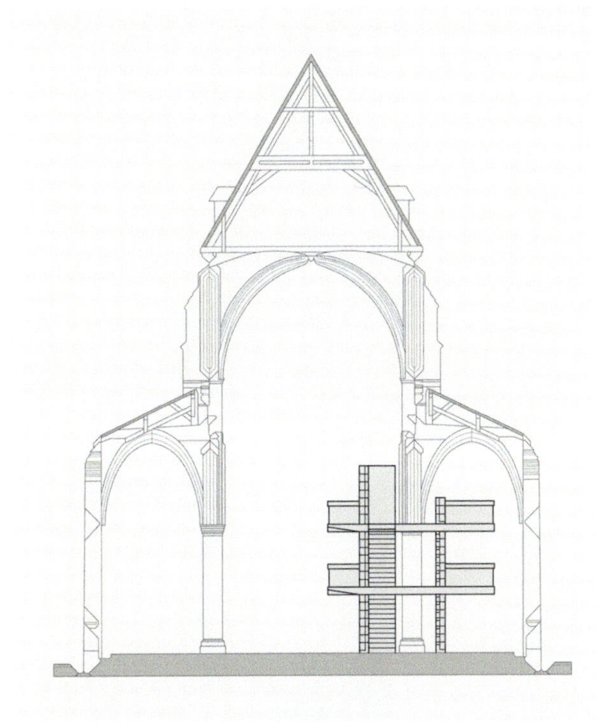

15 A section through the nave of the historic church shows the new armature of the Dominicanen Bookstore (2007) that is arranged asymmetrically to preserve the central axis of the original church.

steel components were brought into the space through the existing doors and erected inside the church with mobile construction equipment. 14 Great care had to be taken with the construction vehicles working within and between the centuries-old stone arches of the historic building.

The new elements in the church are articulated as blackened steel posts, beams, guard rails, and bookshelves, distinct from the warm gray limestone of the Gothic structure. The effect of the new, lightweight, steel structure within the historic stone shell is a remarkable study of contrast and composition. Merkx and Girod's emphasis on the new elements of the bookstore program allowed the original structure to remain largely unmodified.12 Their approach dialogues with the "host" structure by juxtaposing new and old architectural languages, as well as contemporary and historical uses.

12 In many instances of reuse, the "reversibility" of the intervention is considered as a positive feature, so that, in theory, the building could be restored to it original function at a later date.

TEMPORARY SHELTER

One of Charles Rennie Mackintosh's masterpieces, the 1904 Hill House in Helensburgh outside of Glasgow, Scotland, incorporates many of his signature features that have become synonymous with the Art Nouveau movement in architecture and design. The interiors and furnishings, designed with his wife and partner, Margaret Macdonald, include intricate detailing and beautifully crafted elements of wood, glass, and fabric. The exterior massing of the house consists of a composition of distinctive rooflines and chimneys. The exterior is finished with a harling cement finish, which was selected over a more traditional lime render. Unfortunately, the harling cement tends to absorb moisture without allowing the assembly to breathe, leading to accelerated deterioration and weathering. As a result, the National Trust for Scotland and Historic Environment Scotland have undertaken a radical project to protect the house with a purpose-built protective shed while Hill House is allowed to dry out — a process expected to take 15 years. In the interim, researchers and conservationists are studying the original structure and developing an approach for long-term conservation.

The new structure, designed by Kevin Carmody and Andy Groarke, encloses Hill House entirely with a steel-framed long-span structure. 16 The gabled "box" around Hill House protects the landmark from the elements while giving the historic house time to re-acclimate to its new microclimate. It also creates a loose-fitting scaffold around the house that maintains visibility and allows it to continue receiving guests. The steel frame includes a catwalk path around

16 The Hill House Box (2019), designed by Carmody Groarke, creates a temporary shelter for the historic Charles Rennie Mackintosh's Hill House (1902), simultaneously offering protection from the elements, enabling material conservation efforts and ensuring continuity of access for visitors.
17 A section through the Hill House Box (2019) shows how the protective shed allows for ample room around the house as well as a catwalk to observe the house from new vantage points.

18 Digital documentation of the historic Hill House helps produce 2D and 3D documentation, while also helping to survey moisture and visualize thermal bridging and heat loss.

19 A concept model of the Hill House Box shows the house enclosed in a screen mesh to protect it from the elements while the house is studied.

and over the house, providing an unusual itinerary, including vantage points from above. 20 In this way, the architecture makes the conservation process public and visible to visitors.

The architects describe the importance of maintaining access to the house and not obscuring it behind a more conventional scaffolding or construction hoarding. The steel frame structure they designed was prefabricated off site and delivered in large frames that were bolted together. The footings for the new steel structure maintain a distance from the original so as to not impact the historical foundations. The entire structure is clad in a stainless steel chain mail mesh, creating a shimmering screen through which the original Mackintosh gables are still visible. The mesh allows for ventilation while limiting the effects of wind-driven rain. This approach is an architecture-for-the-architecture, a means for extending the life of one building by shrouding it with another.

As with many adaptive reuse, renovation, and preservation projects, new construction requires protection, shoring, and careful sequencing during construction. Often, pre- and post-construction surveys and digital scans are used to maintain a careful record of the original structure. A building is at greatest risk during construction, when equipment may impose concentrated loads on slabs, and safety systems like sprinklers may not yet be operational. For example, the devastating 2019 fire at Notre-Dame Cathedral in Paris was likely caused by electrical issues during its renovation. Similarly, the 2009 high-rise fire in the TVCC hotel in Beijing started due to a fireworks accident while the tower was under construction and the sprinkler system was still being installed.

ADDITIVE AND SUBTRACTIVE

Another category of reuse and transformation consists of large, underutilized industrial structures. For example, the continuous transformation of the shipping industry, its associated equipment, and patterns of logistics make shipping hubs and port facilities susceptible to obsolescence. The last two projects in this category, the Zeitz Museum of Contemporary Art Africa (Zeitz MOCAA) 22 and the Elbphilharmonie Hamburg 21 are significant harborfront buildings that have been radically transformed by new uses. The Zeitz MOCAA by Heatherwick Studio, completed in 2017, employs a subtractive strategy, hollowing out a substantial void in the center of the former grain silo, while the Elbphilharmonie by Herzog & de Meuron, completed

20 A view from the catwalk between the historic Hill House and its new protective shroud.

in 2016, adds a large glass volume perched above an existing brick warehouse.

The host structure of the Zeitz MOCAA is a 1924 concrete grain silo at the tip of a pier in Cape Town harbor. The original silo structure was 190 feet (57m) tall, and was one of the tallest structures in sub-Saharan Africa. The cylindrical shafts of the silos are packed tightly in a six-wide by seven-deep grid. The structures are the centerpiece of the Victoria & Albert Waterfront in Cape Town, which is undergoing a redevelopment process to reposition the industrial harbor with a mix of new uses. The proposal to insert a museum of contemporary art and hotel led Heatherwick to conceive of a subtractive intervention to insert new galleries and other amenities.

The walls of the original concrete silos were only seven inches (180mm) thick and required reinforcement before they could be saw cut. Heatherwick developed an ingenious technique, adding a concrete sleeve to the inside of the original silo and then carving away at both layers to produce a boolean void. 23 The hollowed-out atrium space reveals the original structure of the cylinders, while the galleries are housed in a new structure embedded on the perimeter of the silos. 25 26 The construction sequence of the museum reveals a careful balance of stabilization, surgical demolition, and new construction. Oliver Wainwright, architecture critic at *The Guardian*, describes the museum as "a powerful piece of constructed archaeology,"[13] acknowledging the effort and expense involved in preserving the original silo structures within the complex.

13 Oliver Wainwright, "Mocaa, Cape Town: grain silo reborn as Africa's answer to Tate Modern" *The Guardian*, September 16, 2017.

The original structure of Herzog & de Meuron's Elbphilharmonie project is the Kaispeicher A warehouse, a wedge-shaped 1963 structure sited at the end of a harbor front pier. 1 The building's 66,000-square-foot floor plate made it a viable contender for hosting the 2,100-seat auditorium in the proposed performance venue. Although the finished building appears to simply add a new structure to the historic warehouse, the project involved significant demolition, excavation, subtraction, and structural reinforcement of the original structure.

The Elbphilharmonie includes a concert hall, parking structure, apartments, and a hotel. The concert hall occupies the center of the floor plate in the upper volume, while a primarily single-loaded corridor of hotel rooms occupies the perimeter. 27 The exterior cladding of the upper volume is a unique glass curtain wall with a highly reflective coating. The lower volume has been reused, maintaining the brick facade with punched windows.

21 The Elbphilharmonie (2017), designed by Herzog & de Meuron, radically transforms a historic warehouse into a concert hall and cultural destination.

22 The Zeitz Museum of Contemporary Art Africa (MOCAA), designed by Heatherwick Studio, in Cape Town, South Africa, hollows out a historic industrial building to host a new cultural program.

The process of design and construction of the Elbphilharmonie tells a remarkable story of speculation, advocacy, politics, and adaptive reuse. During the 15-year process, the initiators of the project, Alexander Gérard and Jana Marko, identified the warehouse site, rallied political support, hired architects, developed a design, fundraised, and eventually realized a "once-in-a-lifetime" project in the heart of a major European city.

The process of building the Elbphilharmonie involved the shoring of the brick facades of the existing warehouse while the interior structure was demolished, its foundations were rebuilt, 650 additional piles were added, and the new structure was constructed inside the existing shell. The brick warehouse essentially became the "podium" for the new glass-clad concert hall above. Uses in the podium include a concrete parking structure, recording studios, support functions, and all the circulation elements from the building above, including elevator cores and fire stairs.

The top level of the original building was rebuilt as a public observation deck that also doubles as the sky lobby for the hotel that wraps around the concert hall. A unique, long-span escalator connects the ground-floor lobby to the observation deck, bringing hundreds of visitors to the plaza level. The plaza level also serves as a gap between the brick podium and the glass upper volume above, marking the gap between "old" and "new." [28]

The concert hall volume has the same footprint as the original warehouse, and rises straight up like an extrusion, appearing to duplicate the historic building in its proportions but not in materiality. The silvery reflectivity of the all-glass curtain wall that wraps the upper volume contrasts dramatically with the rather banal brick building of the podium. The concave roofscape of the concert hall, appearing at once like a crown and a geological formation, gives the Elbphilharmonie a distinctive silhouette on the skyline, adding a new icon for the city.

The upper volume's inward-facing concert hall and surrounding outward-facing hotel rooms combine two programs in that they share space and a lobby on level eight, but otherwise function entirely independently. Both need to be completely isolated acoustically. The concert hall's reinforced concrete construction system was chosen specifically for acoustic reasons. Independent mechanical systems for each also help to maintain their acoustic separation. [29]

The complexities of building on top of, and within, an existing masonry shell to house two distinct programs that demand acoustic separation,

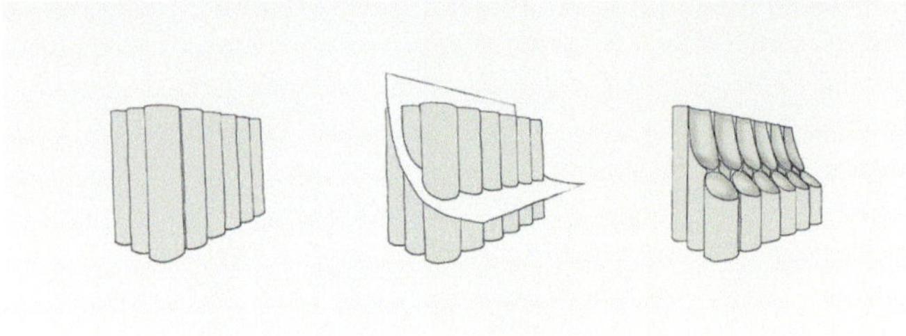

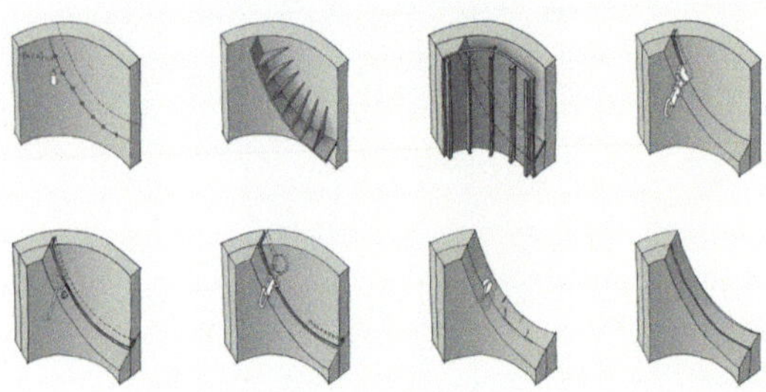

2|3 Conceptual diagrams of the large-scale carving operations performed on the existing silo structure of the Zeitz MOCAA.

2|4 Construction sequence diagrams of the concrete reinforcement and saw cutting operations performed on the existing silo structure of the Zeitz MOCAA.

2|5 Concept model of the Boolean operation for carving an ovoid figure from the cylinders of the existing silo structure of the Zeitz MOCAA.

both with complex mechanical systems, demands, and convoluted circulation and servicing, makes this type of construction extremely expensive. The cost of the project ballooned during development and construction as the complexity of the undertaking became apparent. This raises the question: Was it worth it? Can the reuse of existing buildings for new purposes retain its inherent benefits if the construction cost is so exceedingly high? The premium paid for construction complexity must also be part of the conversation around adaptive reuse.

The public reception of the Elbphilharmonie has been extremely positive and the project is credited with catalyzing the revitalization of the historic harbor area and contributed to the transformation of an entire neighborhood. A public space master plan by Benedetta Tagliabue has created a consistent language of pavers, plantings, street furniture, and amenity structures. New multi-family residential buildings have brought new users to the city's center and spurred a dramatic urban transformation. A visit to the public observation deck on a weekday afternoon finds it packed with both tourists from around the world and locals hoping to catch a glimpse of their city from a new vantage point. The plaza level is a popular venue for school field trips, pre- and post-theater dining, and for taking wedding photos. Public access to the observation deck is free, but ticketed, to prevent overcrowding — a clear indication that the building has become a new cultural landmark in the city.

Returning to the question of whether it is "worth it" to undertake adaptive reuse projects when the complexity and cost of construction are so high, it is important to consider not only cost but also value. By transforming the derelict Kaispeicher A warehouse into the Elbphilharmonie, the designers did more than add a new building on top of an old one. The Elbphilharmonie is an alchemical transformation, akin to turning lead into gold, that employs conceptual design and meticulous execution to create something wholly unique.

UN-PRECIOUS PRESERVATION

Much of historic preservation deals with buildings of significance and distinguished examples of architecture of a specific period. The Hill House by Mackintosh is an exemplary of the Art Nouveau style. The church in Maastricht is a prime example of Gothic architecture. What about buildings that are not significant? When is the mundane building the subject of preservation or reuse?

The following are two examples that illustrate the

6.2 The vertiginous interior atrium carved out of the existing silo structure of the Zeitz MOCAA.

27 **Section model of the Elbphilharmonie (2017), showing the cellular arrangement of the hotel rooms wrapped around the concert hall.**

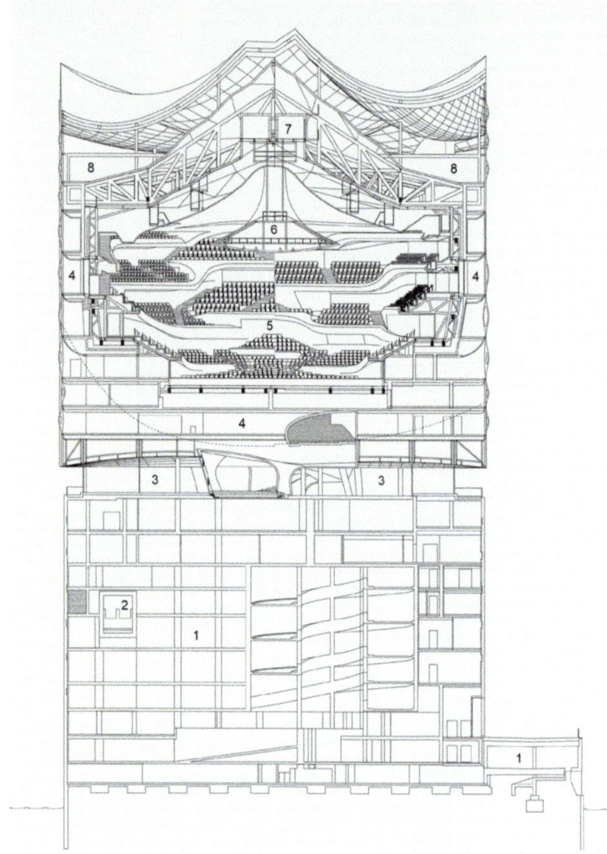

28 **Section of the Elbphilharmonie (2017), showing the concert hall within the new structure, bearing on the completely rebuilt structure of the historic warehouse below.**

adaptive reuse of structures that might be considered "unprecious." The SESC Pompeia community center in São Paulo, designed by Lina Bo Bardi, and completed in 1987, transforms a decommissioned oil drum factory into a vibrant social and cultural hub in the working-class neighborhood of Pompeia. The new design retains the long-span industrial sheds of the original structure for workshops and gallery spaces while adding a vertical sports complex for the community. 30 The Fondazione Prada, in Milan, designed by Rem Koolhaas of OMA, completed in 2018, intervenes in a former gin distillery in Largo Isarco, Milan, to create a new cultural campus in the city. 35 In each case, the existing architecture was deemed "not precious," even "ugly,"[14] but the owners and designers opted to retain rather than demolish. As a result, the two projects develop a new attitude toward design with respect for existing structures — a pragmatic, matter-of-fact approach to preservation and retention. This period-agnostic approach to historical and contemporary structures treats buildings equally, flattening their differences and freeing designers from certain challenges of over-deference.

14 Lina Bo Bardi referred to the sports complex as ugly. "I hope the sports complex will be ugly, uglier than MASP. It is a silo, a bunker, a container, and more, divided by an area where nothing can be built." Zeuler Rocha Mello de Almeida Lima and Barry Bergdoll, *Lina Bo Bardi* (New Haven; London: Yale University Press, 2019), p. 169.

The original oil drum factory in São Paulo was built in 1938 and consists of several bays of asymmetrical, gabled sheds on a gently sloping site. 32 A central north-south street splits the site, with deeper bays on the western side and narrower bays on the east. The original structure consists of a prefabricated, reinforced concrete frame system, developed by pioneering French engineer François Hennebique. The factory moved out after 30 years, and the building was abandoned for five years before it was bought by the SESC institution for conversion into a cultural and recreation center. Initial plans to raze the site and build a new facility were drawn up but were abandoned as too expensive. The idea of reusing the existing structure was proposed by architect Lina Bo Bardi, who was selected thanks to her experience with adaptive reuse.

Bo Bardi's approach to the site was to retain the existing industrial sheds, remove the interior partitions, and expose the original concrete frames and wooden trusses. Bo Bardi reimagines the sheds filled with a new "landscape" of cultural programs like galleries, workshops, a library, and a new theater. In maintaining the low-rise structures for cultural programs, the athletic programs were pushed to the southeast, where they stack to form a vertical tower for sports. Further complicating the site planning, an existing underground stream along the southern edge of the site split the gymnasium into two volumes with the large floor plate pool and gym building

9 2 Aerial photo of the Elbphilharmonie under construction showing the large void of the concert hall surrounded by other programs.

30 SESC Pompeia (1986), designed by Lina Bo Bardi is an adaptive reuse of an existing factory, in the foreground, and the addition of a vertical gymnasium building on the limited remaining area of the site. The gymnasium is accessed via a freestanding tower and a series of concrete catwalk bridges.

31 Inside the vertical gymnasium, Bo Bardi stacked a swimming pool, indoor soccer fields, and a solarium.

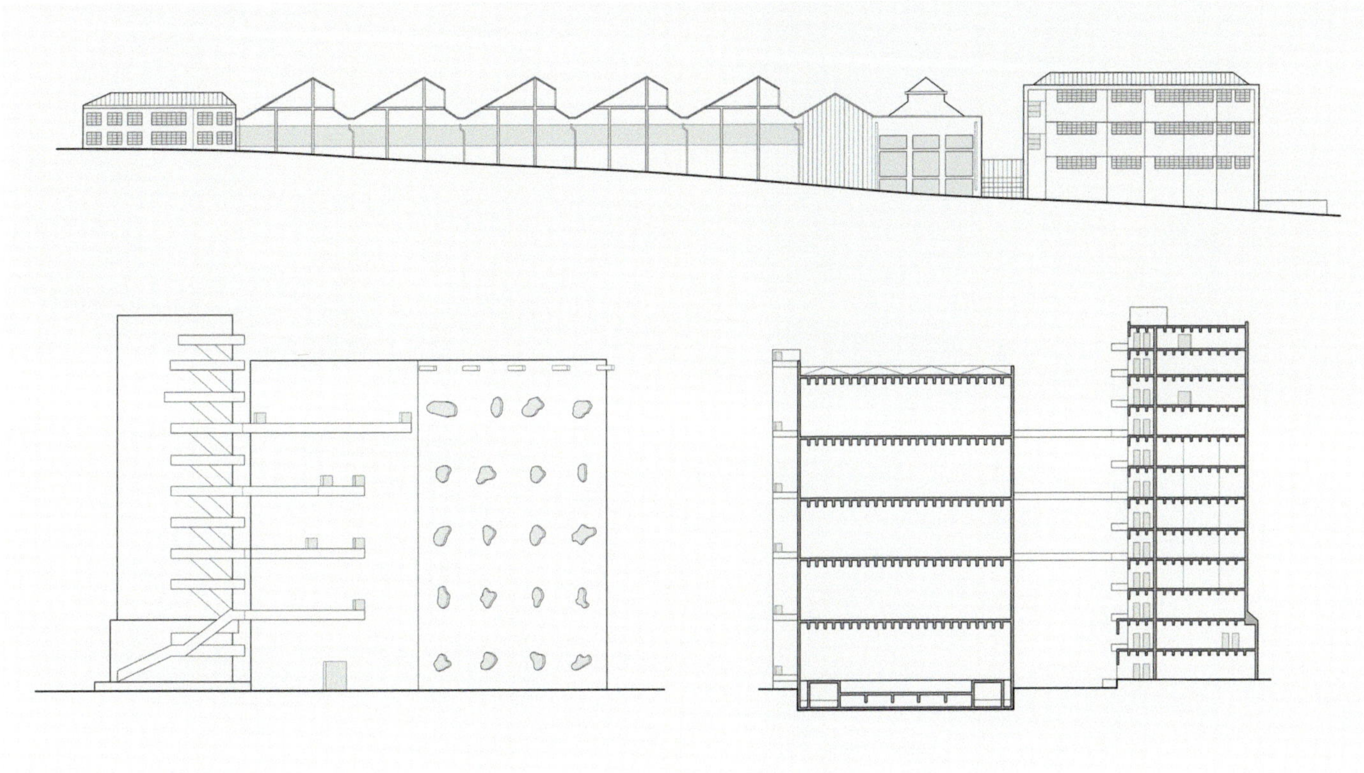

3 2 **Section and elevations of SESC Pompeia show the low rise existing factory buildings and the high rise gymnasium and access tower.**

forming the primary structure and a separate circulation tower connected by a series of diagonal bridges that stitch the two towers together. 3 2

1 5 Zeuler Rocha Mello de Almeida Lima and Barry Bergdoll, *Lina Bo Bardi*, p. 171.

Within the repetitive structural grid of the factory shed, Bo Bardi inserted a mezzanine level to house a library with a new terraced landscape of seating areas. Adjacent bays house a lounge with custom wooden benches, also designed by Bo Bardi. Classrooms and pottery workshops sit adjacent to a large gallery space. A shallow water feature with a curvilinear profile is carved into the floor of the gallery. A narrow bay is used as a foyer space for a new auditorium. The auditorium is inserted into one of the tallest bays. A traditional auditorium with a single rake would require more interior height that was available in the existing structure, so Bo Bardi designed a double-raked auditorium with a stage in the center of the audience. Bo Bardi opted for the unusual symmetrical solution as a creative response to given constraints. Retaining the original buildings imposed numerous limitations, requiring architectural invention.

The new gymnasium, circulation, and water towers occupying the southeast corner are all built of exposed, board-formed reinforced concrete. Bo Bardi sought a rough-textured material to reveal the process of making and celebrate its tactile qualities. 3 0 The cylindrical water tower is made with formwork that highlights the individual concrete pours. Slightly loose formwork allows for the intentional concrete texture of a coarse, ad hoc appearance. For Bo Bardi, traces of the process of making were an important way to remind us of the human hands that made the architecture, and that modernization can coexist with history and tradition. 1 5

The new SESC Pompeia complex combines old and new to make a new social hub in the city. The 245-foot (75m) water tower is visible from the surrounding neighborhood, and sports hall users have panoramic views from the bridges between the towers. Meanwhile, the ground-floor programs are well-utilized by artists, who run workshops and exhibitions, library and cafeteria users, as well as athletes, and sunbathers. The different precincts within the site employ distinct approaches to renovation and new construction but are tied together by Bo Bardi's specific design approach. Concrete floors within the factory are inlaid with rounded pebbles, while the mortar between masonry partition walls is deliberately un-tooled and coarsely articulated. The use of bright colors, specifically the red-painted doors, window shutters, ducts, handrails, and truss gusset plates serve as recurring accents and unifying elements across the campus.

33 The Torre of the Fondazione Prada rises above the original low-rise structures of the former industrial complex, adding to the ad hoc composition of structures, both exceptional and banal.

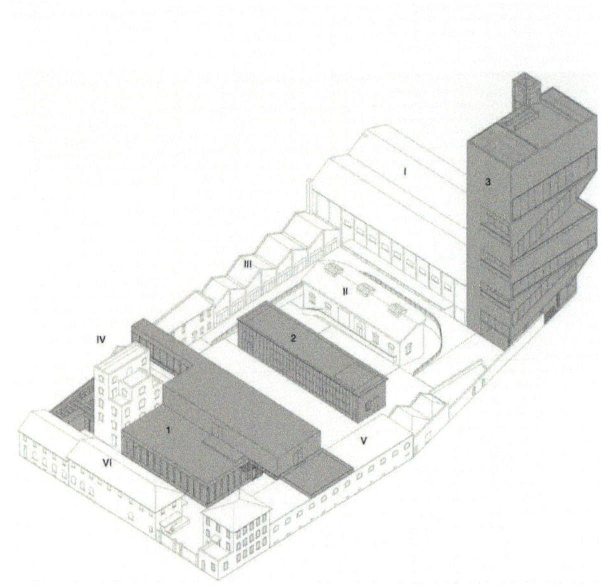

34 Axonometric drawing of the Fondazione Prada cultural complex, highlighting the new structures—Podium Gallery, Beam Gallery, Multimedia Auditorium, and Torre—interspersed with the existing.

Bo Bardi's expansive and wide-ranging design of the SESC Pompeia is inventive at multiple scales and in radically different modes, treating the historic sheds with the surgical precision of insertion and subtraction while simultaneously constructing a bold new structure with a strong presence in the surrounding cityscape. Architect Philippe Blanc describes Bo Bardi's approach as an aggregation: "The context is understood as a ruin, upon which she intervenes. That would be called an 'architectural operation,' making one project from another. In other words, this deals with a construction that is situated between intervention and invention."[16]

Opting to reuse the existing factory buildings, Bo Bardi accepts its constraints and sets in motion several design operations that required inventive solutions throughout. The theater and gallery are insertion strategies where the new inhabits the old, creating a dialog between the two. The athletic tower negotiates the site's constraints to find invention at its limits. Bo Bardi's situational responses to the challenges of the existing structures demonstrate a keen sense of how design across scales can serve multiple purposes, benefiting the social and cultural life of the city, and reanimating a former industrial site at the heart of a neighborhood.

[16] Philippe Blanc, "Lectura de Lina Bo Bardi: Pompeia," in *ARQ*, Ensayos y Documentos, p. 60. "El proyecto se plantea como agregación; el contexto se entiende como ruina sobre la cual se interviene. Es lo se llamaría una "operación de arquitectura" al hacer de un proyecto otro. Dicho de otro modo, se trata de una construcción que se sitúa entre la intervención y la invención." Translated by the author.

[17] Rem Koolhaas statement on the OMA website. See https://www.oma.com/projects/fondazione-prada, accessed on 5.5.2023.

HAUNTED HOUSE

The Fondazione Prada (2018) on the outskirts of Milan employs strategies that preserve, modify, and reconstruct existing parts to produce a new and heterogeneous whole. Much like the SESC Pompeia site, the existing structures at the Largo Isarco industrial complex (1910) were considered architecturally insignificant and unworthy of preservation. Rem Koolhaas' radical approach was to value the difference and diversity over homogeneity. In a provocative statement, Koolhaas decries the expansion of art practices that rely upon a reduction in types of spaces for exhibition: "To apparently everybody's satisfaction, the abandoned industrial space has become art's default preference — attractive because its predictable conditions do not challenge the artist's intentions — enlivened occasionally with exceptional architectural gestures."[17]

Amplifying the diversity of spatial environments, Koolhaas inventoried the buildings at the site and opted to preserve seven, selectively demolishing structures and altering others. He added three new buildings: the Torre, the Podium, and the Cinema. 34 Taken together, the Fondazione Prada offers a varied collection of architectures to house the diverse collection of programs and venues.

3|5 The Podium Gallery and Haunted House structures of Fondazione Prada (2018) mix existing, modified and new construction in a provocative cultural complex.

36 The Podium Gallery of the Fondazione Prada is clad in a modified Miesian curtain wall with articulated mullion caps; however, these mullion caps are not the materialization of a universal grid, but rather the expression of the contingent and combined layers of geometry from the historic building projected onto the modular grid of the new. The result is a mash-up of a modular Miesian facade and a circular arched portal.

Perhaps the most striking element of the entire project is the gold-painted, five-story historic building dubbed the Haunted House. 35 Its architectural ornament suggests that it was once of some cultural significance, perhaps an administrative building capable of surveying the entire complex. Koolhaas' assessment of architectural assets at once preserves the house and transforms it with a layer of gold paint that encapsulates it, like a thin sarcophagus around an architectural body. This gold-embalmed mini tower serves as an exhibition space made up of small rooms that provide a domestic scale for viewing specific artworks. In the Koolhaas taxonomy of display spaces, the gold house is clearly not new, yet it shines brightly as it falls into a "new/historic" category, slightly modified but totally transformed.

Many of the original buildings along the perimeter of the block are two- or three-story structures with punched windows and pitched tile roofs. Their facades are non-descript, finished with weathered white stucco. Koolhaas left the exterior of these buildings as is — as found objects. Though stained and patched, the perimeter buildings were maintained and contrasted with the gold house, and other design elements — for example, the floating Beam Gallery that occupies the central courtyard.

The Beam Gallery is clad in foamed aluminum panels and supported on black-painted cellular castellated beams. Visible from the entrance gate, the building seems to hover over the glass volume of the Podium Gallery. The gallery's glass storefront is articulated with aluminum extrusions attached to vertical mullions that recall a Mies van der Rohe curtain wall. However, the openings in the glass facade are arched and misaligned with the new curtain wall module, and aligned instead with the punched openings in the historic building beyond. 37 The geometry of the semicircular arched portals in the historic building is projected onto the glass facade, creating the improbable hybrid of an arched mullion cap on a contemporary curtain wall. 36

This play of form and material is pervasive across the site. The partially mirror-clad Multimedia Auditorium adjacent to the foamed aluminum-clad Beam Gallery appears to be original. Three of its facades have a projecting cornice at the roof. In contrast, the fourth facade is flat and clad in a mirrored stainless steel panel. Its reflectivity and its tautness are distinctive. The projecting cornices are captured in profile by the mirrored facade that receives them, and it is as if the historic building has been sliced in half, revealing the profile of its mirrored cross section. 38

37 The original circular arched portal, lined in stainless steel and projected onto the curtain wall beyond.

38 The abrupt cut of the historic profiled cornice, capped in a mirrored aluminum plate.

The combination of materials suggests a deliberate play of reference and transference, making for a highly specific, particularly fascinating architectural collage. Architecture critic Aaron Betsky describes the experience of the Fondazione Prada as "old and new, where Koolhaas is employing materials and creating illusions giving the viewer the impression that they are occupying a stage set, as if you were occupying the space behind the scenes, and revealing how effects are created."[18]

Ever polemical, Koolhaas describes the design approach: "The Fondazione is not a preservation project and not a new architecture. Two conditions that are usually kept separate, here confront each other in a state of permanent interaction — offering an ensemble of fragments that will not congeal into a single image, or allow any part to dominate the others."[19]

Koolhaas offers some critical insights into the state of architecture in the early 21st century in his 2009 essay, "Preservation is Overtaking Us." Koolhaas declares that architecture, as we once knew it, as a "serious discipline," is gone.[20] It has been replaced with a system based on luxury goods and starchitects. He argues against the turn toward a commodification of architecture and brands, and points to historic preservation as a kind of refuge.

18 Aaron Betsky, "The Marvels of the Mundane, Rem Koolhaas stages the Prada Foundation as a memory place," in *Architect Magazine*. See https://www.architectmagazine.com/design/the-marvels-of-the-mundane_o, accessed on 5.5.2023.

19 Rem Koolhaas statement on the OMA web site. See https://www.oma.com/projects/fondazione-prada, accessed on 5.5.2023.

20 Rem Koolhaas et al., *Preservation Is Overtaking Us*, GSAPP transcripts (New York: GSAPP Books, 2014), p. 21.

21 Rem Koolhaas et al., *Preservation Is Overtaking Us*, p. 22.

Koolhaas describes preservation as a strategy to escape from "a grotesque, exaggerated form of involuntary newness."[21] He does not mention his design for the Fondazione Prada in this essay, but it is clear how his multi-year engagement with the site and existing buildings could be understood as a means for resisting the prevalent drive toward novelty. In repurposing the old gin factory and its collection of buildings, Koolhaas chose selective demolition — cutting and suturing, attaching and overhanging, cladding and coating — as a radical demonstration of the coexistence of old and new that calls for a rethinking of historical prejudices and disciplinary habits.

MORE LINCOLN CENTER

The legacy of Robert Moses' tabula rasa urbanism consists of bridges and tunnels, as well as cultural institutions and multi-block development projects that replaced neighborhoods designated as slums and slated for "urban renewal." Lincoln Center for the Performing Arts (1962–1969) in New York City is one such cultural megaproject intended to bring together New York's leading performing arts institutions, including the Metropolitan Opera House, the Philharmonic Hall, the New York City Ballet and the

39 The architects of Lincoln Center photographed with a large physical model. From left to right: Edward J. Matthews, Philip Johnson, Jo Mielziner, Wallace Harrison (standing), John D. Rockefeller III, Eero Saarinen, Gordon Bunshaft, Max Abramovitz, and Pietro Belluschi.

Juilliard School of Music. The cultural campus joins 16 acres of land spread across three city blocks. Hundreds of individual buildings were demolished to make way for the four new buildings that replaced them. Lincoln Center could be described as one massive structure articulated as four buildings above a shared podium level and basement.

Lincoln Center was designed by a 1960s dream team of architects, including Wallace Harrison, Eero Saarinen, Gordon Bunshaft, Philip Johnson, Pietro Belluschi, and Max Abramovitz. 39 They used a single plinth to unify a composition of disparate yet materially consistent, monumental structures. The plinth concept gives the buildings a single datum, and the specific arrangement of three massive buildings around a circular plaza alludes to the civic urban space of the Piazza del Campidoglio in Rome. The focus on Lincoln Center's monumental front facades around the plaza simultaneously creates three monumental backs facing outward to the neighboring blocks. The site topography slopes from east to west, resulting in a modest climb on the east face along Columbus Avenue and a 20-foot drop along the west side along Amsterdam Avenue. Ultimately, Lincoln Center became a massive obstacle in the city, presenting large, blank facades and unwelcoming level changes along its edges. Furthermore, the wide pedestrian bridge over 65th Street, connecting to the Juilliard School, turned an otherwise active street into a service zone for the mega-plinth.

The Lincoln Center redevelopment project was launched in 2001 to correct some of the original project's shortcomings and make the complex more welcoming to different audiences. Architects Diller Scofidio + Renfro (DS+R) were selected to lead the public spaces project as well as the addition and renovation of the Juilliard School. The initial pitch, "More Lincoln Center than Lincoln Center," took the buildings and site as an unfinished, ambitious project and saw redevelopment as a means to realize some of the original, but unmet, ambitions.

Lincoln Center raises important questions about the role of institutions and their urban impacts. The inheritance of landmark buildings, both protected and unprotected, and the architectural approaches to modification, addition, and transformation are also interesting problems that arise at this particular site. The original campus designed by Harrison, Johnson, Bunshaft, and Abramovitz was flawed from the outset because it was removed from the street as the travertine-clad buildings turned their backs to the surrounding neighborhoods, and the plaza was disconnected from adjacent sidewalks.

40 The new Lincoln Center, after surgical renovations of the public spaces by Diller Scofidio + Renfro, including the new grand stairs along Columbus Avenue in the foreground, the addition of two large glass canopies linking the colonnade to Columbus Avenue, and a redesigned center fountain in Josie Robertson Plaza.

The DS+R team proposed reconnecting the central plinth to the city through a series of strategic manipulations at its edges. The eastern edge along Columbus Avenue was transformed by creating a set of large, attenuated steps proportioned like landings. The vehicular drop-off was moved below grade to remove conflicts between cars and pedestrians. 40 The risers on the stairs were faced with an LED sign board that displays messages in the tradition of theater marquees.

The northern edge of the campus fronting 65th Street also changed significantly. In the original campus design, a large plaza spanned over the street to connect to the second floor of the Juilliard School. This decked-over approach was a popular solution at the time, effectively turning a three-block plinth into a four-block superblock. The plaza was meant to create a new ground plane but it was rarely active, and the steel level was relegated to back of house actives like and loading docks and dumpsters. The DS+R team proposed the demolition of the wide bridge as part of the 65th Street redesign, as well as the addition of a new two-story restaurant and cafe on the northern edge of the plaza. The restaurant fronts both 65th Street and the plaza, effectively negotiating the change in level indoors with a set of stairs and elevators. 42

The roof of the restaurant was conceived as a green roof on a hypar surface, touching down at the southwest corner, and then twisting over to form an articulated canopy to the north. Adding food and beverage programs to activate both the plaza and the street begged an architectural response. How to integrate a new building among the original stone-clad structures? DS+R opted for an architecture-as-landscape strategy. The restaurant as landform did not compete with the existing monumental buildings, and instead groups with the horizontal reflecting pool, planters, and paving elements. As a patch of green in the entirely hardscaped campus, the lawn is like a surreal fragment floating in a sea of travertine and granite pavers.

Across the Street, at the Juilliard School, the removal of the plaza connection on level two necessitated a new lobby that would, via the old lobby, connect to the street level of the redesigned 65th Street. Alice Tully Hall, the school's main performance space, needed an acoustic upgrade, and 38,500sf (3,575sm) was added to expand the rehearsal spaces and classrooms, and provide additional administrative spaces.

The DS+R team approached the Juilliard School's addition with great attention to the intent of the original building, designed by Pietro Belluschi

4|1 The new Juilliard School of Music (2012) and streetscape design by Diller Scofidio + Renfro. The new addition on the Broadway side includes a new glass enclosed lobby showcasing the volume of Alice Tully Hall within.

4|2 The North Plaza of Lincoln Center for the Performing Arts includes a new reflecting pool framing a Henry Moore sculpture, a reconfigured bosque of trees, and a new grass-topped pavilion containing a restaurant and cafe.

and Eduardo Catalano. Sharing the Roman travertine cladding, the building was cohesive with the others, but unlike the classical allusions made by the colonnades and porticos found on the main campus, the Juilliard School was designed with dramatic horizontal cantilevers and expressed beams. Elizabeth Diller referred to the original building as "a brutalist building in travertine drag."[22]

The new DS+R-designed wing was added to the eastern edge of the original building, extruding the upper volume eastward, undercutting it to expose a taut underbelly, and slicing the face on the Broadway side to reveal the programs within. 4|1 The south facade is coplanar to the original facade and is clad in the same travertine; it extends the same rhythm of recessed windows and articulated piers that mark the original upper-level windows. Because the new programs in the addition do not align with the original slab levels, the windows were shifted, deliberately misaligning with the Belluschi facade pattern. The architects describe the process as akin to a DNA mutation. In falling out of rhythm with the original pattern, the new windows overlap the recesses in unusual ways, producing atypical volumetric conditions referred to as "fish tank" windows. 4|3 These windows register the 3D sculpting of the Belluschi facade, while providing new windows per the new program. The windows employ insulated glass on the interior face and structurally glazed monolithic or laminated glass on the exterior to minimize the outside corners. The misregistration of windows in the Juilliard addition is a visual hint that this is the new wing, not to be confused with the original. 4|4

The Broadway facade presents a taut all-glass face, stretching from soffit to parapet, and from 65th Street to 66th Street. The beveled detailing of the corners in both plan and section allows this facade to appear to turn corners without thickness. [23] 4|5 At the roof parapet, a steel frame is constructed to cantilever the mullion head receptor up and out, providing a knife-edge detail. The sloped fascia panel is fitted around the vertical glass mullions that support the facade. At the base of the facade, a similar tapered detail is employed to minimize the apparent depth of the floor package and the construction of the ceiling soffit below. The transition between the glass face along Broadway and the travertine stone along 65th Street is also achieved with a beveled detail, where the glass is captured in a stainless steel angle, and the stone is tapered and supported with a clip attached to a triangular steel support within. The all-glass facade is hung from the structure above and supported with five-ply laminated glass mullions. The

[22] Diller Scofidio + Renfro et al., eds., *Lincoln Center Inside Out: An Architectural Account*, First edition. (Bologna, Italy: Damiani, 2012), p. 162.

[23] The dimensionless corner articulation was also tested previously at the ICA Museum in Boston (2006), which Elizabeth Diller described as "Saran wrap over a shoe box." Many of the details at the Juilliard School were first tested at the ICA in Boston.

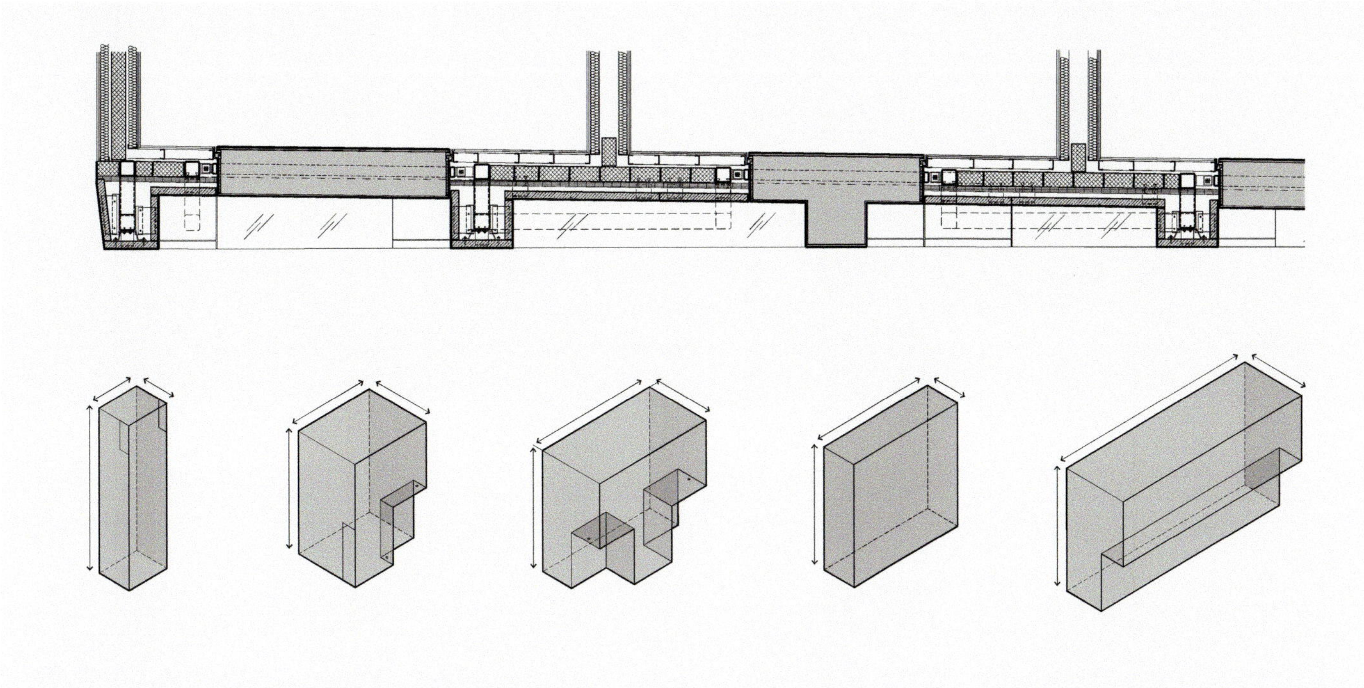

4|3 The new travertine-clad facade of the Juilliard School of Music acknowledges both its continuity with the original, as well as its discontinuity, in the shifted placement of the new windows that intersect the facade in unusual configurations resulting in the "fish tank" windows, that negotiate the deep relief of the facade with the need for new windows in seemingly random locations.

curtain wall assembly consists of laminated and insulated glass to provide the acoustic separation necessary for some of the practice rooms. The tautness of the glass facade is ensured by the fact that there are no projections beyond the glass. In fact, the insulated glass is mechanically secured with a rotating clip on the second lite of glass within the dimension of the insulated glass unit.

Beneath the cantilevered volume of the classrooms and rehearsal rooms, the newly exposed lobby is designed to showcase the volume housing Alice Tully Hall. The original design hid the concert hall behind layers of opaque surfaces and program spaces. The renovation of the Juilliard School proposed a kind of architectural "strip tease," where the interior spaces are placed on display and visible from the street. Enclosing the new lobby is an all-glass, one-way cable net facade that supports the oversized laminated glass lites on high-tension, stainless steel cables. The cable facade is suspended from the steel frame above and tensioned by a spring in the ceiling. Stainless steel patch fittings support the laminated glass to carry the gravity load and resist wind loads. The highly transparent enclosure for the lobby minimizes the presence of glass and highlights the continuity of public space from the sidewalk into the building.

24 Diller Scofidio + Renfro et al., eds., *Lincoln Center Inside Out*, pp. 14–15.

DS+R's approach to the Lincoln Center redevelopment addresses each condition situationally, considering the legacy and intent of the original, and asking how to apply contemporary thinking to questions of access and visibility, display and performance. The interventions across the campus, including signage and lighting, canopies and drop-offs, new planters and water features, facades, and envelopes demonstrate a sophisticated repertoire. In each case, the new is in dialogue with the old, extending a conversation and inviting the public — the audience — to appreciate the art of the architecture.

The architects describe their approach to the project: "Rather than replacing the image of this cultural icon with one alien to it, the design team devised a strategy to work with the ready-made and amplify its most successful features. Preservation in this context had to be understood as a creative act." 24

In response to the technical challenges of working on existing buildings while the buildings remained occupied, the architects worked closely with the construction team to develop a sequence plan to keep spaces protected and minimize disturbance during construction. "In a fashion akin to emergency medicine, LCDP [Lincoln Center

44 A close-up view of the "fish tank" windows in the Juilliard School of Music south facade.

Development Project] masterminded open-heart surgery on a patient that was wide awake, as the work had to be carefully planned, translated, and executed while Lincoln Center's venues remained open."[25]

The Lincoln Center redevelopment project took 12 years to design and execute, divided into various scopes, and spread out into a series of phases. Unlike the original Lincoln Center project that brought together six architects to design a single campus, the redevelopment employed a single team to rework the complex and update it for contemporary audiences. In extending the life of the 1960s campus, DS+R developed and deployed various technical means to apply these life-extending operations and recast Lincoln Center for a new century and new generations.

DEMOLITION VS. REUSE

In contemporary Chinese cities, historic low rise structures that make up the typical urban fabric, have been bulldozed at alarming rates and replaced with modern high-rise apartment buildings. 46 For example, the traditional *hutong* alleys and *siheyuan* courtyard house type is being replaced by higher-density modern structures. Architectural historian Thomas Campanella describes China's breakneck redevelopment and rapid urbanization as the "wholesale reinvention of the city as we know it,"[26] in his book, *The Concrete Dragon, China's Urban Revolution, and What it Means for the World*. As part of the "24/7 frenzy of consumerism and construction,"[27] he also highlights the rampant demolition of the historic fabric in Chinese cities as part of the feverish effort to modernize after prolonged economic stagnation during the Cultural Revolution.

In response to this *tabula rasa* approach, a Chinese historic preservation movement was born to prevent the wholesale destruction of the historic neighborhoods in Beijing and Shanghai. Recognizing that the traditional laneways could be repurposed as retail stores, the Chinese developer Shui On redeveloped the Xin Tian Di neighborhood in Shanghai as an outdoor shopping center. They hired Boston architect Benjamin Wood, who had previously overseen the successful transformation of Faneuil Hall, in Boston, into a festival marketplace in the 1980s, to transform a dilapidated neighborhood into a high-end destination shopping neighborhood in downtown Shanghai. Xin Tian Di established a new model for neighborhood-scale historic preservation through adaptive reuse.

25 Diller Scofidio + Renfro et al., eds., *Lincoln Center Inside Out*, p. 18.

26 Thomas J. Campanella, *The Concrete Dragon: China's Urban Revolution and What It Means for the World*, 1st ed. (New York, NY: Princeton Architectural Press, 2008), p. 15.

27 Thomas J. Campanella, *The Concrete Dragon*, p. 18.

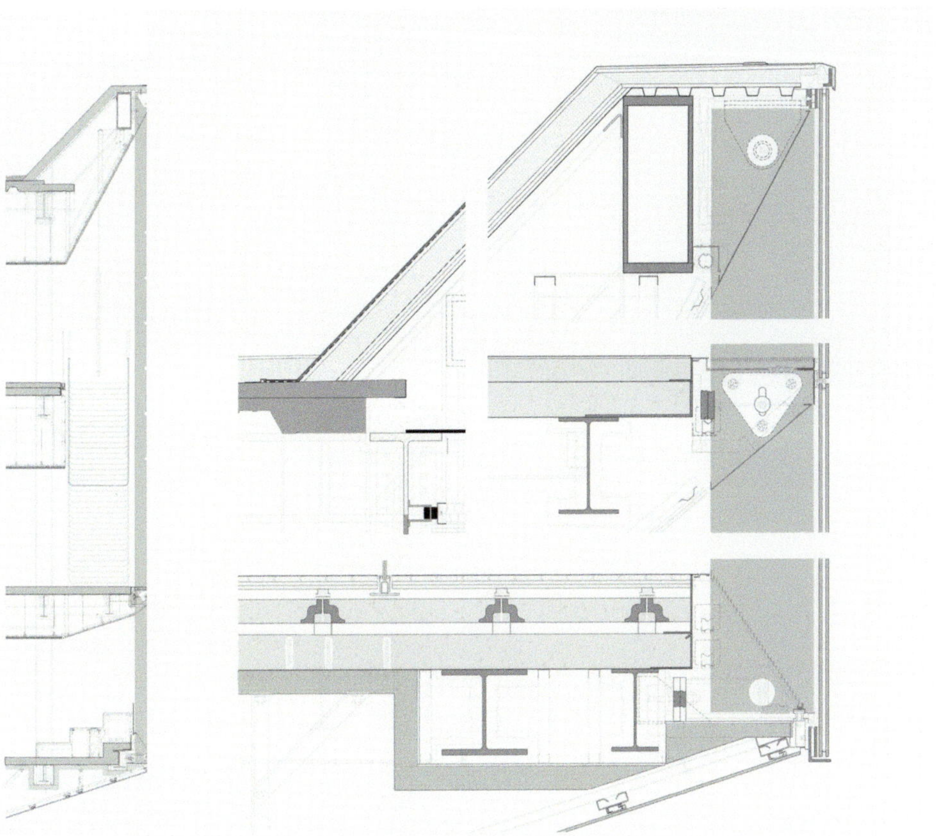

45 The Broadway facade of the Juilliard School of Music is a taut glass plane, in contrast to the stone facade along 65th Street. The beveled edges at the head and sill mask the actual thickness of the structure and floor packages to make the glass surface appear to be stretched taut over the frame, "like Saran Wrap over a shoebox."

Rather than restoring historic buildings and turning them into pristine shopping centers, architects Lyndon Neri and Rosana Hu, of Neri&Hu, developed a different approach for architectural interventions in the historic fabric of Shanghai. The Waterhouse Hotel in the South Bund district transforms a dilapidated 1930s Japanese army building into a boutique hotel. 48 For the project, the architects chose to carefully curate moments of contrast between old and new, calling attention to the patina in the original building while adding moments of striking modernity. Neri and Hu describe the process: "Peeling back layers of finishes was akin to performing an autopsy — uncovering the lives and narratives hidden within each imperfection, an excavation of memories that brings the most intimate moment of inhabitation to public light."28

In addition to selectively restoring elements of the existing building and updating its concrete structural frame, the renovation also includes the addition of a new penthouse. Clad in corten steel, the new structure includes additional rooms and a rooftop terrace. The contrast between the original structure's rough patched concrete and peeling paint and the new steel facade with tall glass "periscope" windows heightens the tension between old and new, between the precious and the mundane.

28 Neri&Hu Design and Research Office, *Neri and Hu Design and Research Office: Works and Projects 2004–2014* (Zurich, Switzerland: Park Books, 2017), p. 34.

Plans and sections produced by Neri&Hu employ a three-tone technique, where black linework is original, gray lines are to be demolished, and red linework denotes new construction. 47 These layered drawings evoke palimpsests, where traces of the past coexist with the new. Detailed wall sections reveal the effort of patching and repairing, sometimes from the inside out, to preserve the original structure.

For the designers, the notion of a "ruin" is elevated as a signifier of persistence and duration. The hotel's elevations reveal a patchwork of finishes and materials that combine old and new, weathered and worn, with hard and crisp. The facade along Maojiayuan Road, includes the new entrance clad in corten steel, and lined with wood, set into a stucco-faced wall marked by staining and peeling paint. Their surgical interventions within historic fabric never strive for a seamless restoration, preferring to allow the building to flicker between ruination and renewal.

SALVAGE AND SPOLIA

In the context of rampant demolition, Chinese architects Wang Shu and Lu Wenyu of Amateur Architecture Studio have also been working to find ways to incorporate the elements of historic buildings

46 An unfortunately all too common scene in Chinese cities: The demolition of an old neighborhood to make way for new development.

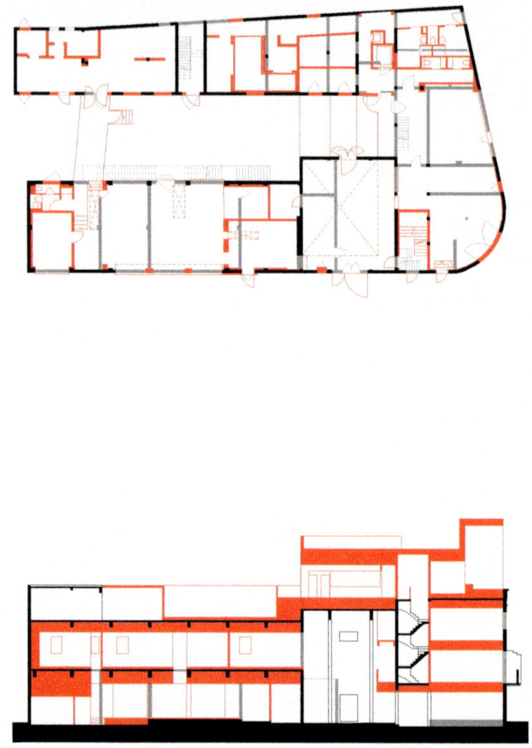

47 "Palimpsest drawings" by Neri&Hu, document the existing, the demolished and the new in one plan and section of the Waterhouse Hotel, in Shanghai.

into their new structures. Rebuilding with salvaged elements is not new in China. *Wapan* construction refers to rapid rebuilding techniques, particularly after a disaster like earthquake or typhoon. Indeed, the process of rebuilding with fragments and components of previous structures is found across cultures. For example, the Latin word *spolia* refers to an architectural fragment displaced and reused in a new context.[29]

Amateur Architecture Studio's design for the Ningbo Museum, constructed in 2008, employs the striking use of wapan construction for its main exterior facades. 49 The massive new concrete structure includes battered walls that are covered in a layer of seemingly randomly placed materials and elements taken from demolished buildings. The new masonry structure incorporates gray stones and bricks as well as roof tiles and stone blocks to create a pattern with "found" materials that give the museum a bricolage-like appearance. 50 51

In the introduction of the Louisiana Museum exhibition catalog the curators write, "Wang Shu personifies the architect as researcher — an archeologist who, by inverse excavation, piles history up as buildings, not to be exhibited in a museum...but to give life to new houses and create a framework around new lives."[30]

29 Another example of the use of spolia in new construction is the paved walkways on the base of the Acropolis in Athens. Greek architect Dimitris Pikionis employed fragments of buildings as pavers for the new landscape plan. The new landscape plan, completed in 1957 incorporates slabs and other construction debris to form a new patchwork pattern of pavers.

30 Shu Wang et al., eds., *Wang Shu Amateur Architecture Studio* (Zürich: Lars Müller Publishers, 2017), p. 9.

31 Charlotte Malterre-Barthes, *A Moratorium on New Construction* (London: Sternberg Press, 2025).

Amateur Architecture Studio is dedicated to preserving traditional methods of construction, craftsmanship, and cultural practices. The reassembled collages of construction materials, neatly arranged into orderly stacks, are immensely beautiful, but also deeply melancholic. The massive scale of the Ningbo Museum and its mountains of rubble recall the 30 villages that were demolished in the city center to make way for the new district. Wang Shu's patchwork masonry seems to mourn the demolished structures. When nothing is left to preserve, and the historic city has been destroyed, then the act of reuse is a powerful way to recall the past and breathe new life into the present.

MORATORIUM

"To build is to destroy,"[31] writes architect and urban scholar Charlotte Malterre-Barthes in her essay *A Moratorium on New Construction*. She argues that all new construction entails harvesting and extracting raw materials, which are themselves acts of destruction. Malterre-Barthes makes a case against building new, or at least for taking a moment to stop and reflect before rushing toward new construction. Her radical call for a pause that disrupts the default, business-as-usual way of design

4.8 A non-descript Japanese army building from the 1930s in Shanghai, China, was renovated into the Waterhouse Hotel (2010) by Neri&Hu while maintaining significant elements of the original and the patina of age and weathering.

4|9 The Ningbo Museum (2008), designed by Amateur Architecture Studio, contains artifacts related to the history of Ningbo, but the building itself is a kind of curated display of the history of the recent past. The architects, Wang Shu and Lu Wenyu, incorporated salvaged building materials into the new building facade to create a composite structure that is part ruin and part new construction, while providing a critical commentary on China's rapid urbanization and its consequences.

and building, of extracting and of mining, refining, fabricating and constructing, is a call to action. Rather than build new, we might instead use what we already have. Rather than purpose-built, perhaps we adapt existing buildings. With her work, Malterre-Barthes calls for architects to design differently, and to inhabit differently — with an appreciation for architecture as an enduring artifact that demands rethinking and an ethics of the existing.

SIMULTANEITY

Approaching the Great Mosque of Córdoba from the narrow cobblestone streets of the dense fabric of the surrounding city, the building appears fortress-like and imposing. Its buttressed limestone walls are entirely closed and topped with crenelated battlements. The entrance is through an arched gate in what is known as the Bell Tower. The forecourt to the Great Mosque is an orange garden, planted in neat rows and dotted with fountains. The courtyard is lined with a colonnade on three sides, save the blank wall of the mosque, which appears to have multiple arches filled in with stone. Construction of the Great Mosque began in 786 and the building was expanded several times over its long history. 5|2 Everywhere, there are traces of architectural elements that speak to the various lives the mosque has lived.

Entering the hypostyle hall of the Great Mosque, one notices the seemingly endless field of columns surmounted by double-tiered arches made of alternating red and tan stone. 5|3 The columns vary in material and some are believed to be salvaged from Roman structures in the region. There are 11 bays in the original footprint that run perpendicular to the *qibla* wall to the south. Over time, the original bays were extended to the south, and eight additional bays were added to the east. Overall, the plan is roughly square, with 19 bays and 35 aisles. The interior space of the mosque is vast and extensive, containing nearly six acres of covered space for worship. It is fascinating to imagine the multitudes of worshipers that have at times, filled the mosque, all oriented toward the qibla to the south.

In the 13th century, Spain was reconquered by the Castilians, and the Great Mosque was consecrated as a church, dedicated to the Virgin Mary, and altered to include a Christian chapel. In the 16th century, Emperor Charles V commanded a Gothic cathedral to be built in the middle of the mosque. Today, the Great Mosque is also the Cathedral of Córdoba. The cathedral occupies nine bays at the center of the mosque's original

50 Wapan construction – the practice of rebuilding after disaster with existing building materials – gives the Ningbo Museum facade a unique patchwork effect.
51 The combination of red terracotta roof tiles, grey bricks, stones and grey roof tiles arranged in irregular patterns evoke the organic patterns of landscapes.

5 2 Aerial photo of the Great Mosque of Córdoba showing the repetitive roof structure of the Islamic mosque and the central figure of the cathedral extending above the roof line.

hypostyle hall. Its central nave rises higher than the adjacent roofs, and heavy buttress walls flank the stone volume. A soaring dome caps the transept, supported on pendentives that spring from bundled piers in the Gothic tradition. The piers of the cathedral correspond to the grid intersections of the original mosque. Spanish architect Hernán Ruiz was entrusted with designing the cathedral within the existing grid of the mosque. The result is an astounding simultaneity of forms and materials. Circulating through a non-directional hypostyle hall the visitor is able to slip through a field of columns in an array of itineraries. As one enters the cathedral nave, the Gothic space takes on an axial itinerary, on route to the transept and altar. 5 4 Upon exiting the cathedral, one reenters mosque, with its non-hierarchical field of columns and arches. 32 5 3

The coexistence of the Islamic Great Mosque and the Catholic cathedral recalls a passage from Sigmund Freud's *Civilization and its Discontents*, where he evokes an image of Rome as an analog for the subconscious, whereby all new structures coexist with old structures, and nothing is ever demolished: "Now let us, by flight of imagination, suppose that Rome is not a human habitation but a psychical entity with a similarly long and copious past — an entity that is to say, in which nothing that has once come into existence will have passed away and that all the earlier phases of development continue to exist alongside the latest one... On the piazza of the Pantheon, we should find not only the Pantheon of today, as it was bequeathed to us by Hadrian, but, on the same site, the original edifice erected by Agrippa; indeed the same piece of ground would be supporting the church of Santa Maria Sopra Minerva and the ancient temple over which it was built. The observer would perhaps only have to change the direction of his glance or his position to call up the one view or the other." 33

For Freud, memories are not forgotten but rather overlaid and superimposed. A change of perspective or position would allow the viewer to perceive the different versions of Rome superimposed on each other. The Great Mosque of Córdoba provides an approximation of that simultaneity. In fact, Hernán Ruiz did demolish a section of the mosque to build the cathedral. But he also retained the mosque and transformed the hypostyle hall into a great antechamber for the cathedral, allowing one to coexist with the other. The mosque and the cathedral offer a striking example of architectural reuse and revitalization, as well as an ethics of co-existing.

32 Stan Allen has written on the field condition of the Great Mosque versus the figure of the Western church. See, Stan Allen, *Points + Lines: Diagrams and Projects for the City*, 1st ed. (New York, NY: Princeton Architectural Press, 1999), p. 93.

33 Sigmund Freud, *Civilization and Its Discontents* (New York, NY: Norton, 2010), p. 17.

5|3 A study of contrasts and coexistence, the Great Mosque of Córdoba is also the Cathedral of Córdoba. The hypostyle hall of the mosque is non-hierarchical and repetitive, allowing the visitor, or the worshiper, to move through the space in many possible ways.

5|4 The Christian church, embedded in the hypostyle hall of the mosque, is distinct in its singular, axial organization, and vertical proportions. The coexistence of the two structures is a unique occurrence and perhaps a model for the future of architecture that will need to contend with the presence of past structures.

1 Mockup panel for Ashen Cabin, designed by HANNAH, for a site in Ithaca, New York.

Chapter 10 Construction Futures

The publicity photo from 1957 shows an interior living room scene with sleek furniture and a smartly dressed couple. [2] He is sitting cross-legged on a couch. She is standing by the window, gesturing to the window shades. The windows span from floor to ceiling, and the seam where the wall meets the ceiling is a tight curve. Both the ceiling and the wall are bowed slightly, seemingly billowing in response to an imaginary gust of wind. The built-in furniture along the wall is curved to match the geometry of the exterior wall, resulting in a keystone-shaped cover for the integrated record player. The sofas are tapered at their edges and cantilevered from their supports, giving the impression of floating. A flat panel mounted to the wall with slightly bowed edges feels like a theater prop, a placeholder for a yet-to-be-invented flat-screen TV. Everything about the scene hints at an imminent future being willed into being through design, materials research, and visual storytelling.

The Monsanto House of the Future (MHOF) was the product of American industry's efforts to find peace-time markets for wartime materials research. A design-build research project with MIT's Department of Architecture, staged by the "Imagineering" team at Disneyland, the MHOF showcased the potential of plastic in residential buildings. For MIT professors Richard Hamilton and Marvin Goody, the house presented an opportunity to study new housing types by offering a futuristic vision of the American home when construction was booming with the GI Bill-fueled single-family suburban neighborhoods. For Disney, the demonstration house was a theme park attraction promising a glimpse into the near future.

The interior geometry of the MHOF living spaces evokes the inside of an airplane fuselage, where nearly all surfaces curve. The house is streamlined and smoothed, seemingly designed to avoid drag like an airfoil. These forms are a necessity in aviation, but in architecture, they are styling. The compound curvatures of the MHOF are meant to showcase the materials from which it is made: materials capable of deformation, elasticity, or, more precisely, what architectural historian Stephen Phillips refers to as "plasticity."[1]

The MHOF was conceived as a demonstration of the inherent qualities of plastic, in particular, its unique ability to be formed into any shape. The floors, walls, and roofs were made of large, prefabricated, fiber-reinforced plastic L-shaped panels called "bents." Fiberglass was developed during World War II and applied to boat building and aircraft design. The lightweight, high-strength, thin shell material could

[1] Stephen Phillips, "Plastics," in Beatriz Colomina, Annmarie Brennan, and Jeannie Kim, eds., *Cold War Hothouses: Inventing Postwar Culture, from Cockpit to Playboy* (New York, NY: Princeton Architectural Press, 2004), p. 113.

[2] The Monsanto House of the Future (1957), designed by MIT professors Richard Hamilton and Marvin Goody, showcased the material possibilities of plastic to produce a compelling prototype of the future, complete with curvilinear walls and streamlined interiors.

be shaped into the compound curvatures needed for the hull of a boat — or, the shell of a futuristic house. The curved interiors of the MHOF were not only evocative of aerodynamic cabins but also marketed as hygienic and easy to clean. Curves and filleted corners eliminated areas where dust could accumulate.

The MHOF designers insisted that the house's form, with its curvature, modules, and joints, was designed "according to the dictates of the material."[2] Even though plastics could assume nearly any shape, the designers opted for the compound curvature due to the inherent structural properties of monocoque shapes. Unlike a skin wrapped over a structural frame, the monocoque shell is both structure and enclosure. The architects made the case for the compound curve based on the fact that "this was a concept that very few structural systems and materials are capable of accomplishing," and the geometry of the house was "the form in which plastic might best exhibit its promise, unique from and exclusive of most other materials."[3]

As a new materially-driven architectural typology, the MHOF could be compared to the drawing of the Maison Dom-Ino by Le Corbusier. The Maison Dom-ino was proposed as a polemical illustration of the capacity of new material systems, like the steel-reinforced concrete frame, to revolutionize architecture and offer a new syntax of form and expression. Similarly, the MHOF was intended to offer a new "plastic aesthetic" and what architecture critic Douglas Haskell called a "'second modern' order... to which today's 'modern' will be just an antecedent."[4]

[2] Quoted in Stephen Phillips, "Plastics," p. 108.

[3] Quoted in Stephen Phillips, "Plastics," p. 111.

[4] Quoted in Stephen Phillips, "Plastics," p. 105.

The MHOF was organized in a cruciform plan with a central core and four cantilevered wings. [3] The core was constructed of concrete and contained the primary structure, mechanical rooms, bathrooms, and kitchen. Cantilevered from the core were four projecting wings, each containing a primary program: entry and dining room, living room, kids' bedroom, and primary bedroom. The cantilevered wings were made of fiber-reinforced plastic, with the floor, end wall, and roof articulated as a single surface. Side walls were infilled with a filleted glass window wall to accentuate the ribbon-like expression of the plastic hull.

Despite the promises of an easy-to-assemble, factory-made kit house, the MHOF was a one-of-a-kind prototype that involved intensive manual labor. Each wing of the house was constructed of four large plastic bent modules, which were 16 feet (4.9m) long and 8 feet (2.45m) wide. The size of the plastic shells was limited by the practical constraints of transportation. During construction and installation

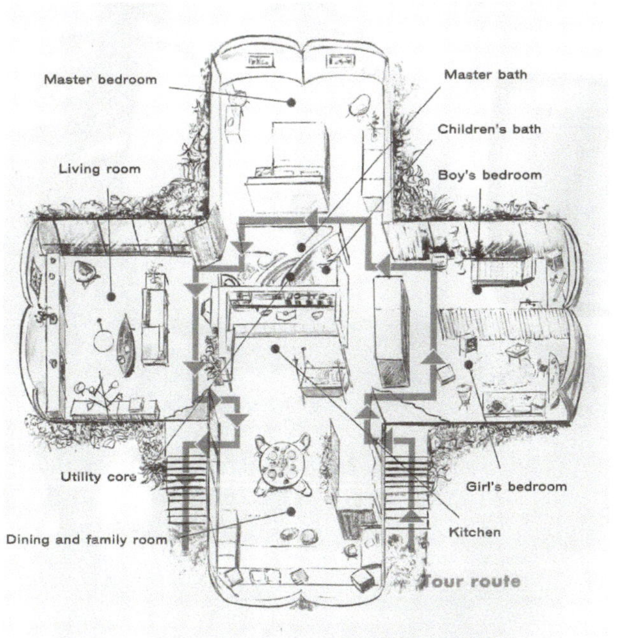

3 The floor plan of the MHOF shows the cruciform configuration with the service core in the middle and cantilevered wings for bedrooms, living, and dining rooms.

4 Construction of the MHOF involved the assembly of large scale prefabricated L-shaped "bents."
5 Once craned into position, the "bents" were bolted together and joints were patched and sanded to produce the streamlined monolithic appearance.

at Disneyland, the fiberglass parts were hoisted into place and supported by temporary shoring. 4 5 Once the parts were bolted together, patched, and sanded, and the scaffolding was removed, the MHOF hovered above the ground, cantilevering dramatically over its landscaped context.

MHOF showcased the potential of plastic to reshape the American domestic realm. The house included furniture pieces by Charles and Ray Eames fabricated with the same fiberglass materials. The kitchen included new appliances and features like a microwave oven, an ultrasonic dishwasher, and retractable storage. Bathrooms featured curved and molded sinks and tubs. Everything from the furniture to the lighting was carefully integrated to create a new domestic stage, whose total effect was evocative of what came to be known as "lifestyle."

The translation from material research, marketing speculation, and theorization to construction was not without its challenges. To achieve the necessary rigidity of the cantilevered wings, the L-shaped bents were formed to create a series of ribs that had to overcome the qualities that make plastic plastic, namely its pliancy and flexibility. The wings were attached to the concrete core with several steel sleeves to achieve the necessary rigidity for the residential live loads, making the house more of a material composite than a pure mono-material.

Ultimately, the MHOF proved too expensive to compete with the conventional stick-built single-family house, and the designers' dreams of a mass-produced kit house that would leverage the efficiencies of the assembly line would not be realized. The plastic house belongs to the long lineage of projects that have promised futures that never arrived. During its 10 years at Disneyland, it attracted millions of visitors and undoubtedly helped shape attitudes about domesticity. It offered a powerful prototype, but like many futurisms, its version of the future became obsolete, and it was eventually demolished in 1967. However, the MHOF premise of large-format prefabricated parts and minimal on-site labor still points to a future architecture that brings together the aspirations of new materials, mass manufacturing, economies of scale, and lifestyle demands.[6]

[6] For more on the history of prefabrication in residential architecture see Barry Bergdoll et al., eds., *Home Delivery: Fabricating the Modern Dwelling* (New York, NY: Museum of Modern Art, 2008).

PAST FUTURES

Some of the most compelling prototypes of the future are world's fairs, where progress and products are displayed for the masses. Many of the projects in this book debuted at world's fairs. Beginning in 1851 with

6 The geodesic sphere, designed by Buckminster Fuller for the Montreal Expo 67, was intended as a showcase American culture and offered visitors a glimpse into a version of a technological future.

7 The Eden Project (2001), designed by Nicholas Grimshaw, for a site in Cornwall, United Kingdom, applies many of the principles of Buckminster Fuller's original geodesic structures, applied to the program of an ecological learning center and rainforest biome.

Joseph Paxton's Crystal Palace, architecture has played an important role in housing these "pageants of progress" within novel structures. The tradition of exhibiting new technologies and architectural experiments continued with the Eiffel Tower in Paris in 1887. The 1970 Osaka World's Fair showcased the work of Japanese Metabolist architects like Kenzo Tange, Kisho Kurokawa, and Kiyonori Kikutake. World's fairs also registered changing attitudes about progress. Spokane's 1974 World's Fair brought attention to the environmental crisis. The Shanghai Expo in 2010, with the theme "Better City, Better Lives," included memorable structures like Thomas Heatherwick's Seed Cathedral.

PROTOTYPES AND PROMISES

Many world's fair pavilion structures are temporary and provisional, but they introduce new concepts and cultural phenomena that often ultimately make their way from the experimental to the everyday. The Crystal Palace translated the light-framed glass shed from the greenhouse to the railway station, and eventually the department store. The Montreal Expo of 1967 showcased engineered lightness with Buckminster Fuller's geodesic dome 6 and Frei Otto's tensile net canopies. Fuller's spherical structure hinted at the premise of a total enclosure capable of containing worlds with their own ecosystems. The expo dome was a partially realized project prefigured by a larger proposal for an unrealized dome over Manhattan. Nicholas Grimshaw's 2000 Eden project for an enclosed ecosystem has an uncanny likeness to Fuller's 1960s geodesic structures. 6 7

6 Grimshaw employed a combination of geodesic domes for their structural efficiency and long span capacity, but also, for their ability to be cropped at the intersection of an arbitrary ground plane, as the superstructure met the site topography of the converted stone quarry.

The articulated space frames of the Osaka Expo realized, at least temporarily, some of the expressive structural logics of the Japanese Metabolist architects, seemingly in dialog with the British Archigram group. Peter Cook's fantastical drawings and diagrams for Plug-In City, published in magazines and periodicals, illustrated a future of continuous assembly and disassembly. 8 The construction cranes depicted on the monumental diagrid frame of Plug-In City were intended to allow users to tailor the city to their needs as a continuously reconfigurable technological utopia. Archigram's imagery precipitated a new cultural sensibility that embraced technology, media, and consumer society. The permanent construction crane, signaling an architecture of transformable systems, would reappear in the work of the British High-Tech movement and Norman Foster's Hongkong and Shanghai Bank, where the facade

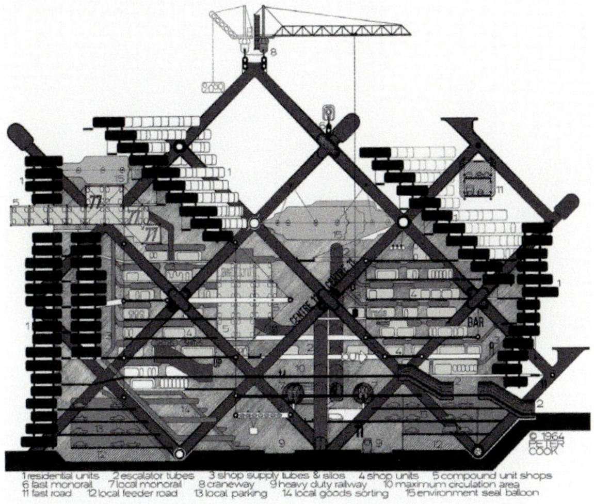

8 The evocative drawings of Peter Cook's Plug-In City (1964) imagined a future urbanism built around reconfigurable megastructures, transportation networks, and modular construction.

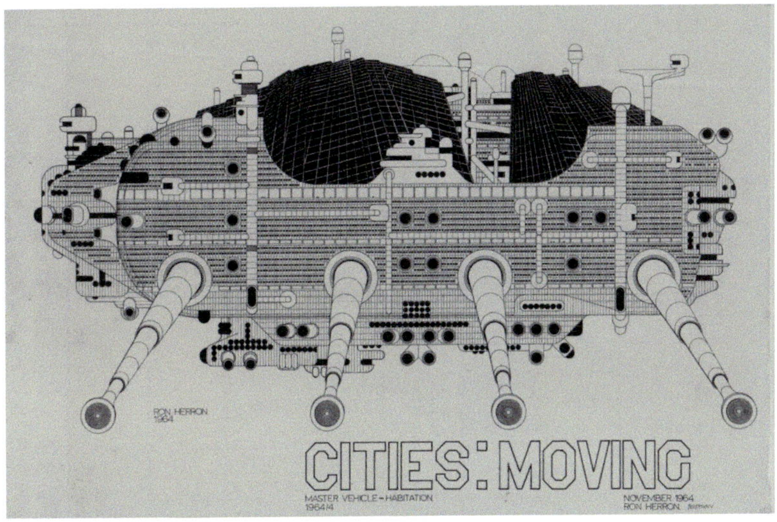

9 Ron Herron's Walking City (1966) project envisioned the city as fleets of mobile structures capable of migration and reconfiguration. Like many of "yesterday's tomorrows," the promises of a technologically enabled future were only partially realized.

maintenance equipment is expressed as four crane-like mechanisms mounted to the corners of the steel frame. 7

Another glimpse of a transportable and reconfigurable mechanical utopia was Ron Herron's 1966 Walking City project. 9 The bodies of the amphibious vessels appear curved and capsule-like: part submarine and part insect. The segmented and articulated supports recall the configurations of jet bridges and multi-directional structural nodes. The compound curvature of Herron's Walking City reappeared in early-2000s experiments with blobs and shells. The Kunsthaus in Graz, designed by Peter Cook, was realized in 2003 as one of the first "blob" buildings of compound curvature by cladding the structure in heat-formed acrylic panels. Also completed in 2003, the Selfridges Department Store, designed by Jan Kaplický of Future Systems, features a compound-curved form with an exterior wall made of bright blue stucco with an added layer of stainless steel discs that defines the implied outer surface of the building form.

7 The kit-of-parts formal aesthetic of the HSBC led to the rumor, soon after it was constructed, that the building was designed to be disassembled and taken back to England before the handover of Hong Kong to China in 1997.

CARTESIAN CURVES

Compound curvature and irregular bulbous forms fall within a certain category of futurism as the complexity of digital form-making in the early 2000s tested strategies for translating digital models into constructed material realities. The Metropol Parasol project constructed in Seville in 2011 hints at a possible future in a few ways. 10 The bulbous geometry of the structure is consistent with the formal experimentation of the time. However, the translation of the compound curvatures into a contoured and intersecting planar lattice is an alternative strategy for approximating the complex curvilinear geometry planar material. 11 For German architect, Jürgen Mayer H., the use of curved and filleted forms are simultaneously signifiers of a forward-looking future, as well as an oblique reference to the 1970s. For Mayer, the futurisms of the 1960s and 70s are also deeply nostalgic reference points.

While the imagined futures of post-war North America and Europe were predicated on synthetic materials like the plastic used in the MHOF prototype, the Metropol Parasol project is realized primarily in engineered lumber. Mayer's competition-winning proposal was not material specific, but in its design development, the idea of a renewable material like glue-laminated timber became a critical part of the narrative. The entire structure was to be assembled of planar sheets of laminated timber, giving the

10 The Metropol Parasol (2011), designed by Jürgen Mayer H., for Seville, Spain, creates an exceptional and curvilinear intervention in the historic city center.
11 The construction of the mushroom-like structures was realized in engineered timber plates arranged in a waffle grid to approximate the compound curved geometry of the pavilion. The choice of wood, as a rapidly renewable biogenic material, is one of the Metropol Parasol's most futuristic qualities.

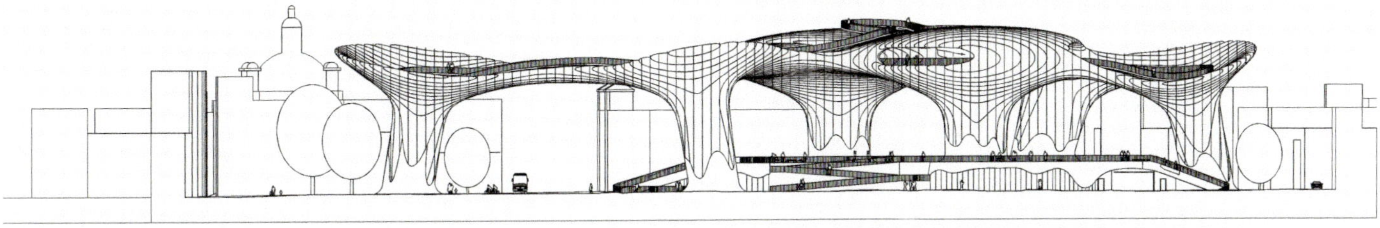

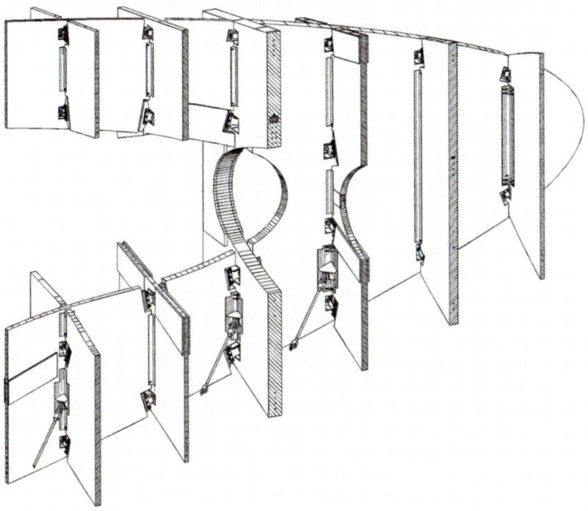

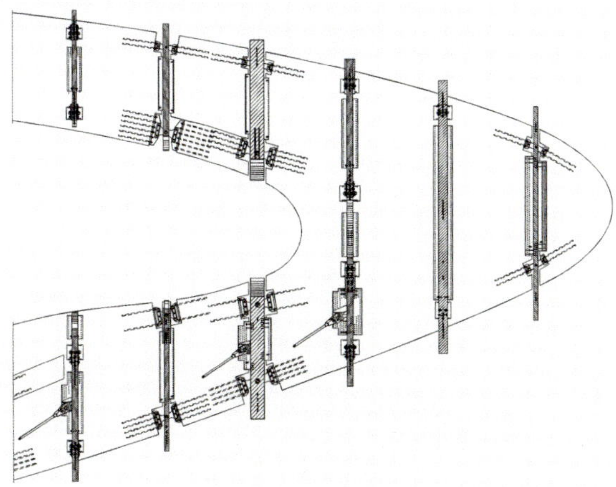

12 The section of Metropol Parasol illustrates the continuous canopy configuration and its gridded planar logic.
13 A partial view of the intersecting panels and the necessary steel joints and cross bracing.

14 The technical means to the surrealist ends, a detail section of the wood panel lattice reveals the amount of steel embeds and mechanical fasteners needed to make the wooden lattice appear to float effortlessly.

structure a surprising sense of warmth and materiality, not typically found in futuristic imagery.

The massive planes of engineered wood are spliced together with steel inserts and cross-bracing to create the uncanny effect of a Cartesian cloud suspended over the city. 13 14 It is planar and rectilinear along the x and y axes but expressively curvilinear in profile. The aggregated profiles imply the volume of the cloud-like canopy, while a distinct grid pattern is visible in the shadows it casts.

The widespread use of engineered lumber in Europe has preceded its use in the United States by decades, where fire codes have been late to recognize the flame-resistant qualities of mass timber. Mayer's surreal *setas*, or mushrooms, as they are called by locals, give expressive form with primarily renewable materials, suggesting a future where complexity can be achieved in tandem with a shift away from carbon-intensive materials and practices. The use of wood as a biogenic material that sequesters carbon is a greater indicator of its futurity than its curvilinear form.

CARBON FORM

Of the many ways we could describe our contemporary moment — from postmodernism to posthumanism — carbon is emerging as one of the most powerful factors impacting design discourse and practice, both of which shape how we think about the built environment in the age of climate crisis. Andreas Malm's writing on global warming and industrialization describes fossil capitalism and carbon modernity, mapping out relationships between fossil fuels and economic systems that define what we call modernity. Malm writes, "The fossil economy has the character of a totality, a distinguishable entity: a socio-ecological structure, on which a certain economic process and a certain form of energy are welded together."[8]

In her essay, "Overcoming Carbon Form," architect Elisa Iturbe enumerates the various manifestations of carbon form: "not only urban and architectural typologies but also the spaces and objects of our everyday lives: fast food chains, roadside motels, a delivery truck full of bottled water idling on an asphalt street, the appliance section of a department store, military installations, the internet, Las Vegas, natural gas pipelines, the Indy 500."[9] Iturbe makes a compelling case for a reading of the present as inextricably bound up in the logics of extraction, refinement, and combustion, which renders all of architecture complicit with carbon form. "In the totalizing framework of carbon modernity, architectural

8 Andreas Malm, *Fossil Capital: The Rise of Steam-Power and the Roots of Global Warming* (London; New York, NY: Verso, 2016), p. 12.

9 Elisa Iturbe, "Architecture and the Death of Carbon Modernity," in Overcoming Carbon Form, *Log* 47 (New York, NY: ANY Corp, 2019), p. 13.

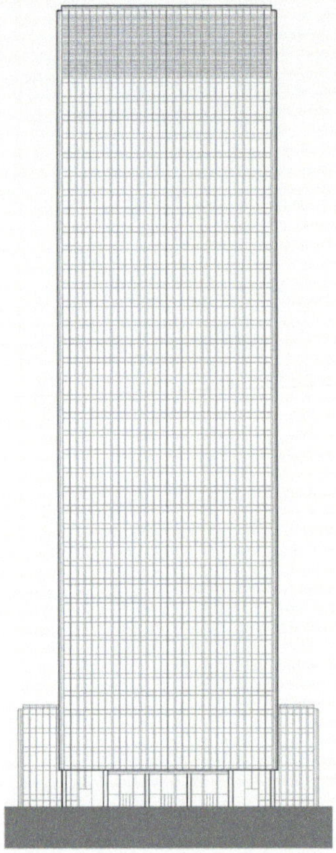

15 The elevation of the Seagram Building (1958), designed by Ludwig Mies van der Rohe, presents architecture as a singular, sculptural, and autonomous object in the city.

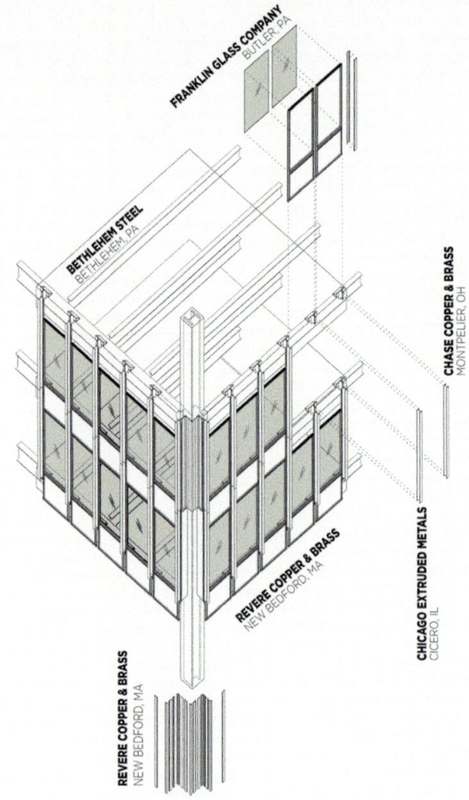

16 Kiel Moe's "terrestrial reading" of the Seagram Building highlights the building as a confluence of material, engery, and labor configured temporarily as an office building in Midtown Manhattan.

thought is constantly subsumed by the ideology of the fossil economy, making it impossible to decouple both the discipline and the practice of architecture from environmental collapse. Carbon form is everywhere, and the possibility of critical architecture is negated by its breadth."[10]

Unlike the discourses of cultural postmodernism and late capitalism, both characterized by immateriality, carbon form has a very tangible quality and offers some very real possibilities for architecture. Fredric Jameson's theories of postmodernism argue that everything falls under the rubric of consumer society, and architecture is the form of cultural production most closely tied to large amounts of capital.[11] If postmodernism was distinguished by its immateriality — it was the invisible and immaterial flows of capital that Jameson remarked upon — carbon form seems to provide a more physical object of inquiry.

As carbon form comes into focus as a critical concern for contemporary architecture and culture today, offering both analytical and material means for engagement, its potential is garnering interest among researchers and practitioners.[12] Matter and material practices consume and embody energy, impacting our climate and, importantly, shaping our lifestyles and perpetuating ways of life that exacerbate global warming. Examining how and with what we build is our best hope for effecting change on many levels. Inescapably, it is largely material that orders and reorders the world around organizing principles, schemas, and forms.

CONSTRUCTION ECOLOGIES

Reading the built environment as a set of material flows profoundly recasts architecture as a sum of all the energy that produced it. Kiel Moe describes buildings as "material in transit." The physical building that stands on the corner of Park Avenue and 53rd Street in New York City is an "exquisite pile of technomass," an "object instance," otherwise known as the Seagram Building. 15 16 Moe's analysis of the modernist icon in his book, *Unless: The Seagram Building Construction Ecology*, shows the building through the lenses of material ecologies, environmental load, and embodied energy.[13]

The title of Moe's book, "Unless," contains a dire warning. "Unless architects begin to describe buildings as terrestrial events, processes, and artifacts, architects will — to professional and collective peril — continue to operate outside the key environmental and political dynamics of this century."[14] Moe argues for a new lens through

10 Elisa Iturbe, "Architecture and the Death of Carbon Modernity," in *Log* 47, p. 19.

11 Fredric Jameson, *Postmodernism, or, The Cultural Logic of Late Capitalism*, 2. print. in paperback, repr. 2008. (London: Verso, 2008).

12 See Antoine Picon's book, *Materiality and Architecture*, (Minneapolis, MN: University of Minnesota Press, 2021).

13 Kiel Moe, *Unless: The Seagram Building Construction Ecology* (New York, NY: Actar Publishers, 2017).

14 Kiel Moe, *Unless*, p. 28.

17 In tracing the material makeup of the Seagram Building, Kiel Moe speculates about the origins of the copper and zinc extracted from an open pit mine in Chuquicamata, Chile.

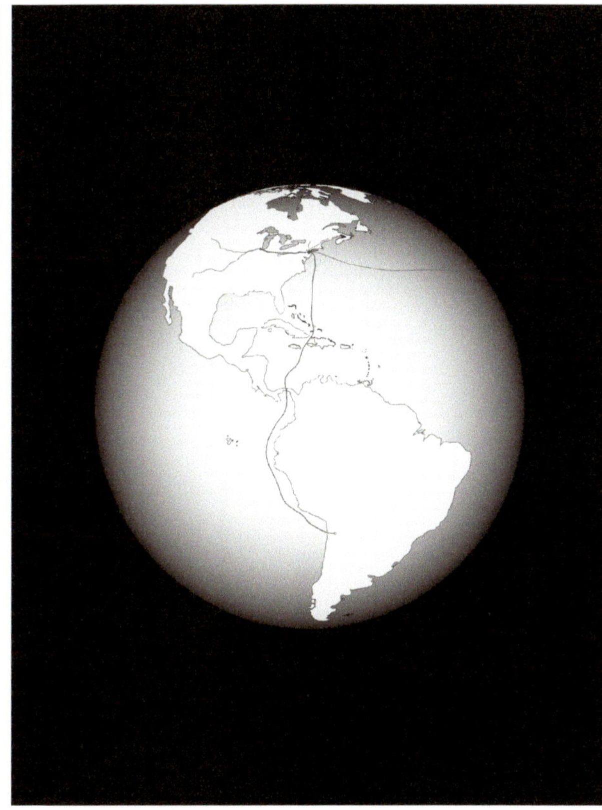

18 Kiel Moe diagrams the material flows and embodied energy of the Seagram Building by mapping the global network of procurement and production that resulted in the modernist icon.

which to assess architecture as a material, geological, and ecological artifact — essentially the substance of which it is made, and the energy and labor it embodies.

Tracing materials back to their sources entails the skills of a scientist and a detective. Moe maps the material flows and itineraries of the bronze (brass) mullions back to the open pit mine in Chuquicamata, Chile, 17 and the bronze tinted glass that was manufactured at the Franklin Glass Company kilns in Butler, Pennsylvania. His research process combines material science and social histories. The investigation of the Seagram Building's material origins is what Moe refers to as a "terrestrial reading." 18 Buildings shift materials and "goods" from their sources to building sites while transferring and accumulating environmental degradation and waste, what Moe calls "bads," to the sites of origin. Moe refers to these impacts as "environmental loads," which he argues are asymmetrically displaced from centers of power and wealth onto the peripheries where these materials come from. Only by understanding the expanded web of materials, their sources, waste factors, tailing, cut-offs, byproducts, and toxicological impacts can we develop an accurate account of the ecological costs of architecture.

15 Kiel Moe, *Unless*, p. 56.

According to Moe, "architects anachronistically still tend to focus on the design of buildings as objects, rather than the design of buildings as an active terrestrial monitor of material, energy, and social transformation."[15] How might we change the propensity to think of architecture as a finished product rather than as a complex set of undeniably material and social processes?

Moe's analysis of the Seagram Building suggests that architects have both tremendous agency and responsibility. Every choice made and every material specified has extended consequences in the world and communities well beyond the building itself. Each design decision is tied to a web of procurement and practices that stretches back to mines, forests, factories, and families. Moe's position on material and construction ecologies is a call to rethink the very substance of architecture, to more deeply consider its expanded footprint, and to better understand and process its larger impact on our terrestrial world.

FOOTPRINT, HANDPRINT

Returning to the concept of material footprint and human handprint touched on in Chapter 2, we see that material and labor remain two of the dominant drivers of design and construction. A growing

19 The Aerial Assemblies project, by Skylar Tibbits and Self-Assembly Lab, MIT, imagines "barrier eliminating" scenarios for construction processes, like the buoyancy scenario of airborne assembly.

20 Demonstrating the principles of design with "low-energy" scenarios, the Aerial Assemblies project allows modules to "dock" with each other in specific predetermined ways.

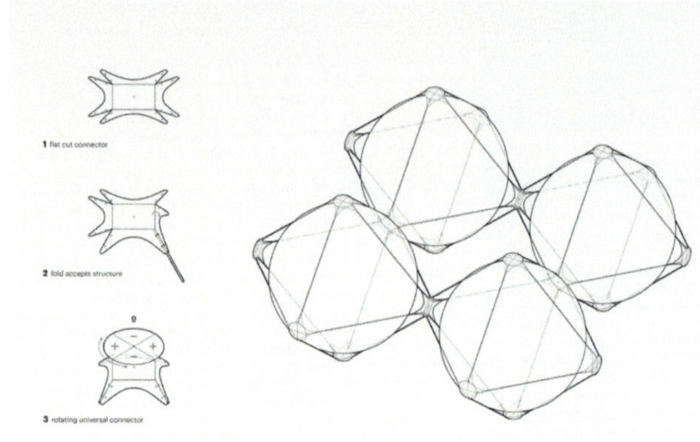

21 The design of the selective joints in the Aerial Assemblies project encourages desired connections while discouraging others.

material consciousness around carbon form and embodied energy, as articulated by Iturbe and Moe, has essentially reset conversations around material in contemporary architecture. Questions of labor and assembly are also undergoing a reevaluation of similar depth and significance.

Labor is, and has always been, an integral part of architecture and its construction. It is a key determinant of construction costs. From a social perspective, the availability of a skilled labor force is an essential part of the construction process and is an integral part of the construction ecosystem.[16] Primary concerns for craft and industrial production that characterize modernity are today redirected toward automation and artificial intelligence. Unlike manufacturing, where the search for cheap labor results in offshore production, architectural labor cannot be offshored. Site specificity means the labor is primarily local.[17]

Social and technological transformations will continue to impact workforce training, fair labor practices, and innovation in construction. Automation and computer-controlled fabrication are transforming the production of components, while digital scanning, surveying, and machine vision are impacting the precision of on-site documentation, installation, and assembly.

16 Recall the efforts undertaken by MASS Design, described in Chapter 2, "Means and Methods," to not only design and build the structures of the Dian Fossey campus, but also to build the capacity and skills of the local labor force.

17 For a discussion of the "offshoring" of architectural labor see, Pedro Fiori Arantes et al., *The Rent of Form: Architecture and Labor in the Digital Age*, Buell center books in the history and theory of American architecture (Minneapolis, MN: University of Minnesota Press, 2019).

SELF-ASSEMBLY

Skylar Tibbits, founder of the Self-Assembly Lab at MIT, has been researching low-energy and low-labor mechanical assembly topics in the built environment since the early 2010s. Tibbits, who belongs to the generation of architects trained entirely on digital tools, recalls working on an installation project in the early 2000s, where designers used digital tools to cut hundreds of unique parts and then found themselves assembling the installation by hand. He wondered why, with all the sophisticated technology, he was still putting the parts together piece by piece. "There's got to be a better way," he recalls thinking.

Digital fabrication at the time required an inordinate amount of manual assembly. Tibbits wondered if parts could be imbued with qualities that would help them fit together more automatically, in other words, if they could even assemble themselves. He founded the Self-Assembly Lab to research smart materials, energy, and labor-saving strategies in architecture and design.

The work of the Self-Assembly Lab investigates how materials and components can be imbued with qualities that eliminate mistakes and assist with precision in assembly. Tibbits describes a mode of design for "error trapping," in which joints

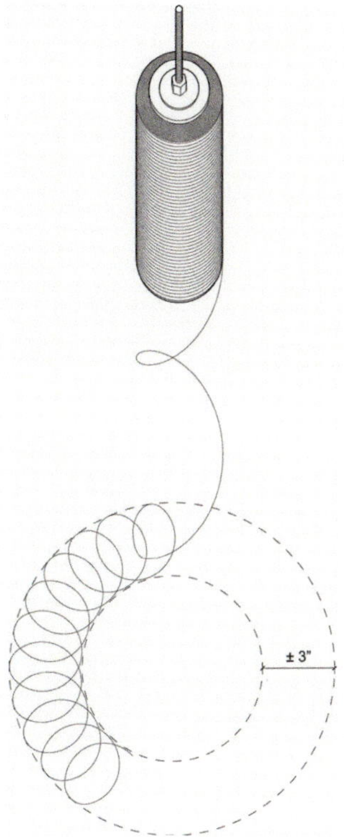

22 The principles of the Rock Print project (2015), designed by Gramazio Kohler Research, ETH Zurich, and the Self-Assembly Lab, MIT, involved reversible, mortar-free, granular construction that leverages the binding tension of string and tendency of granular material to jam.

23 Employing a multi-axis robot with a custom fixture, the designers of Rock Print created a dry aggregation of gravel capable of standing like a column, while also being entirely disassemblable.

are differentiated and only fit together in specific configurations, thus eliminating the possibility of incorrect assembly.

One such demonstration of this concept is a project called Aerial Assemblies, whereby a space frame of triangulated modules is floated in the air with helium-filled balloons, suspending the effects of gravity to allow for the assembly of different joints that form a three-dimensional assemblage. 19 20 For this experiment, 16 fiberglass modules with custom nodes are fitted with 36-inch weather balloons. The geometry of the frames was designed with specific positive and negative connectors to error-correct, and Velcro nodes were limited to a single degree of freedom to allow rotation to facilitate "docking" with other modules. 21

The aerial assembly process reveals how specifically designed modules, coupled with a barrier-eliminating factor, in this case, the helium, could enable the space frame to be assembled with minimal human labor. Tibbits describes the principle of self-assembly: "Most self-assembly systems operate in an open system whereby they can trade energy and entropy with their environment to move toward ordered structures. Thus, we are not going against entropy; rather, we are using entropy to build precise structures that move comfortably from an unordered state to a more organized and functional state."[18]

[18] Skylar Tibbits and Massachusetts Institute of Technology, eds., *Self-Assembly Lab: Experiments in Programming Matter* (New York, NY: Routledge, 2016), p. 140.

It is not hard to see how critical issues in construction, including labor cost, tolerance, and sequencing could benefit from insights gained from the Self-Assembly Lab. If the future of design and construction involves on-site labor and assembly, then design that incorporates error trapping and low-energy state configurations could include elements or principles of self-assembly as part of its conceptualization.

DESIGN FOR DISASSEMBLY

Tibbits also collaborated with the Swiss research group, Gramazio Kohler Research, ETH Zürich, on the Rock Print project to explore the material logics of robotically assisted construction that anticipates its future disassembly. Rock Print was temporarily installed at the Chicago Architecture Biennial in 2015 and consisted of a 12-foot (3.6m) gravel column constructed using a multi-axis robot. The robot was equipped with a custom fixture that allowed it to lay down a continuous pattern of string in a precise looping arrangement. As the string was placed, researchers added layers of gravel. 22 23 The result was a composite mix of loose gravel bound

2|4 Felix Heisel, of the Circular Construction Lab at Cornell University, is piloting projects for salvaging building materials through designed processes of deconstruction.

by 75 miles of string without mortar or adhesives. Tibbits explains that the installation utilized the compressive strength of gravel and the tensile strength of string — two ordinary materials with opposite properties and behavior.

Once the entire column was built, or as the designers say, "printed," within the wooden formwork, the team slowly removed the formwork, allowing excess gravel to fall away. The elements at the center, forming the four-legged column, remained standing. The gravel and string Rock Print column demonstrated their ability to "3D print" a dry stack structure that would achieve its necessary structural capacity and stability by combining the traits of gravel and string in a highly redundant configuration. Tibbits describes the gravel column's operating principle as "jamming," or the behavior of the piece of gravel that is subject to gravity and, in the process of falling, is arrested by other pieces of gravel. The "binding tension" of the string produces enough cohesion in the column to keep the gravel together and allow it to stand.

The "dry cohesion" of the Rock Print project is the material method and technique, but the real promise of the project is its ability to be undone, making it a demonstration of "design for disassembly." After the Rock Print was exhibited in Chicago, the column was dramatically deconstructed. The string was rewound onto a spool, and the column returned to its original state as a pile of gravel on the floor.

Tibbits sees these projects as highly tailored experiments meant to highlight possibilities and constraints within design and construction processes. In each project, he changes the variables and testing scenarios. Aerial Assemblies uses helium to counteract gravity and wind to "mix" the modules until they lock into position. Rock Print imagines the reuse of materials as architectural needs change over time. In each case, Tibbits strives for low-energy states, what he calls "passive processes," allowing the material to "switch" or "jam" with natural processes that reveal new potential.

CIRCULAR CONSTRUCTION

Given shifting demands and the relative persistence of the need to change, transform, repurpose, and often demolish buildings, reconfigurable architecture is highly relevant today. Increasingly, the embodied energy of buildings and the appreciation of their materials as resources that could be repurposed have led designers and researchers to explore means of construction that plan for future disassembly.

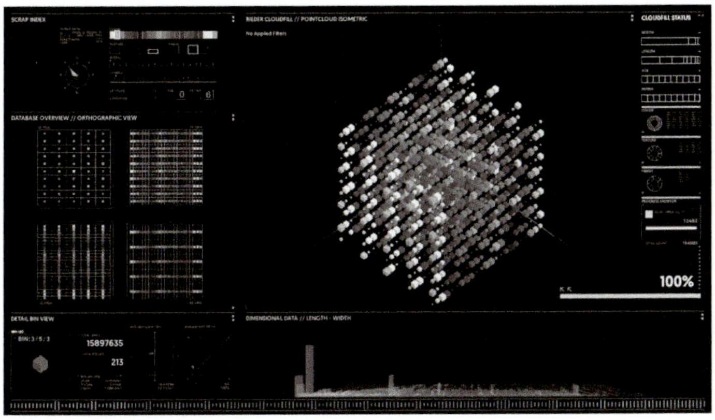

25 **Certain Measures is researching means of cataloging, scanning, and inventorying waste materials with machine vision tools with the intent to develop design processes premised on the availability of waste materials as a basis for design.**

26 **Certain Measures is developing designs based on the availability of waste materials of irregular dimensions, resulting in irregular forms, as depicted in the rendering of a small cabin with "waste" wood siding, from a project called CloudFill.**

Architect Felix Heisel, director of the Circular Construction Lab at Cornell University, has been studying two scenarios: first, the means and methods of disassembling existing buildings, cataloging their salvageable parts, and making them available for incorporation into future buildings, and second, the design of new construction that assembles materials without the use of adhesives and sealants that make it difficult to disassemble them in the future. Heisel and his team of researchers have been working with local officials and developers on small-scale test cases that show how demolition waste can be diverted from landfills by rapidly disassembling structures, cataloging components, and making them available to builders through searchable databases. 24 Heisel's team is testing a circular construction business model that scales up the processes of cataloging and redeploying reused building materials in new projects.

Heisel's second area of research involves design for disassembly, where he develops details that anticipate future uses. Demolition waste often ends up in landfills because the material is damaged beyond reasonable repair. For example, the wood is full of nails, or the glass is marred by silicone sealants. For Heisel, designing for future disassembly involves an increased use of dry assembly methods to begin with, including rubber gaskets that replace silicone seals and other reversible details that allow for parts to be repurposed.

REORDERING DESIGN

Questions of labor practices, risk and its management, costs of materials, waste streams, procurement, and automation will continue to transform the worlds of design and construction. Two speculative projects illustrate the search for new material sources as a way to free architecture of its dependency on raw materials that are high in embodied energy and its complicity with extractive practices. ScrapCrete 28 by Certain Measures and Ashen Cabin 29 by HANNAH both explore the potential of waste streams and digital tools to derive an alternative design vocabulary for a discard-aware future, and, in so doing, reorder the fundamental sequences of design and construction processes.

Designers and researchers at Certain Measures combine the principles of circular construction with machine learning to develop designs based on available waste materials. Rather than develop a design assuming the abundant availability of standard-sized materials, Certain Measures inverts the question and asks: Given a set of non-standard,

27 An inventory of waste materials forms the basis for "design with what you have,"or inventory constrained design. A dry layout of the salvaged off-cuts prior to assembly as a new facade.

28 Certain Measures is testing the ScrapCrete concept on a facade that employs scrap material and off-cuts, aggregated to form a new facade, forming a new aesthetics of the existing.

reclaimed materials of variable lengths, what designs could emerge from a different mode of working? Certain Measures applies these principles and workflows to demonstrate what a waste-source design approach would produce.

This presumption of irregular material modules demands a quick and efficient assessment of available source material. ScrapCrete leverages machine vision to sort and select building elements from waste streams. The design tool scans inventories of waste materials like lumber off-cuts, producing an index of irregular parts as a 3D inventory. 25 The ScrapCrete tool then processes the dimensions of components to generate new facade designs and geometries based on a data set of the shapes available. With irregular modules as a basis for design and a generative approach to "form finding," designers can produce a multitude of possibilities.

Early tests with the method arranged panels and boards of irregular sizes to produce curved geometries and tiled panels. Prototypes like the CloudFill project where Certain Measures generated new versions of datchas from the demolition of old ones. 26 Another ScrapCrete project involves the cladding of a big box warehouse building for the German manufacturer Rieder Facades GmbH, is also made of irregular tile modules diverted from waste streams. The constituent panels comprise a range of sizes, planks, boards, and panels that tile together to form a quilt-like building facade. 27 The range of gray tones in the scrap panels adds to the ad hoc compositional effect. 28

By rethinking assumptions around standardized and modular materials, orthogonal and repetitive patterns, and inverting the design process around resource availability, the designers at Certain Measures practice an alternative architecture where aesthetics are informed by an inventory of the irregular byproducts of construction waste. Inspired by the aim of a zero-waste world, their speculative projects reorder many of the conventions that drive architecture. In the designers' words, ScrapCrete combines the "logic of the quilt with customized shape and pattern detection," resulting in a project that "uses big data to tackle big waste."[19]

19 See https://www.certainmeasures.com/projects/scrapcrete, accessed 5.19.2024.

WASTE

The Emerald Ash Borer is a bright green beetle native to Asia that feeds on ash trees. The ash borer is considered an invasive species in North America and Europe. The beetles burrow under the tree's bark, limiting its ability to transport water and nutrients,

92 The Ashen Cabin (2019), designed by HANNAH, employs robotically sawn lumber from diseased trees ⊃ reorder the processes of material availability and design operations.

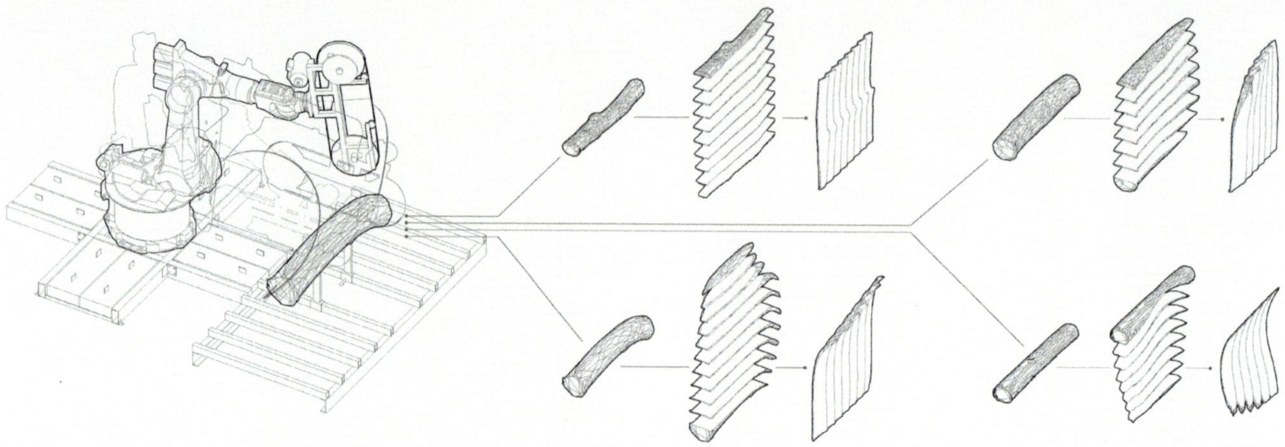

30 Material processing diagram for robotically sawn lumber used for Ashen Cabin (2019). The irregular geometries of the existing logs served as a starting point for producing irregular planks and boards, which in turn produced irregular walls and awnings.

and it eventually dies. Architects Leslie Lok and Sasa Zivkovic, principals of HANNAH, and professors at Cornell University Department of Architecture, have been researching waste lumber sources, including forests devastated by the Emerald Ash Borer. Lok and Zivkovic sought new uses for diseased lumber, and new means of incorporating alternative building materials as a means of reconceiving architecture.

HANNAH's Ashen Cabin is constructed of robotically sawn lumber from dead ash trees. The architects employed 3D scanning and robotic sawing techniques for optimal utilization of the trees, including their curved and branched sections. A band saw attached to a multi-axis robot allows for the "slabbing" of lumber in non-planar units. 30 These bent and curved elements provide the cabin's cladding, and the curvature inherent to the non-planar boards creates projecting windows or canopies. An alternative to the typical process of fabricating irregular trees into standard dimensions that frame architecture, HANNAH develops a non-standard architectural form derived from the trees' natural irregularity. The original branching figure of a tree is inscribed in the curvilinear forms of the cabin. 33

In procuring materials from sources outside of standard material "streams," HANNAH inserts new materials into building stock that the construction industry might otherwise overlook. Reordering the design process, which typically begins with straight and planar elements, to accommodate curvature and embrace non-standard building components further develops a formal language around a material's inherent qualities. Using diseased lumber, a byproduct of an ecosystem of invasive species and environmental degradation, Ashen Cabin makes compelling commentary on our planet's undeniable interconnectedness, and future possibilities for architecture.

SIMPLY

The Bavarian town of Bad Aibling, 35 miles (56km) southeast of Munich, has a population of 19,613 and is known for its mineral water thermal baths. It is an unlikely place to look to the future of architecture, but three modest structures built on the outskirts of town offer clues. The three buildings are called *Forshungshäuser*, or Research Houses, and were completed in 2020. 39 The buildings are part of a research initiative by Technical University of Munich (TUM) professor and architect Florian Nagler. Nagler has noted that despite well-established discourse on sustainable design and emerging technologies

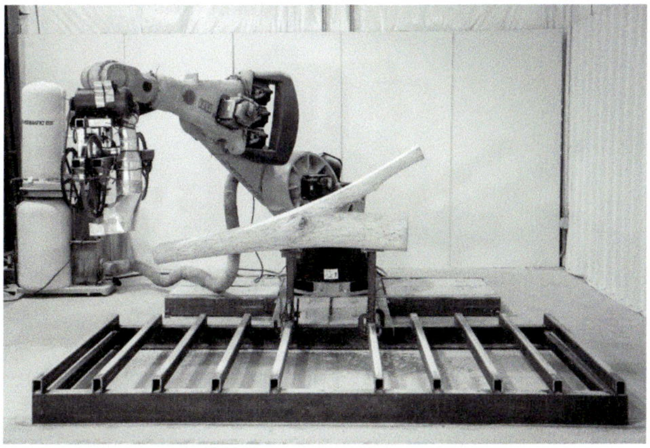

31 The fabrication set up for the robotically sawn lumber used on Ashen Cabin (2019).
32 Irregular boards produced from irregular logs for Ashen Cabin (2019).

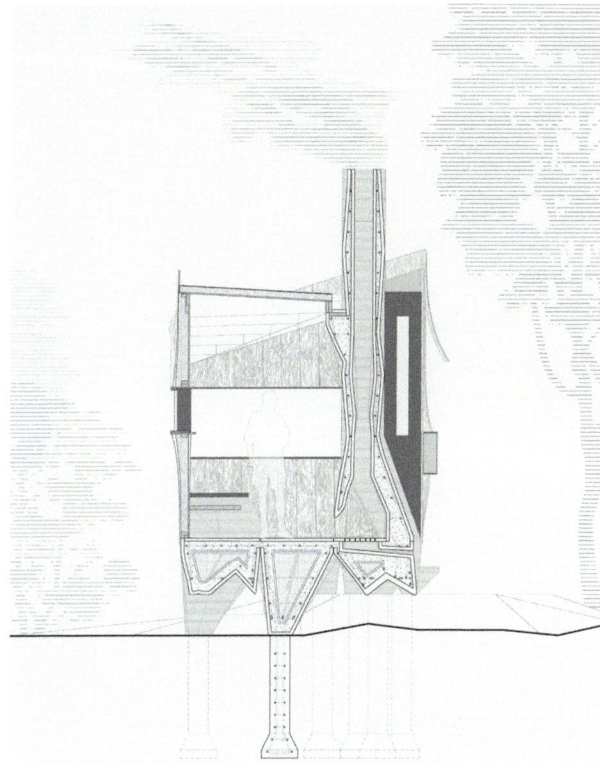

33 A section cut through Ashen Cabin (2019), describes the structure of the foundation as well as the framing and cladding members for the curved geometry of the bay windows and entrance awning beyond.

for climate control, contemporary buildings often do not live up to the expectations and intentions of the designers. Between design, construction, and occupation, something changes, and the post-occupancy conditions of contemporary buildings generally underperform. In parallel, Nagler notes that buildings built in the late 19th century, with thick masonry walls, high ceilings, and large windows, perform relatively well from a thermal comfort and energy consumption point of view.

Nagler launched a research initiative he calls "*Einfach Bauen*," or "Build Simply," and argues for reducing the complexity of architecture's configuration, systems, and construction. Rather than complex multi-layer walls of finishes, frames, sheathing, and membranes, he proposes simple monolithic walls. Instead of large, glazed areas tailored to views and exposures, he proposes a basic grid of punched windows in a uniform arrangement. The three houses in Bad Aibling are a full-scale test of his theory of "Build Simply."

The three buildings on Dietrich-Bonhoeffer-Strasse share the same massing strategy and are perpendicular to the street in a roughly north-south orientation. They are three-story structures with pitched roofs and a single-story projection on the ground floor. They are nondescript, gray in color, and easily overlooked. The differences between the three are subtle. The first building employs semi-circular arched windows, the second has rectangular windows with straight lintels, and the third has windows with a segmental or shallow-arched header. 34 38

Upon closer inspection, it becomes apparent that the three are built of different materials. The first building is fair-faced concrete. The second is wood cladding, stained gray. The third is a gray stucco over a masonry structure. All three utilize load-bearing perimeter walls and the different window geometries hint at the different wall constructions. The segmental-arched header is a masonry detail spanning over a modestly sized opening. The rectangular opening employs a wooden lintel over the windows. The circular arch is used in the concrete wall and was chosen for the concrete structure to make the houses out of singular materials and eliminate rebar. Since the concrete walls have no rebar, the arched form allowed the loads in the wall to transfer around the windows. As Nagler states: "Construction follows material."[20]

[20] Conversation with the author, August 14, 2024.

In a departure from many contemporary design approaches, Nagler designed the buildings with high floor-to-floor dimensions of 10 feet (3.1m) tall punched windows. The window to wall ratio (WWR) is approximately 15 percent, keeping the amount

34 The material and window mockups for the three Research Houses (2020) in Bad Aibling, Germany, designed by Florian Nagler. Each house was designed and built of a different material: brick, mass timber, and lightweight concrete, and each material dictated the shape of the window: segmental arch, straight lintel, and circular arch.

of glazing low to improve thermal performance. The high ceilings and tall windows ensure the penetration of daylight and limit the need for interior lighting, further improving the energy efficiency of this approach. The pivot window also allows hot air to vent from the top while admitting fresh air through the bottom. 34

Nagler's premise was to construct three similar buildings with similar configurations, exposures, and orientations but with one major variable: the material of the exterior walls. 35 In each building, the walls were constructed of a single material in the simplest possible configuration. The concrete walls were built with an aerated lightweight concrete technique to introduce air pockets into the wall, improving its thermal performance. 37 The wood house has load-bearing CLT walls and exterior wood cladding. 36 The masonry house is built with an extruded brick module containing air cavities and finished with exterior plaster. All the buildings have wood-framed floors and roofs. The buildings are equipped with the same basic mechanical systems, but not with additional insulation. The intent was to use the thermal mass inherent to the materials themselves.

The Research Houses form a control set for the experiment, where the variables are limited and carefully monitored. 38 Unlike many research projects, the three buildings are constructed to be lived in by actual residents. In addition to collecting data on energy utilization, the researchers can also interview users about their experiences living in the different houses. Post-occupancy evaluations are increasingly part of the expanded design process that seeks to better understand the impacts of design decisions after the building is completed. In the case of the Forshungshäuser, feedback from occupants proved surprising. While a calculation of thermal performance might evaluate the three wall sections and determine that one is higher performing than the others due to its better R-value, the expanded assessment of the houses from occupant interviews revealed that although the wood house was not the highest performer based on calculations, the users in the wood-walled house generally kept their apartments colder than the other houses. Thomas Auer, from the environmental engineering firm Transsolar, attributes this to the psychological effect of the wood interior material: "The residents feel warmer in a wood house, even though the temperature reading on the thermostat is objectively lower than the other houses."[21] The Research House's occupant interviews yielded different data points and insights that would not have

[21] Conversation with the author, April 14, 2023.

37 The material basis of design for the Research Houses (2022) in Bad Aibling: lightweight concrete, extruded terracotta block, and an engineered lumber with air pockets to improve insulation value.

35 Construction photo of the three Research Houses (2022) in Bad Aibling, each employing a different material system to compare environmental performance.
36 The mass timber Research House under construction.

been possible with straightforward energy flow calculation, nor in a simulated lab environment.

Nagler is a firm believer in the "Build Simply" ethic, and that this approach to design and construction produces benefits on many levels. The cost of construction is lower, contributing to a more affordable supply of housing, and it limits building materials to a few sources, potentially leading to a lower-embodied-energy house. The building systems are also reduced to the simplest workable systems. "If we do not build it, it won't need to be maintained in the future,"[22] says Nagler. Keeping synthetic products out of the project benefits indoor air quality and wellness, and also makes the house easier to demolish or repurpose in the future. Lastly, the house as a lived-in research project yields insights that can be applied to future design projects, like the benefits of the psychological material factor.

22 Conversation with the author, August 14, 2024.

For Nagler, the future of architecture will not look like the space capsules of Peter Cooke's Plug-In City or the high tech of Norman Foster's Hongkong and Shanghai Bank. Nor will it look like the enclosed worlds of Buckminster Fuller or Nicholas Grimshaw. The tropes and motifs of post-war visions of future worlds have not delivered on the questions of comfort, cost-effectiveness, ease of construction, or the possibility of circularity. These criteria lead to a surprising conclusion: the future of architecture will look more like the past. "Build Simply," the straightforward three-story building with a pitched roof and punched windows performs well, does not require lots of energy, and is comfortable for its inhabitants. For Nagler, simple design and simple construction are good enough. 39

SHIFTS AND CONSTANTS

Revisiting past utopian imaginings, taking stock of the current state of design and construction, and searching for clues about trends and patterns to detect the outlines of possible futures, this chapter has identified shifting concerns as well as intractable constants. Despite technological advances in design tools, fabrication equipment, telecommunications, logistics, digital surveying, and smart building systems, which are all revolutionizing the way buildings are conceptualized and constructed, there remain constraints like material sizes, on-site labor, and transportation that are highly resistant to change.

Looking ahead — peering into the future — some of the key drivers for design and construction include sophisticated design tools, artificial intelligence, and machine learning, which will undoubtedly

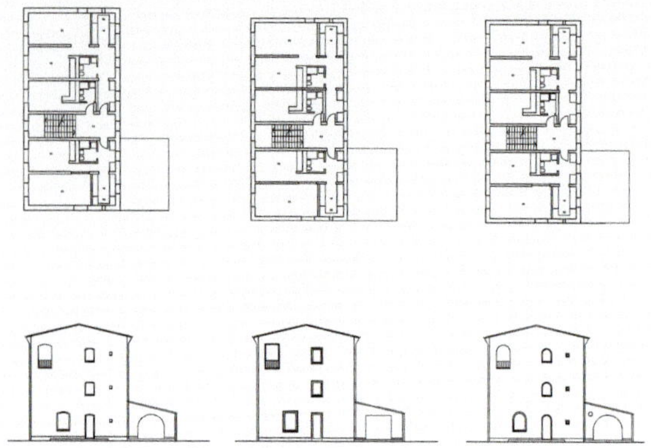

38 Plans and elevations of the three Research Houses (2022) in Bad Aibling, which reveal the consistency of organization and facade fenestration, isolating the variables to just the material makeup of the wall construction.

39 From the exterior, the three Research Houses in Bad Aibling, appear unremarkable, and their differences are manifested primarily in the shapes of windows. The simplicity of their designs and construction conceals a sophisticated design research project at the intersection of material science, post-occupancy studies, and human behavior.

impact how architects conceptualize, develop, and document their projects. Concerns with climate, energy use, material sourcing, adaptive reuse, and embodied energy will continue to be important considerations. More careful accounting of the true costs of materials — their embodied energy lifecycle assessment (LCA) will inform design decisions, and their procurement on the construction site will require more tracing of provenance. Labor will continue to be an important factor in construction, both on and off site. The creative procurement of materials from atypical and renewable sources will encourage material imagination in design development. Despite trends toward the automation of cutting and fabricating components, the labor required to assemble architecture from parts will persist. Thinking through questions of assembly and sequencing, as well as transportation and pre-fabrication, will offer new ways of conceiving architecture as a carefully assembled whole.

Future buildings will undoubtedly continue to integrate sophisticated systems for monitoring and control, adapting building systems to environmental and occupant conditions. Despite the celebrated arrival of the "Internet of things," with networked thermostats and smart appliances, the performance of the buildings, their envelope, and thermal mass will likely return to basic passive principles of thermal inertia and daylighting. Energy codes will dictate the thermal performance of building envelopes and limit the window-to-wall ratios of buildings. Despite the apparent divergence between high-tech and low-tech approaches to building systems, designers will make choices informed by ever greater amounts of information: site information, energy simulations, material procurement, and performance analysis.

Historian Mario Carpo declared that "every canon is a data set" and "every data set is a canon," combining two critical contemporary issues. Current reevaluations of the architectural canon and the decolonization of knowledge bring to light the many ways the built environment and the discipline of architecture inscribe power structures and perpetuate exclusive practices. Simultaneously, the rise of AI and the superabundance of information impacting design and construction render so much of the world as data to be sifted, sorted, analyzed, and applied. Carpo's assertions highlight how these two conditions are conflated in contemporary design, as images and precedents, both historical and contemporary, increasingly become data sets in machine learning and automation processes. The presence of past designs will be amplified by AI tools that sample enormous data sets of

40 The dendriform column structural load test (1937), designed by Frank Lloyd Wright for the Johnson Administrative Building, Racine, Wisconsin, represents a means of design research in materials and structural performance.

41 The Rock Print column installation (2015) by Gramazio Kohler Research, ETH Zurich, and Self-Assembly Lab, MIT, illustrates a contemporary approach to design research in the areas of design for disassembly and robotic fabrication.

projects, impacting the design process amidst a consciousness of similar or comparable designs. The influences of this data set will inevitably change what it means to design and what it means to imagine.

TESTING 1 2 3

Recalling Frank Lloyd Wright's legendary 1937 experiment for his SC Johnson Administrative Building design, where the unusual geometry of the dendriform column was subjected to a structural load test to prove Wright's intuition about material behavior, we might ask about the contemporary state of material testing and experimentation. Might the Johnson Adminsitrative building column find its contemporary counterpart in the gravel and string Rock Print column tested by Gramazio Kohler Research, and the Self-Assembly Lab? Where Wright found a sponsor in a private client to test his unusual design for implementation in the proposed project, the Rock Print column has no specific client nor project application in mind. Contemporary material experiments are often carried out at academic institutions, in open-ended formats, where outcomes are understood as cumulative. Insights gained from one project on dry and reversible construction techniques can be used for subsequent projects with different parameters and applications.

These examples highlight the fact that architecture, as a discipline, is a mode of knowledge production. Beyond a single project, practice, or institution, the making of architecture involves marshaling existing practices, experiences, and knowledge to transform matter and material, which ultimately, and almost inevitably, produces new knowledge. As we struggle to imagine future worlds in light of challenges and constraints amplified by climate change, resource scarcity, energy awareness, and social and political upheaval, we can understand architecture as a practice of design and research that produces both new knowledge and new tools to act on the future.

DESIGN RESEARCH

The projects shared in this chapter, from the testing of entropy and self-assembly to the reordering of procedures of material procurement, are examples of design and research projects occurring within institutional frameworks where risk is managed, scenarios are structured, and outcomes are published and shared. And at the same time, these experiments conducted by researchers at MIT,

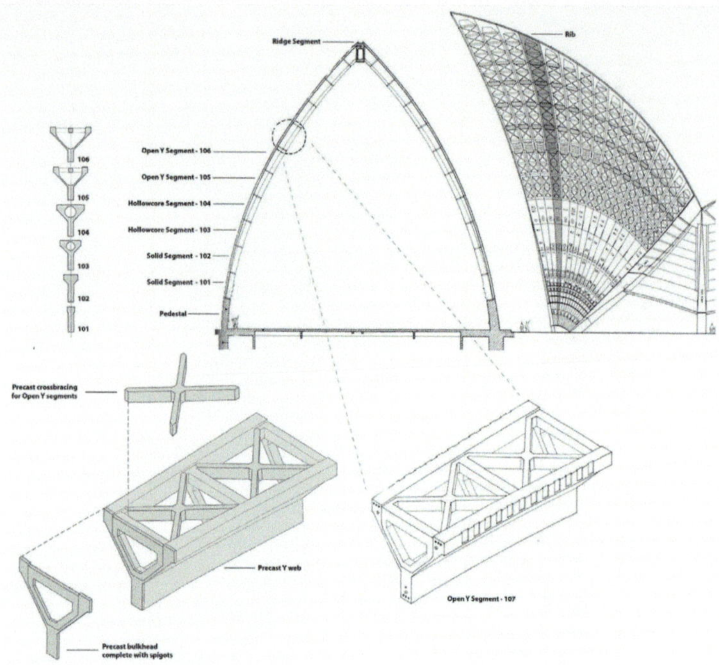

42 Sections and isometric drawings of the precast concrete ribs of the Sydney Opera House (1973).

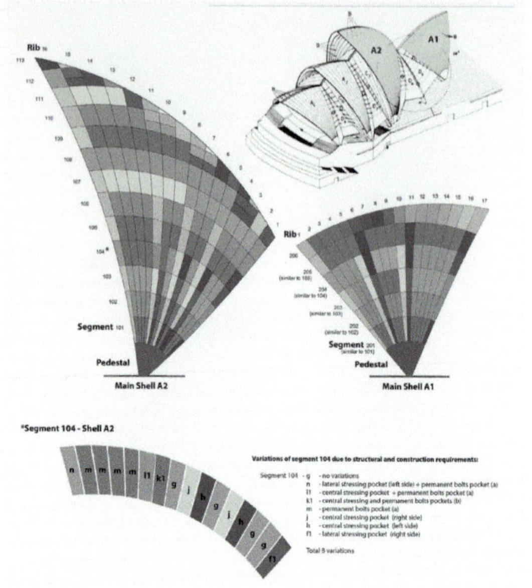

43 Analytical drawing of the unique precast rib segments used to construct the Sydney Opera House (1973).

Cornell, ETH, and TU Munich offer prototypes and glimpses of possible futures.

How do we bring these alternative futures into being? How do we shift from the practices of speculation and forecasting to projection and building? What tools are necessary to move past imagination and on to implementation? And, how to execute the unprecedented in a world of risk aversion and inertia for normative practices?

OUGHT TO BE

American social scientist Herbert A. Simon wrote that "science is the study of the world as it is," but that "design is the study of the world as it ought to be."[23] Design seeks to transform the world from how it is, to how it ought to be. Simon's definition speaks to the anticipatory nature of design and its utopian aspirations. As a discipline and a profession, architecture aims to transform the world. The work of architecture is a proposition — inextricably linked to imagining alternative futures. Much like science fiction, architecture and design are speculative practices. Speculative fiction scholar Sherryl Vint describes Science Fiction (SF) as "a vision of the world made otherwise and the possibilities that might flow from such change."[24] She notes that

23 Herbert A. Simon, *The Sciences of the Artificial*, 3rd ed. (Cambridge, MA: MIT Press, 1996).

24 Sherryl Vint, *Science Fiction*, The MIT Press essential knowledge series (Cambridge, MA: The MIT Press, 2021), p. 4.

25 Quoted in Sherryl Vint, *Science Fiction*, p. 11.

political theorist Fredric Jameson has described science fiction as an effort to recast the present as the past, not to predict the future, but to "defamiliarize and restructure our experience of the present."[25] Where science fiction imagines worlds and scenarios beyond the present, architecture, too, is anticipatory, speculative, and projective. Jameson's reframing of the present as past suggests that we have the agency in the present, to imagine and enact alternative versions of the world as it ought to be. To design is to imagine and act on a "world made otherwise." 44

IMAGINATION AND AGENCY

When we look around at the world today, we see rapidly growing cities as well as obsolete office buildings in our downtowns, landscapes that bear traces of human activity — always already brownfields — and infrastructure that embodies an immense amount of energy and labor. The built environment is the sum total of intent. It is all the actions, decisions, policies, and practices that have resulted in the cities and buildings that we call our homes, neighborhoods, workplaces, and institutions. We turn to history to explain the complex factors, including political, economic, and cultural, that led to these outcomes. It is the inertia of the everyday.

44 The Memorial to Enslaved Laborers at the University of Virginia, designed by Höweler + Yoon. Two hundred years after its founding, the University of Virginia inaugurated a memorial dedicated to the enslaved laborers that worked to construct the university. The memorial is an acknowledgement of the unjust forced labor of individuals in the distant past, with a new structure in the present.

Titled *Design for Construction*, this book asks: How should we equip ourselves to better act on the world? And, following Simon, How do we act on the world to make it more like it "ought to be"? To design for construction is to apply an imagination for change. It is to claim the agency to build thoughtfully. It is an imperative of improvement. The term "tectonic imagination" refers to an applied imagination and an instrumental way of thinking. It is a mode of applied knowledge and knowledge production. Given the unprecedented availability of information and our access to emerging technologies, it is our responsibility to ask: In what way should we build? And, when seemingly anything is possible, *what* and *how* should we design? In arguing that architecture is a verb, design puts into practice processes and methods informed by an expanded understanding of the pre-conditions for design, namely material and labor, and the post-conditions for design, its impact and consequences.

Architecture is risk. It is often unprecedented and unknown. In the face of the new, the alternative, and the "world made otherwise," design and research offer processes to prototype futures through structured encounters with the new, be it new materials and systems, or new means of implementation. Recalling the 15-year process of the Sydney Opera House's design and construction that started with an audacious imagined future, and resulted in an unprecedented structure in the present, the collective labor of the architects, engineers and contractors who undertook the difficult processes of translation, is highly instructive. 42 Design and invention did not end after the architect's drawings were completed. The years of work by the engineering team and the contractor were essential in the realization of the unique structure. 43

By conceptualizing design and construction together as part of a continuous process that begins with concepts and ends with constructs, I hope to show how these processes are intertwined and codependent, and connected by feedback loops that are generative of new knowledge. It is through this messy process of design that architects assume agency and act on the world, transforming what is into something perhaps closer to what "ought" to be.

Bibliography

Ábalos, Iñaki, and Juan Herreros. *Tower and Office: From Modernist Theory to Contemporary Practice*. Cambridge, MA: MIT Press, 2002.

Addis, William, *Building: 3000 Years of Design Engineering and Construction, Building History Series*. London; New York, NY: Phaidon Press, 2007.

Adelson, Sheldon and ORO Editions, eds. *Reaching for the Sky: The Marina Bay Sands, Singapore*. 1st ed. Novato, CA: ORO Editions, 2013.

Albers, Josef. *Interaction of Color. 50th anniversary edition; 4th edition*. New Haven, CT: Yale University Press, 2013.

Alberti, Leon Battista, Joseph Rykwert, and Leon Battista Alberti. *On the Art of Building: In Ten Books*. 7. print. Cambridge, MA; London: MIT Press, 1997.

Ali, Mir M., and Fazlur Rahman Khan. *Art of the Skyscraper: The Genius of Fazlur Khan*. 1. publ. New York, NY: Rizzoli, 2001.

Allen, Edward, Wacław Zalewski, and Joseph Iano. *Form and Forces: Designing Efficient, Expressive Structures*. Hoboken, NJ: John Wiley & Sons, Inc, 2010.

Allen, Stan. *Points + Lines: Diagrams and Projects for the City*. 1st ed. New York, NY: Princeton Architectural Press, 1999.

Allen, Stan and Princeton University, eds. *Landform Building: Architecture's New Terrain*. Baden, Schweiz: Lars Müller, 2011.

Arantes, Pedro Fiori, Adriana Kauffmann, Reinhold Martin, and Pedro Fiori Arantes. *The Rent of Form: Architecture and Labor in the Digital Age*. Buell Center Books in the History and Theory of American Architecture. Minneapolis, MN: University of Minnesota Press, 2019.

Asgaard Andersen, Michael. *Jørn Utzon: Drawings and Buildings*. First English edition. New York, NY: Princeton Architectural Press, 2014.

Bacigalupi, Paolo. *The Water Knife*. First edition. New York, NY: Alfred A. Knopf, 2015.

Balmond, Cecil, Jannuzzi Smith, and Christian Brensing. *Informal*. Munich; New York, NY: Prestel, 2007.

Banham, Reyner. *The Architecture of the Well-Tempered Environment*. 2nd ed. Chicago, IL: University of Chicago Press, 1984.

Barber, Daniel A. *A House in the Sun: Modern Architecture and Solar Energy in the Cold War*. New York, NY: Oxford University Press, 2016.

———. *Modern Architecture and Climate: Design before Air Conditioning*. Princeton, NJ: Princeton University Press, 2020.

Barkow, Frank, Regine Leibinger, Barkow Leibinger Architekten, and Architectural Association (Great Britain), eds. *Barkow Leibinger: An Atlas of Fabrication*. London: Architectural Association, 2009.

Beal, Justin. *Sandfuture*. Cambridge, MA: MIT Press, 2021.

Beck, Ulrich. *Risk Society: Towards a New Modernity. Theory, Culture & Society*. London; Newbury Park, CA: Sage Publications, 1992.

Bell, Michael, and Craig Buckley, eds. *Solid States: Concrete in Transition*. 1st ed. Columbia Books on Architecture, Engineering, and Materials. New York, NY: Princeton Architectural Press, 2010.

Bell, Michael, and Jeannie Kim, eds. *Engineered Transparency: The Technical, Visual, and Spatial Effects of Glass*. 1st ed. New York, NY: Princeton Architectural Press, 2009.

Benfey, Christopher E. G. *Red Brick, Black Mountain, White Clay: Reflections on Art, Family, and Survival*. [1st U.S. pbk. ed.]. New York, NY: Penguin Books, 2013.

Bergdoll, Barry, Peter Christensen, Ron Broadhurst, and Museum of Modern Art (New York, N.Y.), eds. *Home Delivery: Fabricating the Modern Dwelling*. New York, NY: Museum of Modern Art, 2008.

Bernstein, Phillip G., *Machine Learning: Architecture in the Age of Artificial Intelligence*. London: RIBA, 2022.

Betsky, Aaron, ed. *Scanning: The Aberrant Architectures of Diller + Scofidio*. New York: Whitney Museum of American Art; New York, NY: Distributed by Harry N. Abrams, Inc, 2003.

Billington, David P. *The Tower and the Bridge: The New Art of Structural Engineering*. Princeton Paperbacks, Science/Architecture. Princeton, NJ: Princeton University Press, 1985.

Bonner, Jennifer, Hanif Kara, and Jennifer Bolande, eds. *Blank: Speculations on CLT*. First edition. Novato, CA: Applied Research and Design Publishing, an imprint of ORO Editions, 2021.

Brook, Timothy. *Vermeer's Hat: The Seventeenth Century and the Dawn of the Global World*. Paperback ed. New York, NY; Berlin: Bloomsbury, 2009.

Brown, G. Z. *Sun, Wind, and Light: Architectural Design Strategies*. New York, NY: Wiley, 1985.

Buck-Morss, Susan. *The Dialectics of Seeing: Walter Benjamin and the Arcades Project*. 8. print. Cambridge, MA: MIT Press, 1999.

Byles, Jeff. *Rubble: Unearthing the History of Demolition*. 1st pbk. ed. New York, NY: Three Rivers Press, 2006.

Campanella, Thomas J. *The Concrete Dragon: China's Urban Revolution and What It Means for the World*. 1st ed. New York, NY: Princeton Architectural Press, 2008.

Carpo, Mario. *Being Post-Digital: Design and Automation at the End of Modernity*. Cambridge, MA: MIT Press, 2023.

———. *The Second Digital Turn: Design Beyond Intelligence*. Writing Architecture. Cambridge, MA: MIT Press, 2017.
Ching, Francis D. K., Mark Jarzombek, and Vikramaditya Prakash. *A Global History of Architecture*. Third edition. Hoboken, NJ: Wiley, 2017.
Cobb, Henry Nichols, Mark Pasnik, and Emma Cobb. *Henry N. Cobb: Words & Works 1948-2018: Scenes from a Life in Architecture*. New York, NY: The Monacelli Press, 2018.
Cohen, Jean-Louis. *Le Corbusier and the Mystique of the USSR: Theories and Projects for Moscow, 1928–1936*. Princeton, NJ: Princeton University Press, 1992.
Colomina, Beatriz, Annmarie Brennan, and Jeannie Kim, eds. *Cold War Hothouses: Inventing Postwar Culture, from Cockpit to Playboy*. New York, NY: Princeton Architectural Press, 2004.
Cruvellier, Mark, Bjørn Normann Sandaker, and Luben Dimcheff, *Model Perspectives: Structure, Architecture and Culture*. London; New York, NY: Routledge, Taylor & Francis Group, 2017.
Davidson, Justin, et al., eds., *What Is a Museum Now? Snøhetta and the San Francisco Museum of Modern Art*. Zürich, Switzerland: Lars Müller Publishers, 2017.
Diller Scofidio + Renfro, Elizabeth Diller, Ricardo Scofidio, Iwan Baan, Matthew Monteith, FXFOWLE Architects, and Beyer Blinder Belle, Architects & Planners, eds. *Lincoln Center Inside Out: An Architectural Account*. First edition. Bologna, Italy: Damiani, 2012.
Drew, Philip, and Jørn Utzon, eds., *Sydney Opera House: Jørn Utzon, Repr., Architecture in detail*. London: Phaidon, 2003.
Dudley, George A. A *Workshop for Peace: Designing the United Nations Headquarters*. New York, NY: Cambridge, MA: Architectural History Foundation; MIT Press, 1994.
Easterling, Keller. *Subtraction*. Critical Spatial Practice 4. Berlin: Sternberg Press, 2014.
Eiffel, Gustave, and Bertrand Lemoine. *The Eiffel Tower: The Three-Hundred-Metre Tower*. Köln: TASCHEN, 2021.
Eisenman, Peter, and Greg Lynn. *Archaeology of the Digital: Peter Eisenman, Frank Gehry, Chuck Hoberman, Shoei Yoh*. Accompany the Exhibition Archaeology of the Digital Presented at the CCA from 7 May to 13 October 2013. Berlin: Sternberg Press, 2013.
Eisenman, Peter, and Michael Graves, Charles Gwathmey, John Hejduk, Richard Meier, Colin Rowe, and Kenneth Frampton. *Five Architects: Eisenman, Graves, Gwathmey, Hejduk, Meier*. New York, NY: Oxford University Press, 1975.
Evans, Robin. *Translations from Drawing to Building and Other Essays*. London: AA Publications, 2023.
Fannon, D., M. Laboy, and P. Wiederspahn. *The Architecture of Persistence: Designing for Future Use*. New York, NY: Routledge, 2021.
Ferro, Sérgio, *Architecture from Below: An Anthology*, ed. Silke Kapp and Mariana Moura, trans. Ellen Heyward and Ana Naomi de Sousa, First edition. London: MACK, 2024.
Fierro, Annette. *Architectures of the Technopolis: Archigram and the British High Tech*. London: Lund Humphries Publishers Ltd, 2023.
———. *The Glass State: The Technology of the Spectacle, Paris, 1981-1998*. Cambridge, MA: MIT Press, 2003.
Flury, Aita, ed., *Cooperation: The Engineer and the Architect*. Basel: Birkhäuser, 2012.
Ford, Edward R. *The Details of Modern Architecture. Vol. 1*. 1st paperback ed. Vol. 1. Cambridge, MA: MIT Press, 2003.
Fortmeyer, Russell Martin, and Charles D. Linn, *Kinetic Architecture: Designs for Active Envelopes*. Mulgrave, Victoria: Images Publishing Group, 2014.
Foster, Hal, ed. *The Anti-Aesthetic: Essays on Postmodern Culture*. New York, NY: New Press, 2002.
———. *The Art-Architecture Complex*. London; New York, NY: Verso, 2011.
Frampton, Kenneth. *Modern Architecture: A Critical History*. 3rd ed., rev. Enlarged. World of Art. London: Thames and Hudson, 1992.
———. *Studies in Tectonic Culture: The Poetics of Construction in Nineteenth and Twentieth Century Architecture*. 3. Aufl. Cambridge, MA: MIT Press, 2001.
Freud, Sigmund. *Civilization and Its Discontents*. New York, NY [u.a.]: Norton, 2010.
Geiser, Reto, ed. *House Is a House Is a House Is a House Is a House: Architectures and Collaborations of Johnston Marklee*. 2. edition. Basel: Birkhäuser, 2017.
Gerrewey, Christophe Van, ed. *OMA/Rem Koolhaas: A Critical Reader*. Basel: Birkhäuser, 2019.
Goldberger, Paul. *Building Art: The Life and Work of Frank Gehry*. First Vintage Books edition. New York, NY: Vintage Books, 2017.
Gramazio, Fabio, Matthias Kohler, and Michael Hanak, eds. *Digital Materiality in Architecture: Gramazio & Kohler*. Baden: Müller, 2008.
Grigsby, Darcy Grimaldo. *Colossal: Engineering the Suez Canal, Statue of Liberty, Eiffel Tower, and Panama Canal Transcontinental Ambition in France and the United States during the Long Nineteenth Century*. 1st ed. Pittsburgh, PA: Periscope publ, 2012.
Guinness, Desmond, and Julius Trousdale Sadler. *Mr. Jefferson, Architect*. A Studio Book. New York, NY: Viking Press, 1973.
Gutierrez, Laurent, ed. *HK Lab 2: an exploration of Hong Kong interior spaces*. Hong Kong: MAP Book Publ, 2005.
Hagberg Fisher, Eva, Aline B. Saarinen, and Eero Saarinen, *When Eero Met His Match: Aline Louchheim Saarinen and the Making of an Architect*. Princeton, NJ: Princeton University Press, 2022.
Harriss, Joseph. *The Tallest Tower: Eiffel and the Belle Epoque*. Boston, MA: Houghton Mifflin, 1975.
Hays, K. Michael, Andrew Holder, and Harvard University, eds. *Inscriptions: Architecture before Speech*. Cambridge, MA: Harvard University, Graduate School of Design, 2021.
Heisel, Felix, Dirk E. Hebel, and Ken Webster. *Building Better-Less-Different: Circular Construction and Circular Economy: Fundamentals, Case Studies, Strategies*. 1st ed. Boston, MA: Birkhäuser, 2022.
Hix, John. *The Glasshouse*. London; New York, NY: Phaidon, 2005.
Hofmeister, Sandra and OMA - Office for Metropolitan Architecture, eds. *OMA - Architektur Und Baudetails: = OMA - Architecture and Construction Details. 1*. Auflage. Edition DETAIL. München: Edition DETAIL, 2019.
Holden, Kimberly J., ed. *SHoP: Out of Practice*. 1st ed. New York, NY: Monacelli Press, 2012.
Höweler, Eric, and J. Meejin Yoon. *Expanded Practice: Höweler + Yoon Architecture / My Studio*. 1st ed. New York, NY: Princeton Architectural Press, 2009.
———. *Verify in Field: Projects and Conversations Höweler + Yoon*. Zürich: Park Books, 2021.

Hughes, Francesca. *The Architecture of Error: Matter, Measure, and the Misadventures of Precision*. Cambridge, MA: MIT Press, 2014.

Idenburg, Florian, Jing Liu, and Ilias Papageorgiou. *Solid Objectives: Order, Edge, Aura*. Zürich: Lars Müller publishers, 2017.

Ingels, Bjarke, ed. *Formgiving*. Köln: TASCHEN, 2020.

Ingels, Bjarke, BIG Bjarke Ingels Group, National Building Museum (U.S.), and Taschen (Firm), eds. *Hot to Cold: An Odyssey of Architectural Adaptation*. Köln: Taschen, 2015.

Ivy, Robert Adams. *Fay Jones: The Architecture of E. Fay Jones, FAIA*. New York, NY: McGraw-Hill, 2001.

Jarzombek, Mark. *Architecture Constructed: Notes on a Discipline*. London; New York, NY: Bloomsbury Visual Arts, 2023.

———. *Designing MIT: Bosworth's New Tech*. Cambridge, MA: MIT Press, 2017.

Jencks, Charles. *The Language of Post-Modern Architecture*. New York, NY: Rizzoli, 1977.

Johnston, George Barnett. *Assembling the Architect: The History and Theory of Professional Practice*. London: Bloomsbury Visual Arts, 2020.

Joselit, David. *After Art*. Point. Princeton, NJ: Princeton University Press, 2013.

Joyce, Nancy, and Frank O. Gehry. *Building Stata: The Design and Construction of Frank O. Gehry's Stata Center at MIT*. Cambridge, MA: MIT Press, 2004.

Kallipoliti, Lydia. *The Architecture of Closed Worlds: Or, What Is the Power of Shit?* 1. Auflage. Zürich: Müller, Lars, 2018.

Knowles, Ralph L. *Sun Rhythm Form*. Cambridge, MA: MIT Press 1981.

Koch, Natalie. *Arid Empire: The Entangled Fates of Arizona and Arabia*. London; Brooklyn, NY: Verso, 2022.

Kolarevic, Branko, and Kevin R. Klinger, eds. *Manufacturing Material Effects: Rethinking Design and Making in Architecture*. New York, NY: Routledge, 2008.

Koolhaas, Rem, *Delirious New York: A Retroactive Manifesto for Manhattan*. New ed. New York, NY: Monacelli Press, 1994.

Koolhaas, Rem, Bruce Mau, Jennifer Sigler, Hans Werlemann, and Office for Metropolitan Architecture, eds. *Small, Medium, Large, Extra-Large: Office for Metropolitan Architecture, Rem Koolhaas, and Bruce Mau*. New York, NY: Monacelli Press, 1995.

Koolhaas, Rem, and Brendan McGetrick, eds. *Content: Triumph of Realization*. Köln: Taschen, 2004.

Koolhaas, Rem, James Westcott, Stephan Petermann, Ben Davis, Stephan Trüby, Tom Avermaete, Anna Shefelbine, and Harvard University, eds. *Elements of Architecture*. Köln: Taschen, 2018.

Koolhaas, Rem, Mark Wigley, Jordan Carver, and Jorge Otero-Pailos. *Preservation Is Overtaking Us*. GSAPP Transcripts. New York: GSAPP Books, 2014.

Kuma, Kengo. *Kengo Kuma: Complete Works*. Rev. and Expanded. New York, NY: Thames & Hudson, 2018.

Le Corbusier. *Towards a New Architecture*. New York, NY: Dover Publications, 1986.

Le Corbusier, and Tim Benton. *Precisions: On the Present State of Architecture and City Planning*. Reprint of the original American edition, MIT Press, 1991. Zurich: Park Books, 2015.

Le Corbusier, and Pierre Jeanneret. *Oeuvre complète. 1: Oeuvre complète 1910 - 1929* / W. Boesiger / O. Stonorov (ed.). Edited by Willy Boesiger and Oscar Stonorov. 16. éd. Basel; Boston; Berlin: Birkhäuser, 2006.

Lepik, Andres. *Small Scale, Big Change: New Architectures of Social Engagement* [Exhibition, Museum of Modern Art, New York, October 3, 2010-January 3, 2011. New York, NY: Museum of Modern Art, 2010.

Lepik, Andres, Jürgen Mayer H., Cristina Inês Steingräber, Kunstbibliothek Berlin, Spanien, and Staatliche Kunstbibliothek Berlin, eds. *J. Mayer H. Metropol Parasol: Redesigning of Plaza de La Encarnación in Sevilla*; [on the Occasion of the Exhibition Metropol Parasol, Organized by the Kunstbibliothek, Staatliche Museen Zu Berlin, in Cooperation with the Embassy of Spain in Germany, Atrium of the Kulturforum Potsdamer Platz, Kunstbibliothek from October 30th - November 6th 2005]. Berlin: Staatliche Museen zu Berlin, 2005.

Leslie, Thomas. *Chicago Skyscrapers, 1871-1934*. Urbana, IL: University of Illinois Press, 2013.

Lewis, Philippa, and Adrian Forty. *Stories from Architecture: Behind the Lines at Drawing Matter*. Cambridge, MA: MIT Press, 2021.

L'Heureux, Erik G., and Grit Vltavsky. *Deep Veils*. First edition. Novato, CA: ORO Editions, 2014.

Machado, Rodolfo, Rodolphe El-Khoury, Detlef Mertins, and Heinz Architectural Center, eds. *Monolithic Architecture*. Munich; New York, NY: Prestel, 1995.

Mack, Gerhard, et al., *Herzog & de Meuron Elbphilharmonie Hamburg*. License edition. Basel; Boston: Birkhäuser Verlag GmbH, 2018.

Malm, Andreas. *Fossil Capital: The Rise of Steam-Power and the Roots of Global Warming*. London; New York, NY: Verso, 2016.

Mark, Robert. *Light, Wind, and Structure: The Mystery of the Master Builders*. McGraw-Hill Edition. New Liberal Arts Series. New York, NY: McGraw-Hill Pub. Co, 1990.

Mathewson, Casey C. M., and Frank O. Gehry. *Frank O. Gehry: Selected Works: 1969 to Today*. Buffalo, NY: Firefly Books, 2007.

May, John. *Signal. Image. Architecture: Everything Is Already an Image*. New York, NY: Columbia Books on Architecture and the City, 2019.

McInnis, Maurie Dee, and Louis P. Nelson, eds. *Educated in Tyranny: Slavery at Thomas Jefferson's University*. Charlottesville, VA: University of Virginia Press, 2019.

McKean, John. *Crystal Palace: Joseph Paxton and Charles Fox*. Architecture in Detail. London: Phaidon, 1994.

McVicar, Mhairi. *Precision in Architecture: Certainty, Ambiguity and Deviation*. London; New York, NY: Routledge, Taylor & Francis Group, 2019.

Mehrotra, Rahul, and Rajesh Vora. *Working in Mumbai: RMA Architects*. Edited by Kate Cahill. Berlin: ArchiTangle, 2020.

Mitchell, William J. *Imagining MIT: Designing a Campus for the Twenty-First Century*. Cambridge, MA: MIT Press, 2007.

Moe, Kiel. *Insulating Modernism: Isolated and Non-Isolated Thermodynamics in Architecture*. Basel, Switzerland; Boston, MA: Birkhäuser, 2014.

———. *Unless: The Seagram Building Construction Ecology*. New York, NY: Actar Publishers, 2020.

Morton, Timothy. *Hyperobjects: Philosophy and Ecology after the End of the World*. Posthumanities 27. Minneapolis, MN: University of Minnesota Press, 2013.

Mostafavi, Mohsen, and David Leatherbarrow. *On Weathering: The Life of Buildings in Time*. Cambridge, MA: MIT Press, 1993.

Murphy, Douglas. *Last Futures: Nature, Technology and the End of Architecture*. London; New York, NY:

Verso, 2016.
Murray, Peter. *The Saga of the Sydney Opera House: The Dramatic Story of the Design and Construction of the Icon of Modern Australia*. London; New York, NY: Spon Press, 2004.
Murray, Scott. *Contemporary Curtain Wall Architecture*. 1st ed. New York, NY: Princeton Architectural Press, 2009.
Nagler, Florian, ed. *Building Simply: Guidelines*. Boston, MA: Birkhäuser, 2021.
Neri&Hu Design and Research Office. *Neri and Hu Design and Research Office: Works and Projects 2004-2014*. Zurich, Switzerland: Park Books, 2017.
Neri, Lyndon, and Rosanna Hu. *Neri&Hu*. New York, NY: Thames & Hudson Inc, 2021.
Nielsen, David. *Bruno Taut's Design Inspiration for the Glashaus*. Routledge research in architecture. London; New York, NY: Routledge, 2016.
Nordenson, Guy. *Guy Nordenson: Reading Structures: 39 Projects and Built Works, 1983-2011*. Zurich: Lars Müller Publishers, 2016.
Olgyay, Victor, and Aladar Olgyay, eds. *Solar Control & Shading Devices*. 1. Princeton paperback print. Princeton, NJ: Princeton Univ. Pr, 1976.
Olgyay, Victor, Aladar Olgyay, Donlyn Lyndon, Victor W. Olgyay, John Reynolds, and Ken Yeang. *Design with Climate: Bioclimatic Approach to Architectural Regionalism*. New and Expanded edition. Princeton, NJ: Princeton University Press, 2015.
Pelkonen, Eeva-Liisa, Kevin Roche, Kathleen John-Alder, Olga Pantelidou, and David Sadighian. *Kevin Roche: Architecture as Environment*. New Haven, CT: Yale University Press: In association with Yale School of Architecture, 2011.
Peter Gluck & Partners, Architects. *The Modern Impulse*. Edited by Oscar Riera Ojeda and Peter L. Gluck. Philadelphia, PA: ORO Ed, 2008.
Pfammatter, Ulrich. *The Making of the Modern Architect and Engineer: The Origins and Development of a Scientific and Industrially Oriented Education*. Basel; Boston, MA: Birkhauser-Publishers for Architecture, 2000.
Picon, Antoine, and Abigail Grater. *The Materiality of Architecture*. Minneapolis, MN: University of Minnesota Press, 2020.
Robertson, Leslie E. *The Structure of Design: An Engineer's Extraordinary Life in Architecture*. First American edition. New York, NY: The Monacelli Press, 2017.
Ronan, John, Carlos Jiménez, Sean Keller, and Clare Lyster. *Out of the Ordinary: The Work of John Ronan Architects*. New York, NY: Actar Publishers, 2022.
Rowe, Colin. *The Mathematics of the Ideal Villa and Other Essays*. 1. MIT Press paperback ed., 10. print. Cambridge, MA: MIT Press, 1995.
Rudofsky, Bernard. *Architecture without Architects: A Short Introduction to Non-Pedigreed Architecture*. Albuquerque, NM: University of New Mexico Press, 1987.
Saarinen, Eero, Eeva-Liisa Pelkonen, and Donald Albrecht. *Eero Saarinen: Shaping the Future*. New Haven, CT: New York, NY: Helsinki: Washington, DC: New Haven, CT: Yale University Press ; In association with The Finnish Cultural Institute in New York ; The Museum of Finnish Architecture ; The National Building Museum ; Yale University School of Architecture, 2006.
Saint, Andrew, *Architect and Engineer: A Study in Sibling Rivalry*. New Haven, CT; London: Yale University Press, 2007.
Sandaker, Bjørn Normann, Arne Petter Eggen, and Mark Cruvellier. *The Structural Basis of Architecture*. Third edition. London; New York, NY: Routledge, 2019.
Scheerbart, Paul, and John A. Stuart. *The Gray Cloth: Paul Scheerbart's Novel on Glass Architecture*. Cambridge, MA: MIT Press, 2001.
Scheerbart, Paul, and Bruno Taut. *Glass Architecture and Alpine Architecture*. London: November, 1972.
Schröpfer, Thomas, and James Carpenter. *Material Design: Informing Architecture by Materiality*. Basel: Birkhäuser, 2011.
Shigematsu, Shohei, Jason Long, Christopher Hawthorne, Lisa Phillips, Massimiliano Gioni, Taryn Simon, Iris van Herpen, et al., eds. *OMA NY: Search Term*. New York, NY: Rizzoli, 2021.
Simha, O. Robert. *MIT Campus Planning, 1960-2000: An Annotated Chronology*. Cambridge, MA: MIT, Office of the Executive Vice President, 2001.
Simon, Herbert A. *The Sciences of the Artificial*. 3rd ed. Cambridge, MA: MIT Press, 1996.
Soleri, Paolo, ed. *Arcology: The City in the Image of Man*. 2. print. paperback ed. Cambridge, MA: MIT Press, 1977.
Space Caviar and V-A-C Foundation, eds. *Non-Extractive Architecture: On Designing without Depletion*. Moscow: Berlin: V-A-C Press; Sternberg Press, 2021.
Tibbits, Skylar. *Things Fall Together: A Guide to the New Materials Revolution*. Princeton, NJ: Princeton University Press, 2021.
Tibbits, Skylar and Massachusetts Institute of Technology, eds. *Self-Assembly Lab: Experiments in Programming Matter*. New York, NY: Routledge, 2016.
Trachtenberg, Marvin. *The Statue of Liberty*. Rev. ed. New York, NY: Penguin Books, 1986.
Vint, Sherryl. *Science Fiction*. The MIT Press Essential Knowledge Series. Cambridge, MA: MIT Press, 2021.
Wang, Shu, Kenneth Frampton, Iwan Baan, Michael Juul Holm, Amateur Architecture Studio, and Louisiana Museum of Modern Art, eds. *Wang Shu Amateur Architecture Studio*. Zürich: Lars Müller Publishers, 2017.
Werkbund Bayern and Museum Villa Stuck, eds. *Frei Otto, Bodo Rasch: Finding Form: Towards an Architecture of the Minimal*: The Werkbund Shows Frei Otto, Frei Otto Shows Bodo Rasch: Exhibition in the Villa Stuck, Munich, on the Occasion of the Award of the 1992 Deutscher Werkbund Bayern Prize to Frei Otto and Bodo Rasch. Stuttgart: Axel Menges, 1995.
Weston, Richard. *Materials, Form, and Architecture*. New Haven, CT: Yale University Press, 2003.
Wigginton, Michael. *Glass in Architecture*. London: Phaidon, 1996.
Wigley, Mark. *White Walls, Designer Dresses: The Fashioning of Modern Architecture*. Cambridge, MA: MIT Press, 1995.
Willis, Carol. *Building the Empire State*. New York, NY: W.W. Norton, 2008.
———. *Form Follows Finance: Skyscrapers and Skylines in New York and Chicago*. 1st ed. New York, NY: Princeton Architectural Press, 1995.
Wong, Liliane. *Adaptive Reuse: Extending the Lives of Buildings*. Basel: Birkhäuser, 2017.
Yoshida, Nobuyuki, ed. *CCTV by OMA*. Kenchiku-to-toshi Bessatsu = Special issue, 2005,7. Tokyo: a+u Publ. Co, 2005.
Zaera-Polo, Alejandro. *The Ecologies of the Building Envelope: A Material History and Theory of Architectural Surfaces*. Print edition. New York, NY: Actar Publishers, 2019. Bibliography

Index

Image Credits

Introduction
Agency and Imagination

1 Arne Magnussen, Courtesy of the collections of the State Library of New South Wales
2–3 Museums of History New South Wales - State Archives Collection
4 Arne Magnussen Courtesy of the collections of the State Library of New South Wales
5 Mitchell Library, State Library of NSW, ON 562/ nos. 744–814
6 Trevor Patt
7 Courtesy of Eisenman Architects
8 Digital Image © The Museum of Modern Art/ Licensed by SCALA / Art Resource, NY / Artists Rights Society (ARS), New York / VG Bild-Kunst, Bonn
9 Courtesy of Greg Lynn Form
10 Zaha Hadid Architects
11 DESIGN EARTH, After Oil (Das Island), 2016
12 Diller Scofidio + Renfro
13 Monte French Design Studio
14 NAARO, Courtesy of Jennifer Bonner / MALL 2018
15 Stephanie Braconnier/ Shutterstock
16 Christian Richters
17 Louis-Émile Durandelle, Digital image courtesy of Getty's Open Content Program
18 OMA CCTV/OMA Rem Koolhaas and Ole Scheeren, image courtesy of OMA
19 Photo by author
20 Courtesy of HANNAH

Chapter 1
Stocktaking

1 Photo by author
2 Bettmann Collection via Getty Images
3 Courtesy of The Skyscraper Museum
4 Architectural Forum 52 (1930): 772
5 Library of Congress, Prints & Photographs Division, Gottscho-Schleisner Collection LC-G612-42061
6 Andy Ryan
7 Bettmann Collection via Getty Images
8 Courtesy of Iron Workers Locals 40 & 361
9 Drawing by author
10 Safdie Architects
11 Dietmar Rabich / Wikimedia Commons / "Singapore (SG), Marina Bay Sands Hotel—2019—4462" / CC BY-SA 4.0
12–13 Safdie Architects
14 © Frank O. Gehry
15 © Ad Meskens / Wikimedia Commons
16 University of Washington Libraries, Special Collections, John Stamets, UW42536
17 ABruszik / Wikimedia Commons
18 Courtesy of Magnusson Klemencic Associates
19–20 Courtesy of Frank O. Gehry & Gehry Partners, LLP
21–22 Zahner
23 Charles Mayer Photography
24 Courtesy of Frank O. Gehry & Gehry Partners, LLP
25 Philip Greenspun
26 Photo by author
27 Dave Burk © SOM
28–30 © SOM
31–32 Diller Scofidio + Renfro
33–34 Photo by author
35 Photo: Iwan Baan
36–40 Höweler + Yoon Architecture
41 Courtesy of the MIT Museum
42 Kyle Klein Photo
43–45 Courtesy of the MIT Museum
46 © Massachusetts Institute of Technology
47 Sayamindu Dasgupta / Wikimedia Commons

Chapter 2
Means and Methods

1 Rafael Gamo
2–4 Félix Candela architectural records and papers, 1950–1984, Avery Architectural & Fine Arts Library, © Columbia University in the City of New York
5–9 © Young & Ayata
10 © Nic Lehoux
11–15 NADAAA
16–17 Peter MacCallum, Courtesy of The Daniels Faculty
18 Drawing by Author
19 Joern Sackermann / Alamy Stock Photo
20 Ensamble Studio
21 Roland Halbe
22 Photo: Iwan Baan
23 SO-IL
24 Photo: Iwan Baan
25 Photo by author
26–28 SO-IL
29–30 Courtesy of GLUCK+
31 © Paul Warchol
32–33 Courtesy of GLUCK+
34 © Paul Warchol
35 Photo: Iwan Baan
36–41 MASS Design Group

Chapter 3
Material Processes

1 Photo by author
2–3 Special Collections, University of Virginia Library, Charlottesville, Va.
4–5 Photo by author
6–7 Archi-Union Architects
8 Edmund Sumner-VIEW / Alamy Stock Photo
9 Photo by author
10 Yihuai Hu, Courtesy of Höweler + Yoon Architecture
11–13 Höweler + Yoon Architecture
14 © The Josef and Anni Albers Foundation / Artists Rights Society (ARS), New York, 2024
15 Acroterion / Wikimedia Commons
16–18 Photo by author
19–20 Morris Adjmi Architects
21 Photo: Iwan Baan
22–23 SO-IL
24 Lorena Darquea
25 Photo by author
26 Thomas Wrede VG Bild-Kunst, Bonn
27 behet bondzio lin architekten GmbH & Co. KG
28–29 Stefano Mori
30 © Studio Anna Heringer
31 Ricola Group AG
32 © Herzog & de Meuron
33 Ricola Group AG
34 © Joe Ladhou
35 Lina Ghotmeh —Architecture
36 Drawing by author

37 Timothy Soar
38 GROUPWORK
39 ClassicStock/Archive Photos via Getty Images
40 Photo: Iwan Baan
41–43 Sou Fujimoto Architects
44–45 © Michael Moran/ OTTO
46 Shigeru Ban Architects
47 Photo: Iwan Baan
48–49 Barkow Leibinger
50–51 nARCHITECTS
52 © Ueyakato Landscape Photo: Yujiro Sagami
53 Photo by author
54 © OMA, by Charlie Koolhaas
55 Photo by author
56–57 SITE
58 Photo by author
59 claudiodivizia via iStock
60 Photo by author
61 Everson Museum of Art Construction View, Collection of Everson Museum of Art
62–69 Photo by author
70–71 The Utzon Archives / The Utzon Center
72 Photo by author
73 BCHO Architects Associates, Mark West
74 Yongkwan Kim
75 Fisac Foundation
76 Lee Bey
77 Studio Gang
78–79 Henrik Kam Photography, Inc.
80–82 Kreysler & Associates

Chapter 4
Part to Whole

1 Roland Halbe
2 John Horner Photography
3 Courtesy of Holly Fairbank and MIT Libraries
4 Takumi Ota
5 DAICI ANO
6 Kengo Kuma & Associates
7 Ron Ellis / Shutterstock
8 Special Collections, University of Arkansas, Fayetteville
9 Drawing by author
10 Special Collections, University of Arkansas, Fayetteville
11 Wirestock, Inc. / Alamy Stock Photo
12 Andreas 2309 / Wikimedia Commons
13 Museo Nacional Thyssen-Bornemisza / Scala / Art Resource, NY
14 Picture Partners / Alamy Stock Photo
15 Roland Halbe
16–19 Ensamble Studio
20 Sorin Vidis / Shutterstock
21 Defears / Wikimedia Commons
22 Photo: Drouin, Serge ©RPBW - Renzo Piano Building Workshop Architects
23–24 ©RPBW - Renzo Piano Building Workshop Architects
25 Daniel Schwen / Wikimedia Commons
26 Tom Harris Photography
27–28 Photo by Author
29 Studio Gang
30–31 Diller Scofidio + Renfro
32 Photo by author
33 SHoP Architects
34 Chuck Choi, Courtesy of Höweler + Yoon Architecture
35–36 Höweler + Yoon Architecture
37–38 Courtesy of Jennifer Bonner / MALL 2018
39 Philippe Ruault, Courtesy of Atelier Jean Nouvel
40 Roland Halbe
41 Scott Murray

Chapter 5
Monoliths

1 Hiromitsu Morimoto
2 Drawing by author
3 Architekturzentrum Wien, Collection, photo: Margherita Spiluttini
4 © Herzog & de Meuron
5 Pedro Pegenaute
6 Tina Nandi Photography
7 Courtesy of RMA Architects
8–9 El Equipo Mazzanti
10 Alejandro Arango
11 The Lyda Hill Texas Collection of Photographs in Carol M. Highsmith's America Project, Library of Congress, Prints and Photographs Division
12 Photo: Iwan Baan
13 Aaron Thompson
14 Eduard Hueber/© archphoto
15 Lydia Gould/© archphoto
16 Estudio Gustavo Frittegotto
17 Johnston Marklee
18–19 Jean NOUVEL & Associés (Emmanuel BLAMONT, Jean-Marc IBOS, Myrto VITART). Drawing © Vincent Lafont
20 Tim Van de Velde
21–22 Tim Hursley
23 Philip Lange / Shutterstock
24–25 Barozzi Veiga
26 © Tim Griffith
27–28 Courtesy of KPF
29 C.TORALDO DI FRANCIA ©
30 FOA, Courtesy of Alejandro Zaera-Polo
31 Edward Caruso Photography
32 Francisco Nogueira

Chapter 6
From Window to Envelope

1 Photo by author
2 Städel Museum, Frankfurt am Main
3–6 © F.L.C. / ADAGP, Paris / Artists Rights Society (ARS), New York 2024
7 Edward R. Ford, The Details of Modern Architecture, Volume 1, figures 8.31 and 8.36, reprinted courtesy of The MIT Press
8 © F.L.C. / ADAGP, Paris / Artists Rights Society (ARS), New York 2024
9 Photo by author
10–12 Diller Scofidio + Renfro
13–14 Photo by author
15 SANAA
16 Photo: Iwan Baan
17 © Frank Dinger
18 Top: SANAA / Bottom: Drawing by author
19 Guy Nordenson and Associates
20–21 Andy Ryan
22 © F.L.C. / ADAGP, Paris / Artists Rights Society (ARS), New York 2024
23 Bettmann Collection via Getty Images
24 Cervin Robinson, Historic American Buildings Survey, Library of Congress, Prints and Photographs Division
25 Architectural Record, Vol. IV, January–March 1895, No.3
26 J.W. Taylor, Historic Architecture and Landscape Image Collection, Library of Congress, Prints and Photographs Division
27 Drawing by author
28 Tony Hisgett / Wikimedia Commons
29 Photo: Jules Verne Times Two / julesvernex2.com / CC-BY-SA-4.0 / https://creativecommons.org/licenses/by-sa/4.0/deed.en
30 Studio Gang
31 Nagel Photography / Shutterstock
32 Studio Gang
33 Drawing by author
34 Historic American Buildings Survey, Library of Congress, Prints and Photographs Division
35 © F.L.C. / ADAGP, Paris / Artists Rights Society (ARS), New York 2024
36 Atlantide Phototravel/ Corbis Documentary via Getty Images
37 Courtesy of Kevin Roche John Dinkeloo and Associates
38 Christian Richters
39 © JongOh Kim
40–42 Barkow Leibinger
43–45 Atelier Jean Nouvel
46 Photo by author
47 Ank Kumar / Wikimedia Commons
48 nj.ap / Shutterstock
49 Courtesy of KPF
50 Photo: Iwan Baan
51–52 REX
53 Tim Waltman for Evan Joseph Studios
54–55 Josef Gartner GmbH

Chapter 7
Force, Form and Gesture

1 Arup
2 Morphart Creation / Shutterstock
3 saai | Archiv für Architektur und Ingenieurbau
4 © Jorge Royan / http://www.royan.com.ar / CC BY-SA 3.0
5 ILEK
6–7 Poleni, Giovanni, 1748 ETH-Bibliothek Zürich, Rar 1229
8 C messier / Wikimedia Commons
9 ILEK
10–11 Eero Saarinen Collection (MS 593). Manuscripts and Archives, Yale University Library

12 Bettmann Collection via Getty Images
13 The State Historical Society of Missouri, Photograph Collection
14 Top: Eero Saarinen Collection (MS 593). Manuscripts and Archives, Yale University Library Bottom: Historic American Buildings Survey, Library of Congress, Prints and Photographs Division
15–17 Courtesy of the MIT Museum
18 Drawing by author
19 Eero Saarinen Collection (MS 593). Manuscripts and Archives, Yale University Library
20 Photo: Iwan Baan
21 Arthur Drooker
22 Google. Design by BIG and Heatherwick Studio
23–24 Arthur Drooker
25 Google. Design by BIG and Heatherwick Studio
26–27 Photo: Iwan Baan
28 Albert Fernique, Library of Congress, Prints and Photographs Division
29 Collection Musee des Artes et Metiers Paris/ Fonds Bartholdi
30 Pierre Petit, Library of Congress, Prints and Photographs Division
31 Jet Lowe, Historic American Building Survey, Library of Congress, Prints and Photographs Division
32–33 Craig Strong, Historic American Building Survey, Library of Congress, Prints and Photographs Division
34 Illus. in: Le genie civil, 1883 Aug. 1, p. 468
35 Compte rendu des travaux de la Société des Ingénieurs Civils Vol. 43, 1885 / CNAM
36 Linley Sambourne, Illustration for Punch, vol. 96, p. 324 (June 29, 1889), Library of Congress, Prints and Photographs Division
37 Bildarchiv ETH Bibliothek
38 Drawing by author
39 Paul Brady Photography / Shutterstock
40 McShane Fleming Studios Courtesy SOM
41 Orlando R. Cabanban Courtesy SOM
42 John Nye
43 LERA Consulting Structural Engineers
44 Pei Cobb Freed & Partners
45 Dreamworld / Wikimedia Commons
46 Drawn by BPR for Foster + Partners
47 ©Wolfgang Hoyt/Esto
48–49 Aiden Li
50 SOM
51 LERA Consulting Structural Engineers
52–53 Photo by author
54–57 Guy Nordenson and Associates
58 OMA, Photo by Alex Yang
59 © Tim Griffith
60 Front
61 Arup
62 Photo: Iwan Baan
63–66 Courtesy OMA
67 Drawing by author

Chapter 8
Climate and Conditioning

1 Photo by author
2 Cosanti Foundation
3 Seth Winslow, Courtesy of Cosanti Foundation
4 Michael E. Brown, Courtesy of Cosanti Foundation
5 © F.L.C. / ADAGP, Paris / Artists Rights Society (ARS), New York 2024
6 Cyrille Weiner
7 Photo by author
8 © F.L.C. / ADAGP, Paris / Artists Rights Society (ARS), New York 2024
9 Guy Gillette, Collier's, October 26, 1956
10 Drawing by author
11 Kenneth Choo
12–13 Pencil Office
14–15 Jason O'Rear
16–17 Jason O'Rear courtesy of Kreysler & Associates
18–19 Gensler
20 Roland Halbe
21 © Morphosis, Credit: Eui-Sung Yi
22 ©Morphosis, Credit: Jasmine Park
23 HGEsch Photography, Courtesy of Christoph Ingenhoven Architects
24 Christoph Ingenhoven Architects
25 HGEsch Photography, Courtesy of Christoph Ingenhoven Architects
26–27 Christoph Ingenhoven Architects
28 Iaian Masterton / Alamy Stock Photo
29 Phillip Bond / Alamy Stock Photo
30 Pio3 / Shutterstock
31 Atelier Jean Nouvel
32 Rudi Sebastian / Alamy Stock Photo
33–34 Christian Richters
35–36 Arup
37 Hugh Ferriss, Cooper Hewitt, Smithsonian Design Museum
38 Studio Gang
39 ©Nic Lehoux
40 Nick Fochtman
41 Photo by author
42 Laurian Ghinițoiu
43 Arch-Exist Photography
44–45 Front
46 Bjarke Ingels Group
47 Courtesy of Harvard Fine Arts Library, Digital Images & Slides Collection
48 © Victoria and Albert Museum, London
49 J. Russell & Sons
50–51 Bildarchiv ETH Bibliothek
52–53 Courtesy of MIT Museum
54–55 Charles Downes. With scientific description by Charles Cowper. London, J. Weale, 1852
56 Photograph by Dell and Wainwright for "The Architectural Review," RIBA
57 © Royal Collection Enterprises Limited 2024 | Royal Collection Trust
58 Messrs. Nash, Haghe, and Roberts, R.A., London: Dickinson Brothers, 1854. Yale Center for British Art, Paul Mellon Collection
59 Katja Schulz / Wikimedia Commons
60 Hbarrison / Wikimedia Commons
61 Akademie der Künste, Image courtesy of Special Collections, Fine Arts Library, Harvard University
62 City of Cologne Historical Archive with Rhenish picture archive
63 Wasmuths Monatshefte fur Bankunst (WMfB)
64 Digital Image © The Museum of Modern Art/ Licensed by SCALA / Art Resource, NY / Artists Rights Society (ARS), New York / VG Bild-Kunst, Bonn
65 Aryan Khalighy
66–67 Trevor Patt
68–71 Front
72–73 Drawing by author
74 Trevor Patt
75 Carol M. Highsmith Archive, Library of Congress, Prints and Photographs Division
76–77 © F.L.C. / ADAGP, Paris / Artists Rights Society (ARS), New York 2024
78 Sergey Norin / Wikimedia Commons
79 Patricia Layman Bazelon
80 Barbara Elliot Martin
81 Cannon Design
82 Annette Kisling
83–84 Sauerbruch Hutton
85–86 HGEsch Photography
87–88 HGEsch Photography, Courtesy of Christoph Ingenhoven Architects
89 Chuck Choi
90 Robert Benson Photography, Courtesy of William Rawn Assoc.
91 William Rawn Assoc.
92 Grimshaw Architects
93 Jürgen Matern / Wikimedia Commons
94 © Hufton+Crow/VIEW
95 FotoFree / Wikimedia Commons
96 © Steve Hall
97 Diego Delso, delso.photo, License CC BY-SA
98 Drawing by author
99 John Ronan Architects
100–101 Drawing by author
102 Eduard Hueber, archphoto © Baumschlager Eberle Architekten

Chapter 9
Afterlives

1 Photo by author
2 HUD Policy Development and Research
3 Walter P. Reuther Library, Archives of Labor and Urban Affairs, Wayne State University
4 © Michael Moran/OTTO
5 Archive Photos / Stringer via Getty Images
6 The Asahi Shimbun via Getty Images

7 Bob Masters / Alamy Stock Photo
8–9 EMBT Architects
10 Rick Ligthelm / Wikimedia Commons
11 Laurie Noble via Getty Images
12 Philippe Ruault
13 Roos Aldershoff
14 Merkx + Girod
15 Drawing by author
16 Johan Dehlin
17 Carmody Groarke
18 Historic Environment Scotland
19 Richard Davies
20 Johan Dehlin
21 Hackercatxxy / Wikimedia Commons
22 Photo: Iwan Baan
23–25 Heatherwick Studio
26 Photo: Iwan Baan
27–28 © Herzog & de Meuron
29 Michael Zapf
30 © Markus Lanz
31 Stefano Monterosso
32 Drawing by author
33 Riccardo Sala / Alamy Stock Photo
34 Image courtesy OMA
35–38 Photo by author
39 Arnold Newman Collection via Getty Images
40 © Michael Moran/OTTO
41–42 Photo: Iwan Baan
43 Diller Scofidio + Renfro
44 Matthew Monteith
45 Diller Scofidio + Renfro
46 Forrest Anderson / The Chronicle Collection via Getty Images
47 Neri&Hu Design and Research Office
48 Tuomas Uusheimo
49 Photo: Iwan Baan
50–51 ©Laksana Studio, Courtesy of Amateur Architecture Studio
52 Toni Castillo Quero / Wikimedia Commons
53 Luis Rogelio HM / Wikimedia Commons
54 Manuel Cohen

Chapter 10 Construction Futures

1 Courtesy of HANNAH
2 "The Future Won't Wait", Monsanto Magazine, 1960
3 "House of the Future", Monsanto Magazine, Summer of 1957
4–5 Courtesy of MIT Museum
6 Naquib Hossain / Flickr
7 Grimshaw Architects
8 © Archigram 1964, Image supplied by the ARCHIGRAM ARCHIVES 2019
9 Digital Image © The Museum of Modern Art/ Licensed by SCALA / Art Resource, NY / Artists Rights Society (ARS), New York / DACS, London
10 Fernando Alda
11 Photo by author
12–14 J.Mayer H. and Partner
15–16 Kiel Moe
17 robas / E+ via Getty Images
18 Kiel Moe
19–21 Self-Assembly Lab, MIT + Autodesk Inc.
22–23 Gramazio Kohler Research, ETH Zurich, and Self-Assembly Lab, MIT
24 Felix Heisel
25–26 Certain Measures
27–28 Courtesy of Rieder Facades GmbH, Photography by Ditz Frejer
29–33 Courtesy of HANNAH
34 Sebastian Schels
35 Max Kratzer
36 Sebastian Schels
37–38 Florian Nagler Architekten
39 Sebastian Schels
40 SC Johnson
41 Gramzio Koehler Research, ETH Zurich, and Self Assembly Lab, MIT
42–43 Courtesy of Paolo Tombesi, Paolo Stracchi, and Luciano Cardellicchio
44 Höweler + Yoon Architecture

Every reasonable attempt has been made to identify owners of copyright. Errors or omissions will be corrected in subsequent editions.

Acknowledgements

My interest in the processes and sequences whereby an idea is conceptualized, developed, and refined to realize a built work of architecture could also be applied to the process of writing a book — "book, verb." Through the process of book making, I am grateful to numerous people. First, to the individuals that helped me conceptualize the book, test the layouts and secure the image permissions, and assemble the book: Chandler Caserta, Alex Xiaotan Yang, Paul Wood, Sia Chen, and Dhruv Mehta. Thanks to my wonderful copy editor Edith Fikes. Thanks also to my super talented graphic designers and collaborators, Yoonjai Choi and Ken Meier from Common Name. I am grateful to everyone at Höweler + Yoon for their great work, and for their patience with the book project.

This book was inspired by the classes that I have been teaching at the Harvard Graduate School of Design, both Cases in Contemporary Architecture, Materials Construction Processes, as well as the seminar Adaptive Architectures. And, in the spirit of placing design studios and building technology courses into productive tension, the design studios that I have been teaching, including Core 3 Studio, Integrate. I am grateful to all the students, Teaching Assistants, Teaching Fellows, for the opportunity to organize my thoughts during the weekly, and early morning, lectures. I thank my inspiring colleagues and co-instructors: Carl Solander, Sal Craig, Mark Mulligan, Jonathan Levi, and Kiel Moe. Guest speakers in my class have shared their work, dived deep, and nerded out on details: Alex Anmahian, Kutan Ayata, Daniel Barber, Frank Barkow, Stella Betts, Marlon Blackwell, Eve Blau, Jennifer Bonner, Billie Faircloth, Ben Gilmartin, Peter Gluck, Lisa Gray, Felix Heisel, Mimi Hoang, Mark Husser, Mariana Ibañez, Florian Idenburg, Sharon Johnston, Sheila Kennedy, Michael Hays, Erik L'Heureux, Grace La, Dominic Leong, Jing Liu, Debora Mesa, Kiel Moe, Toshiko Mori, Farshid Moussavi, Michael Murphy, John Ochsendorf, Neri Oxman, Mark Pasnik, Ben Pell, Thomas Phifer, Anthony Piermarini, Joshua Ramus, Nina Rappaport, Carlo Ratti, Miriam Rich, John Ronan, Paxton Sheldahl, Dennis Sheldon, Marc Simmons, Nader Tehrani, Skylar Tibbits, Frano Violich, Victor Vizgatis, Weston Walker, Mark West, Beth Whittaker, and Dan Wood. A special guest speaker, Les Robertson, traveled from New York to Boston severals years in a row, to deliver the first guest lecture in my course. Les was a constant and an inspiration.

A special thanks to all the architects, designers, and photographers, whose work is included in the book. The work inspires me. Book making involves reading as well as writing, listening and talking. I am grateful to the numerous individuals who took the time to discuss their work with me including: Anthony Birchler, Byong Cho, Mark Cruvellier, Cynthia Davidson, Craig Dykers, Rodolphe el-Khoury, Peter Eisenman, Ed Ford, Tom Gluck, Emily Goldenberg, Roger Goldstein, Patricia Gruits, Louisa Hutton, Colin Koop, William Kreysler, George Johnston, Sam Lasky, Bethia Liu, Meaghan Lloyd, Rodolfo Machado, Kiel Moe, Florian Nagler, Matthias Sauerbruch, Manan Shah, Shohei Shigematsu, Marc Simmons, Elisabeth Spieker, Mark Steyer, Stacy Wong, Joshua Zabel, and Andrea Zanni. Previous versions of some texts were included in essays, editorials, book reviews and conference proceedings, including *Renovating Carbon Form*, *The Plan* magazine, *Architectural Record*, and *Verify in Field: Projects and Conversations: Höweler + Yoon*. The book was supported by the Harvard Graduate School of Design, Faculty Research Grants. A special thanks to Mark Cruvellier, Daniel Barber and Ed Ford, for reading parts of the book and offering comments and feedback.

This book would not have been possible without the many conversations over years around architecture and the discipline with incredibly insightful and supportive interlocutors, colleagues, mentors, and friends: Lucia Allais, Amale Andraos, Thomas Auer, Martin Bechtold, Ann Beha, Tatiana Bilbao, Will Bruder, Eric Bunge, Kyle Coburn, Scott Cohen, Milton Curry, Thomas de Monchaux, Liz Diller, Jim Durham, David Fixler, Jonathan Fournier, Elle Gerdeman, Ben Gilmartin, Lisa Haber-Thomson, David Hacin, Michael Hays, Bob Heintges, Chuck Heuer, Gary Hilderbrand, Mimi Hoang, Andrew Holder, Tom Holzman, Mariana Ibañez, Florian Idenburg, Sungwoo Jang, Mark Jarzombek, J Jih, Hauke Jungjohann, Hanif Kara, Paul Katz, Jerold Kayden, Joe Khoury, Simon Kim, Grace La, Amanda Reeser Lawrence, Mark Lee, Andrea Leers, Ming Leung, Paul Lewis, Jing Liu, Jon Lott, John May, Pat McCafferty, Rahul Mehrotra, Michael Meredith, Ana Miljački, Josephine Minultillo, Lee Moreau, Mohsen Mostafavi, Erika Naginski, Angela Pang, Mark Pasnik, Bill Pedersen, Antoine Picon, Bill Rawn, Charles Renfro, Hilary Sample, Holly Samuelson, Ian Schmellick, Ric Scofidio, Mack Scogin, Ashley Schafer, Jorge Silvetti, Flavio Stigliano, Nader Tehrani, Lily Tsai, Jamie von Klemperer, Charles Waldheim, Sarah Whiting, Beth Whittaker, Mabel O. Wilson, Andrew Witt, Dan Wood, Kim Yao, Ed Young, and Philip Yuan. Thanks to my family for a lifetime of love and support: J. Meejin Yoon and Kaya Miae Höweler, Michael Höweler and Kim Dupcak, Jason (Heung-Iho) Yoon and Hannah (Young-ai) Yoon, and Reinhardt H. Höweler. This book is dedicated to my mother, Ichun Jane Höweler Chang, who first taught me to see the world.